Lecture Notes in Computer Science 1367

Edited by G. Goos, J. Hartmanis and J. van Leeuwen

Springer
Berlin
Heidelberg
New York
Barcelona
Budapest
Hong Kong
London
Milan
Paris
Santa Clara
Singapore
Tokyo

Ernst W. Mayr   Hans Jürgen Prömel
Angelika Steger (Eds.)

Lectures on
Proof Verification
and Approximation
Algorithms

 Springer

Series Editors

Gerhard Goos, Karlsruhe University, Germany
Juris Hartmanis, Cornell University, NY, USA
Jan van Leeuwen, Utrecht University, The Netherlands

Volume Editors

Ernst W. Mayr
Angelika Steger
Institut für Informatik, Technische Universität München
D-80290 München, Germany
E-mail: {mayr,steger}@informatik.tu-muenchen.de

Hans Jürgen Prömel
Institut für Informatik, Humboldt-Universität zu Berlin
D-10099 Berlin, Germany
E-mail: proemel@informatik.hu-berlin.de

Cataloging-in-Publication data applied for

Die Deutsche Bibliothek - CIP-Einheitsaufnahme

Lectures on proof verification and approximation algorithms /
Ernst W. May ... (ed.). - Berlin ; Heidelberg ; New York ; Barcelona ;
Budapest ; Hong Kong ; London ; Milan ; Paris ; Santa Clara ;
Singapore ; Tokyo : Springer, 1998
(Lecture notes in computer science ; 1367)
ISBN 3-540-64201-3

CR Subject Classification (1991): F.2, F.1.3, D.2.4, G.1.2, G.1.6, G.3, I.3.5

ISSN 0302-9743
ISBN 3-540-64201-3 Springer-Verlag Berlin Heidelberg New York

© Springer-Verlag Berlin Heidelberg 1998
Printed in Germany

Typesetting: Camera-ready by author
SPIN 10631845   06/3142 – 5 4 3 2 1 0   Printed on acid-free paper

Preface

Proof Verification and Approximation Algorithms – Hardly any area in theoretical computer science has been more lively and flourishing during the last few years. Different lines of research which had been developed independently of each other over the years culminated in a new and unexpected characterization of the well-known complexity class $\mathcal{NP}$, based on probabilistically checking certain kinds of proofs. This characterization not only sheds new light on the class $\mathcal{NP}$ itself, it also allows proof of non-approximability results for optimization problems which, for a long time, had seemed to be out of reach. This connection, in turn, has motivated scientists to take a new look at approximating $\mathcal{NP}$-hard problems as well – with quite surprising success. And apparently, these exciting developments are far from being finished.

We therefore judged "Proof Verification and Approximation Algorithms" an ideal topic for the first in a new series of research seminars for young scientists, to be held at the International Conference and Research Center for Computer Science at Schloß Dagstuhl in Germany. This new series of seminars was established by the German Society for Computer Science (Gesellschaft für Informatik, GI) with the aim of introducing students and young scientists to important new research areas and results not yet accessible in text books or covered in the literature in a comprehensive way.

When we announced our seminar we encountered considerable interest and received numerous responses. We were able to select 21 qualified doctoral students and postdocs. Each participant then was requested to give a lecture, usually based on several research articles or technical reports, and to submit, in preliminary form and before the workshop began, an exposition of the topic assigned to him/her. The actual workshop then took place April 21–25, 1997 at Schloß Dagstuhl. All participants were very well prepared and highly motivated. We heard excellent talks and had many interesting and stimulating discussions, in the regular sessions as well as over coffee or some enlightening glass of wine after dinner.

This volume contains revised versions of the papers submitted by the participants. The process of revision involved, among other things, unifying notation, removing overlapping parts, adding missing links, and even combining some of the papers into single chapters. The resulting text should now be a coherent and essentially self-contained presentation of the enormous recent progress facilitated by the interplay between the theory of probabilistically checkable proofs and approximation algorithms. While it is certainly not a textbook in the usual sense, we nevertheless believe that it can be helpful for all those who are just starting out to learn about these subjects, and hopefully even to those looking for a coherent treatment of the subject for teaching purposes.

Our workshop was sponsored generously, by Special Interest Group 0 (Fachbereich "Grundlagen der Informatik") of the German Society for Computer Science (GI) and by the International Conference and Research Center for Computer Science (Internationales Begegnungs- und Forschungszentrum für Informatik, IBFI) at Schloß Dagstuhl. We owe them and the staff at Schloß Dagstuhl many thanks for a very successful and enjoyable meeting.

München, Berlin   Ernst W. Mayr
September 1997   Hans Jürgen Prömel
                                                  Angelika Steger

Prologue

Exam time. Assume you are the teaching assistant for some basic course with s students, s very large. The setup for the exam is as follows:

(1) The exam consists of q yes/no questions.

(2) A student passes *if and only if* he or she answers all questions correctly.

You assume that, on average, you'll need at least half a second to check the correctness of each answer. Since you expect the number of students to be close to one thousand (it is a very popular basic course!) and since the number of questions will be several hundred, a rough estimate shows that you are going to spend almost a whole week grading the exam. Ooff.

Is there a faster way?

Certainly not in general: in the worst case you really might have to look at all $s \cdot q$ answers in order to rule out a false decision. But what if we relax the second condition slightly and replace it by

(2') A student definitely passes the exam if he or she answers all questions correctly. A student who does not answer all questions correctly may pass only with a small probability, say $\leqslant 10^{-3}$, independently of the answers he or she gives.

Now you suddenly realize that the grading can actually be done in about $45s$ seconds, even regardless of the actual number q of questions asked in the exam. That is, a single day should suffice. Not too bad.

How is this possible? Find out by reading this book! And enjoy!

Table of Contents

Introduction

During the last few years we have seen quite spectacular progress in the area of approximation algorithms. For several fundamental optimization problems we now actually know matching upper and lower bounds for their approximability (by polynomial time algorithms).

Perhaps surprisingly, it turned out that for several of these problems, including the well-known MAX3SAT, SETCOVER, MAXCLIQUE, and CHROMATICNUMBER, rather simple and straightforward algorithms already yield the essentially best possible bound, at least under some widely believed assumptions from complexity theory. The missing step for tightening the gap between upper and lower bound was the improvement of the lower or non-approximability bound. Here the progress was initiated by a result in a seemingly unrelated area, namely a new characterization of the well-known complexity class $\mathcal{NP}$. This result is due to Arora, Lund, Motwani, Sudan, and Szegedy and is based on so-called probabilistically checkable proofs. While already very surprising and certainly interesting by itself, this result has given rise to fairly general techniques for deriving non-approximability results, and it initiated a large amount of subsequent work.

On the other hand, as if this so-to-speak "negative" progress had inspired the research community, the last few years have also brought us considerable progress on the "positive" or algorithmic side. Perhaps the two most spectacular results in this category are the approximation of MAXCUT using semidefinite programming, by Goemans and Williamson, and the development of polynomial time approximation schemes for various geometric problems, obtained independently by Arora and Mitchell.

These notes give an essentially self-contained exposition of some of these new and exciting developments for the interplay between complexity theory and approximation algorithms. The concepts, methods and results are presented in a unified way that should provide a smooth introduction to newcomers. In particular, we expect these notes to be a useful basis for an advanced course or reading group on probabilistically checkable proofs and approximability.

Overview and Organization of this Book

To be accessible for people from different backgrounds these notes start with three introductory chapters. The first chapter provides an introduction to the world of complexity theory and approximation algorithms, as needed for the subsequent treatment. While most of the notions and results from complexity theory that are introduced here are well-known and classical, the part on approximation algorithms incorporates some very recent results which in fact reshape a number of definitions and viewpoints. It also includes the proof by Trevisan [Tre97] that MAX3SAT is $\mathcal{APX}$-complete.

The second chapter presents a short introduction to randomized algorithms, demonstrating their usefulness by showing that an essentially trivial randomized algorithm for MAXE3SAT (the version of MAX3SAT in which all clauses have exactly three literals) has expected performance ratio 8/7. Later on, in Chapter 7, this ratio will be seen to be essentially best possible, assuming $\mathcal{P} \neq \mathcal{NP}$.

Concluding the introductory part, the third chapter describes various facets and techniques of derandomization, a term coined for the process of turning randomized algorithms into deterministic ones. Amongst other things in this chapter it is shown that the algorithm for MAXE3SAT is easily derandomized.

Chapters 4 to 10 are devoted to the concept of probabilistically checkable proofs and the implications for non-approximability. Chapter 4 introduces the so-called PCP-Theorem, a new characterization of $\mathcal{NP}$ in terms of probabilistically checkable proofs, and explains why and how they can be used to show non-approximability results. In particular, the nonexistence of polynomial time approximation schemes for $\mathcal{APX}$-complete problems and the non-approximability of MAX-CLIQUE are shown in detail. A complete and self-contained proof of the PCP-Theorem is presented in Chapter 5. Chapter 6 is devoted to the so-called Parallel Repetition Theorem of Raz [Raz95] which is used heavily in subsequent chapters.

At the 1997 STOC, Håstad [Hås97b] presented an exciting paper showing that the simple algorithm of Chapter 2 for approximating MAXE3SAT is essentially best possible. Chapter 7 is devoted to this result of Håstad's. The chapter also introduces the concept of long codes and a method of analyzing these codes by means of discrete Fourier transforms. These tools will be reused later in Chapter 9.

Chapter 8 surveys the new reduction techniques for optimization problems using gadgets, a notion for the first time formally introduced within the framework of approximation algorithms by Bellare, Goldreich, and Sudan [BGS95].

MAXCLIQUE cannot be approximated up to a factor of $n^{1-\epsilon}$ unless $\mathcal{NP} = \mathcal{ZPP}$. This result, also due to Håstad [Hås96a], is based on a version of the PCP-Theorem using so-called free bits. This concept, as well as Håstad's result, are described in Chapter 9.

As the final installment in this series of optimal non-approximability results, Chapter 10 presents the result of Feige [Fei96] stating that for SETCOVER the approximation factor of $\ln n$ achieved by a simple greedy algorithm is essentially best possible unless $\mathcal{NP} \subseteq \mathrm{DTIME}(n^{\mathcal{O}(\log \log n)})$.

The last three chapters of these notes are devoted to new directions in the development of approximation algorithms. First, Chapter 11 surveys recent achievements in constructing approximation algorithms based on semidefinite programming. A generalization of linear programming, semidefinite programming had been studied before for some time and in various contexts. However, only a few years ago Goemans and Williamson [GW95] showed how to make use of it in order to provide good approximation algorithms for several optimization problems.

While the PCP-Theorem implied that no $\mathcal{APX}$-complete problem can have a polynomial time approximation scheme unless $\mathcal{NP} = \mathcal{P}$, it is quite surprising that many such problems nevertheless do have such approximation schemes when restricted to (in a certain sense) dense instances. Chapter 12 exemplifies a very general approach for such dense instances, due to Arora, Karger, and Karpinski [AKK95a].

The final chapter then presents one of the highlights of the work on approximation algorithms during recent years. It is the development of polynomial time approximation schemes for geometrical problems like the Euclidean traveling salesman problem, independently by Arora [Aro96, Aro97] and Mitchell [Mit96].

Notations and Conventions

Areas as lively and evolving as proof verification and approximation algorithms naturally do not have a standardized set of definitions and notations. Quite often the same phrase has a slightly different meaning in different papers, or different symbols have identical meaning. In these notes, we have striven for uniform notation and concepts. We have tried to avoid any redefinition of terms, and thus we sometimes had to choose between two (or more) equally well established alternatives (e.g., should approximation ratios be taken to always be $\geqslant 1$, or $\leqslant 1$, or depending on the type of approximation problem?).

We have also tried to avoid phrases like "reconsidering the proof of Theorem x in Chapter y we see that it also shows that ...". Instead, we have attempted to prove all statements in the form in which they'll be needed later on. We hope that in this way we have been able to make the arguments easier to follow and to improve readability of the text.

Finally, we want to explicitly add some disclaimers and an apology. The intention of these notes certainly is *not* to present a survey, detailed or not, on the *history* of research in proof verification or approximation algorithms. This means, in particular, that more often than not only the reference to a paper with the best bound or complexity is given, omitting an entire sequence of earlier work without which the final result would appear all but impossible. Of course, numerous citations and pointers to work that had a major impact in the field are given, but there are doubtlessly many omissions and erroneous judgments. We therefore would like to apologize to all those whose work does not receive proper credit in these notes.

Acknowledgments

This volume, and in particular its timely completion, has only been possible by the joint effort of all participants. We would like to thank them all. Editing this volume in its various stages has been a great pleasure. For the numerous technical details special thanks are due to Katja Wolf for taking care of the bibliography, Stefan Hougardy for preparing the index, and, last but not least, to Volker Heun, who really did an excellent job in solving the thousand and more remaining problems.

1. Introduction to the Theory of Complexity and Approximation Algorithms

*Thomas Jansen**

1.1 Introduction

The aim of this chapter is to give a short introduction to the basic terms and concepts used in several of the following chapters. All these terms and concepts are somehow standard knowledge in complexity theory and the theory of optimization and approximation. We will give definitions of widely used models and complexity classes and introduce the results that are most important for our topic. This chapter is definitely not intended to give an introduction to the theory of complexity to someone who is unaware of this field of computer science, there are several textbooks fulfilling this purpose, see e.g. the books of Bovet and Crescenzi [BC93], Garey and Johnson [GJ79], Wegener [Weg93], and Papadimitriou [Pap94]. Instead we wish to repeat the main definitions and theorems needed in the following text to establish a theory concerned with the task of optimization and approximation.

It will turn out that it is sufficient to make use of four different though closely related machine models, namely the Turing machine and three variants of it: the nondeterministic Turing machine, the oracle Turing machine, and the randomized Turing machine. The first three models are needed to build a theory of problems, algorithms and the time complexity of solving problems. The randomized Turing machine is well suited to introduce the concept of randomness allowing – compared with the other three models – new and quite different types of algorithms to be constructed.

The task we are originally interested in is optimization; for a given problem we are seeking a solution that is in some well defined sense optimal. Since it will turn out that the problem of optimization is too hard to be solved exactly in many cases we will settle with approximation of these hard problems: we will be satisfied if we are able to "come near to the optimum" without actually finding it. Related with the idea of not guaranteeing to find an optimal solution but one that is not too far away is the idea of building algorithms that do not guarantee to find a solution but which have the property, that the probability of finding a solution is not too small. The later chapters will show in detail how the

* This author was supported by the Deutsche Forschungsgemeinschaft (DFG) as part of the Collaborative Research Center "Computational Intelligence" (SFB 531).

relationship of these two different concepts can be described. We will introduce the concept of randomized algorithms that do compute solutions with certain probabilities. We will establish several different levels of confidence that can be related with the output of randomized algorithms. Furthermore, we will show the relationship between solving a problem randomized or deterministically so it will be clear how far solving a problem probabilistically can be easier than solving it deterministically.

1.2 Basic Definitions

In order to have well defined notions as bases for the following sections we give some basic definitions here for very well known concepts like Turing machines and the classes $\mathcal{P}$ and $\mathcal{NP}$. We make some special assumptions about the machines we use which differ from the widely used standard but make some proofs much easier. All assumptions are easy to fulfill by making some simple modifications of what may be considered as standard Turing machine.

Definition 1.1. *A Turing machine M is a tuple $(Q, s, q_a, q_r, \Sigma, \Gamma, \delta)$ where Q denotes a finite set of states, $s \in Q$ is the state in which M starts its computations, $q_a \in Q$ is the accepting state, $q_r \in Q$ (different to q_a) is the rejecting state, Σ is a finite set that builds the input alphabet, $\Gamma \supset \Sigma$ is a finite set that defines the tape alphabet and $\delta : Q \times \Gamma \to Q \times \Gamma \times \{L, R, S\}$ is the transition function.*

A Turing machine M can be imagined as a computation unit that has an infinitely long tape partitioned into cells each containing exactly one letter from Γ. It has a head that is capable of reading exactly one letter from and writing exactly one letter to the position it is at in the moment; the head can be moved to the left or to the right one cell (or stay) according to the transition function δ. The Turing machine M is run on an input $w \in \Sigma^*$ and does its computations in steps. At each moment the Turing machine is in one state $q \in Q$. At the beginning the tape is filled with a special symbol $B \in \Gamma - \Sigma$ (the blank symbol), the letters of the input w are written on consecutive cells of the tape, and the head of M is positioned over the first letter of w. At each step it reads the current symbol $a \in \Gamma$ and according to $\delta(q, a) = (q', a', m)$ it writes a', moves according to m, and changes its state to q'. The computation stops if M reaches the state q_a or the state q_r, and the machine accepts or rejects the input according to the final state. The computation time is the number of steps until M reaches the halting state. The Turing machine we defined is deterministic in the following sense. We know exactly what a Turing machine M will do in the next step if we are informed about the state it is currently in, the content of the tape and the position of the head. We call this triple of information a configuration of M. The start configuration is well defined when the input w is given; since δ

defines exactly what M will do in the next step the next configuration is well defined and by induction we get for each input w a well defined sequence of configurations.

A Turing machine is capable of doing more than just accepting or rejecting a given input. Since it can write to the tape it is possible to use a Turing machine to compute function values, as well. In this case we consider the symbols that are written to the tape and that are different to the blank symbol to be the computed function value if the Turing machine halted in an accepting state. The Turing machine is a very simple model but it is believed to be capable of computing any function that is computable in an intuitive sense. This belief is commonly referred to as the Church-Turing Thesis. What is even more important, it is widely believed that any function that can be computed efficiently on any reasonable computer can be computed efficiently by a Turing machine, given that the term "efficiently" is defined adequately: by "efficiently" we mean in polynomial time as it is defined in the following definition.

Definition 1.2. *We say that a Turing machine M is polynomially time bounded if there exists a polynomial p such that for an input w the computation of M on w takes at most $p(|w|)$ steps ($|w|$ denotes the length of w). The class $\mathcal{P}$ (polynomial time) consists of all languages $L \subseteq \Sigma^*$ for which there exists a polynomially time bounded Turing machine M_L that accepts all inputs $w \in L$ and rejects all inputs $w \notin L$.*

We identify problems which we wish to be solved by a Turing machine with languages $L \subseteq \Sigma^*$. We say a Turing machine solves a problem if it accepts all $w \in L$ and rejects all $w \notin L$ if the considered problem is described by L. Since the answer to an instance w of a problem can either be "yes" or "no" and nothing else we are dealing with so called decision problems. If we are interested in getting more complicated answers (computing functions with other values) we can use Turing machines and consider the output that can be found on the tape as the result of the computation as mentioned above. By Definition 1.2 we know that a problem belongs to $\mathcal{P}$ if it can be solved by a deterministic algorithm in polynomial time. We consider such algorithms to be efficient and do not care about the degree of the polynomial. This might be a point of view that is too optimistic in practice; on the other hand we are not much too pessimistic if we consider problems that are not in $\mathcal{P}$ not to be solvable in practice in reasonable time. We explicitly denote two types of superpolynomial running times: we call a running time of $\Omega\left(2^{n^\varepsilon}\right)$ for some constant $\varepsilon > 0$ an exponential running time. And we call subexponential but superpolynomial running times of the type $\Theta\left(n^{p(\log n)}\right)$ for a polynomial p quasi-polynomial running times. The class of all languages which have a Turing machine that is quasi-polynomially time bounded is denoted by $\tilde{\mathcal{P}}$. Analogous definitions can be made for any class $\mathcal{C}$ that is defined by polynomial running time. The analogous class with quasi-polynomial running time is called $\tilde{\mathcal{C}}$.

We can assume without loss of generality that M stops after exactly $p(|w|)$ steps, so that Definition 1.2 yields the result that a polynomially time bounded Turing machine has exactly one well defined sequence of configurations of length exactly $p(|w|)$ for an input $w \in \Sigma^*$.

An important concept in complexity theory is non-determinism. It can be introduced by defining a nondeterministic Turing machine.

Definition 1.3. *A nondeterministic Turing machine is defined almost exactly like a Turing machine except for the transition function δ. For a nondeterministic Turing machine $\delta : (Q \times \Gamma) \times (Q \times \Gamma \times \{L, R, S\})$ is a relation.*

The interpretation of Definition 1.3 is the following. For a state $q \in Q$ and a letter $a \in \Gamma$ there may be several triples (q', a', m); at each step a nondeterministic Turing machine M may choose any of these triples and proceed according to the choice. Now we have a well defined start configuration but the successor is not necessarily unique. So we can visualize the possible configurations as a tree (we call it computation tree) where the start configuration is the root and halting configurations are the leaves. We call a path from the root to a leaf in this tree a computation path. We say that M accepts an input w if and only if there is at least one computation path with the accepting state at the leaf (which is called an accepting computation path), and M rejects w if and only if there is no accepting computation path. We can assume without loss of generality that each node has degree at most two. The computation time for a nondeterministic Turing machine on an input w is the length of a shortest accepting computation path. For $w \notin L$ the computation time is 0.

Definition 1.4. *We say that a nondeterministic Turing machine M is polynomially time bounded if there exists a polynomial p such that for an input $w \in L$ the computation of M on w takes at most $p(|w|)$ steps. We can assume without loss of generality that M stops after exactly $p(|w|)$ steps if $w \in L$. The class $\mathcal{NP}$ (nondeterministic polynomial time) consists of all languages $L \subseteq \Sigma^*$ for which a polynomially time bounded nondeterministic Turing machine exists that accepts all inputs $w \in L$ and rejects all inputs $w \notin L$.*

The choice of polynomial bounded computation time leads to very robust complexity classes; a lot of natural problems can be solved in polynomial time and, what is more, polynomial computation time on Turing machines corresponds to polynomial computation time on other computation models and vice versa. But of course, other functions limiting the computation time are possible, so the classes $\mathcal{P}$ and $\mathcal{NP}$ can be viewed as special cases of more general complexity classes as will be described now.

Definition 1.5. *The class $\text{DTIME}(f)$ consists of all languages L that have a Turing machine M_L which is $\mathcal{O}(f)$ time bounded.*

The class NTIME(f) *consists of all languages L that have a nondeterministic Turing machine M_L which is $\mathcal{O}(f)$ time bounded.*

As a consequence of Definition 1.5 we get $\mathcal{P} = $ DTIME(poly) $= \bigcup_p$ DTIME(p) and $\mathcal{NP} = $ NTIME(poly) $= \bigcup_p$ NTIME(p) for polynomials p.

Nondeterministic Turing machines are obviously at least as powerful as deterministic ones. Any deterministic Turing machine M with a function δ can be taken as a nondeterministic Turing machine where δ is seen as a relation. Therefore, DTIME(f) $\subseteq$ NTIME(f) holds for all f which implies that $\mathcal{P} \subseteq \mathcal{NP}$ holds. Whether $\mathcal{P} = \mathcal{NP}$ holds is probably the most famous open question in computer science; the discussion of this question was the starting point for the theory of $\mathcal{NP}$-completeness based on polynomial reductions which we define next.

Definition 1.6. *A problem $A \subseteq \Sigma^*$ is said to be polynomially reducible to a problem $B \subseteq \Delta^*$ ($A \leqslant_p B$) if and only if there exists a function $f : \Sigma^* \to \Delta^*$, computable in polynomial time, such that $\forall w \in \Sigma^* : w \in A \Leftrightarrow f(w) \in B$. The function f is called* polynomial reduction.

We call a problem A $\mathcal{NP}$-hard if and only if all problems $B \in \mathcal{NP}$ are polynomially reducible to A. We call a problem A $\mathcal{NP}$-complete if and only if A is $\mathcal{NP}$-hard and belongs to $\mathcal{NP}$.

Obviously, if any $\mathcal{NP}$-complete problem is in $\mathcal{P}$ then $\mathcal{P} = \mathcal{NP}$ follows. On the other hand, if any $\mathcal{NP}$-complete problem is provably not in $\mathcal{P}$, then $\mathcal{P} \neq \mathcal{NP}$ follows. There are a lot of problems which are known to be $\mathcal{NP}$-complete, for a list see e.g. the book of Garey and Johnson [GJ79]. The historically first and perhaps most important $\mathcal{NP}$-complete problem is the satisfiability problem SAT.

Sᴀᴛ

Instance: A Boolean formula Φ in conjunctive normal form.
Question: Does a satisfying assignment for Φ exist?

Cook [Coo71] proved that Sᴀᴛ is $\mathcal{NP}$-complete by showing that for any problem $A \in \mathcal{NP}$ it is possible to code the behavior of the nondeterministic Turing machine M_A for A on an input w in a Boolean formula in conjunctive normal form Φ such that Φ is satisfiable if and only if M_A accepts w. This result is called Cook's theorem and will be used later in this chapter.

The key idea of the proof is that it is possible to describe the computation of a nondeterministic Turing machine by a Boolean expression Φ. The first step is to show that some special assumptions about nondeterministic Turing machines can be made without loss of generality. We assume that the input alphabet is $\Sigma = \{0, 1\}$, the tape alphabet is $\Gamma = \{0, 1, B\}$, the states are $Q = \{q_0, \ldots, q_n\}$ with $s = q_0$, $q_a = q_1$, and $q_r = q_2$. Furthermore, we assume that M stops on an input w after exactly $p(|w|)$ steps for a given polynomial p. Finally, all nondeterministic steps are assumed to happen at the beginning of the computation: M starts by producing a string of 0s and 1s of appropriate length by nondeterministic steps. After this phase all following steps are deterministic. We define $|Q|\,(p(|w|)+1)$ variables $v_q(q,t)$, $2\,|\Gamma|\,(p(|w|)+1)^2$ variables $v_t(c,l,t)$, and $2(p(|w|)+1)^2$ variables $v_h(c,t)$ with $q \in Q$, $t \in \{0,\ldots,p(|w|)\}$, $c \in \{-p(|w|),\ldots,p(|w|)\}$, and $l \in \Gamma$. The interpretation of the variables is the following. The variable $v_q(q,t)$ is true iff M is in state q after step t, $v_t(c,l,t)$ is true iff the tape at position c contains the letter l after step t, and $v_h(c,t)$ is true iff the head of M is over the cell c after step t. We want to build a set of clauses such that all clauses can only be satisfied simultaneously if and only if M accepts the input w. First, we make sure that the start configuration is defined adequately. Second, we have to make sure that for each t the Turing machine is in one well defined state, the head has one well defined position, and each cell contains one well defined letter of Γ. Third, we must assure that the configuration after step t is a proper successor of the configuration after step $t-1$. Fourth, we have to state that the final state must be the accepting state. All conditions are either of the form "after step t exactly one of the following possibilities is true" or "if A is the case after step $t-1$ then B is the case after step t". For the first type of condition we consider the state of M after the first step as an example. The clause $v_q(q_0,1)\vee\cdots\vee v_q(q_n,1)$ is satisfiable if and only if M is in at least one state after the first step. The $(n+1)n/2$ clauses $(\neg v_q(q_i,1)) \vee (\neg v_q(q_j,1))$ with $0 \leqslant i \neq j \leqslant n$ are only satisfiable simultaneously if and only if M is in at most one state after the first step; so the $1+(n+1)n/2$ clauses can only be satisfied together if and only if M is in exactly one state after step 1. For the second type of condition we only mention that the Boolean formula "$A \Rightarrow B$" is equivalent to "$\neg A \vee B$". It is easy to see, how all conditions of valid computations can be expressed as clauses. Since the sizes of Q and Γ are constant it is obvious that there are only polynomially many variables and that only polynomially many clauses are formulated. So, the reduction can be computed in polynomial time. Given an assignment τ that satisfies all clauses simultaneously we not only know that the Turing machine M accepts the input w. We know the exact computation of M, too. For problems $L \in \mathcal{NP}$ we can assume that some "proof" for the membership of the input w in L of polynomial length can be stated such that a deterministic Turing machine can verify this proof in polynomial time. So, assuming that we use nondeterministic Turing machines that "guess" a proof of polynomial length nondeterministically and verify it deterministically in polynomial time implies that – given τ – we are able to reconstruct the proof for the membership of w in L in polynomial time.

For any class $\mathcal{C}$ of languages we define co-$\mathcal{C}$ as the set of all languages $\bar{L} = \{w \in \Sigma^* \mid w \notin L\}$ such that $L \subseteq \Sigma^*$ belongs to $\mathcal{C}$. Obviously $\mathcal{P} = \text{co-}\mathcal{P}$ holds, while $\mathcal{NP} = \text{co-}\mathcal{NP}$ is widely believed to be false.

We will now introduce another type of Turing machine that can be even more powerful than the nondeterministic one. We will use the so called oracle Turing machine to define an infinite number of complexity classes that contain the classes $\mathcal{P}$ and $\mathcal{NP}$ and put them in a broader context.

Definition 1.7. *An* oracle Turing machine M *for an oracle* $f : \Sigma^* \to \Sigma^*$ *is a (deterministic or nondeterministic) Turing machine that has two more states* q_o *and* q_a, *and a separate tape called oracle tape. The machine M can write questions* $v \in \Sigma^*$ *to the oracle on the oracle tape and then switch to state q_o. Then in one step the word v is erased from the tape, an answer $f(v)$ is written (ε if no answer exists) and the head of the oracle tape is positioned over the first letter of $f(v)$. Finally, M switches to the state q_a.*

We now use this definition to construct an infinite number of complexity classes. We start by defining how oracle Turing machines can be used to define new complexity classes based on the definitions of $\mathcal{P}$ and $\mathcal{NP}$.

Definition 1.8. *For a class $\mathcal{C}$ of languages we denote by* $\mathcal{P}(\mathcal{C})$ *($\mathcal{NP}(\mathcal{C})$) the class of all languages L, such that there is a language $H \in \mathcal{C}$ so that there is a polynomially time bounded (nondeterministic) oracle Turing machine M_L^H with oracle H for L.*

Definition 1.9. *We define* $\Delta_0 := \Sigma_0 := \Pi_0 := \mathcal{P}$. *For any $k \in \mathbb{N}$ we define* $\Delta_k := \mathcal{P}(\Sigma_{k-1})$, $\Sigma_k := \mathcal{NP}(\Sigma_{k-1})$, $\Pi_k := co\text{-}\Sigma_k$. *The classes Δ_k, Σ_k, and Π_k form the* polynomial hierarchy $\mathcal{PH} := \bigcup_{k \geqslant 0} \Sigma_k$. *The classes Δ_i, Σ_i, and Π_i are said to form the i-th level of $\mathcal{PH}$.*

We note that $\Delta_1 = \mathcal{P}$, $\Sigma_1 = \mathcal{NP}$, and $\Pi_1 = co\text{-}\mathcal{NP}$ hold. It is believed though not proven that no two classes of this hierarchy are equal. On the other hand, if there is an i such that $\Sigma_i = \Pi_i$ holds, then for all $j > i$ one gets $\Sigma_j = \Pi_j = \Delta_j = \Sigma_i$. This event is usually referred to as the collapse of the polynomial hierarchy; it collapses to the i-th level in this case.

We finally define another variant of Turing machines that allows us to introduce the concept of randomness.

Definition 1.10. *A randomized Turing machine M is defined as a nondeterministic Turing machine such that $\forall q \in Q, a \in \Sigma$ there are at most two different triples (q', a', m) with $(q, a) \times (q', a', m) \in \delta$. If there are two different triples M chooses each of them with probability $\frac{1}{2}$. There are three disjoint sets of halting states A, R, D such that computations of M stop if a state from $A \cup R \cup D$ is reached. If the halting state belongs to A (R), M accepts (rejects) the input. If the halting state belongs to D then M does not make a decision about the input.*

For the randomized Turing machine the computation time is declared in analogy to the deterministic Turing machine and not in the way it is defined for the nondeterministic Turing machine. Since the computation time depends on the random choices it is a random variable. While for deterministic and nondeterministic Turing machines where the classes $\mathcal{P}$ and $\mathcal{NP}$ were the only complexity classes we defined we will define a number of different complexity classes for randomized Turing machines which can all be found in the following section.

1.3 Complexity Classes for Randomized Algorithms

The introduction of randomized Turing machines allows us to define appropriate complexity classes for these kind of computation; we will follow Wegener [Weg93] in this section. There are many different ways in which a randomized Turing machine may behave that may make us say "it solves a problem". According to this variety of choice there is a variety of complexity classes.

Definition 1.11. *The class $\mathcal{PP}$ (probabilistic polynomial time) consists of all languages L for which a polynomially time bounded randomized Turing machine M_L exists that for all inputs $w \in L$ accepts and for all inputs $w \notin L$ rejects with probability strictly more than $\frac{1}{2}$ respectively.*

$\mathcal{PP}$-algorithms are called Monte Carlo algorithms.

Definition 1.12. *The class $\mathcal{BPP}$ (probabilistic polynomial time with bounded error) consists of all languages $L \in \mathcal{PP}$ for which a constant ε exists so that a polynomially time bounded Turing machine M_L for L exists that accepts all inputs $w \in L$ and rejects all inputs $w \notin L$ with probability strictly more than $\frac{1}{2} + \varepsilon$ respectively.*

Obviously, $\mathcal{BPP}$ is a subset of $\mathcal{PP}$. Assume there are a $\mathcal{PP}$-algorithm A_1 and a $\mathcal{BPP}$-algorithm A_2 for a language L. Our aim is to increase the reliability of these algorithms, that means to increase the probability of computing the correct result. We try to clarify the difference in the definitions by considering an input w of length $|w|$ for the $\mathcal{BPP}$-algorithm A_2 first. There is a constant ε such that the probability that A_2 computes the correct result is $p > \frac{1}{2} + \varepsilon$. For the input w we use A_2 to create a more reliable algorithm by running A_2 on w for t times (t odd) and accept if and only if A_2 accepted at least $\lceil \frac{t}{2} \rceil$ times.

The probability that A_2 gives the correct answer exactly i times in t runs is
$c_i = \binom{t}{i} p^i (1-p)^{t-i} = \binom{t}{i} (p(1-p))^i (1-p)^{2(\frac{t}{2}-i)} < \binom{t}{i} \left(\frac{1}{4} - \varepsilon^2\right)^i \left(\frac{1}{4} - \varepsilon^2\right)^{\frac{t}{2}-i}$
$= \binom{t}{i} \left(\frac{1}{4} - \varepsilon^2\right)^{\frac{t}{2}}$. It follows that the probability of our new algorithm to give the
correct answer is $1 - \sum_{i=0}^{\lfloor \frac{t}{2} \rfloor} c_i > 1 - 2^{t-1} \left(\frac{1}{4} - \varepsilon^2\right)^{\frac{t}{2}} \geqslant 1 - \frac{1}{2}\left(1 - 4\varepsilon^2\right)^{\frac{t}{2}}$.

So if we want to increase the probability of answering correctly to $1 - \delta$ we can run A_2 $\frac{2 \log 2\delta}{\log(1-4\varepsilon^2)}$ times to achieve this. If δ and ε are assumed to be constants the running time is polynomial.

If we do the same thing with the $\mathcal{PP}$-algorithm A_1 we can use the calculations from above if we manage to find a value for ε. For an input w with length $|w|$ there can be $2^{p(|w|)}$ different computation paths each with probability $2^{-p(|w|)}$. So the probability to give the correct answer for the input w is $\geqslant \frac{1}{2} + 2^{-p(|w|)}$. Notice that ε is not a constant here, it depends on the length of the input. Since for $-1 < x < 1$ we have $\ln(1+x) < x$ we know that $\frac{2 \log 2\delta}{\log(1-4\varepsilon^2)} > \frac{-\log 2\delta}{2}\varepsilon^{-2}$.

Since ε^{-1} grows exponentially with $|w|$ we need exponentially many runs of A_1 to reduce the error probability to δ.

There exist natural probabilistic algorithms that may make an error in identifying an input that belongs to the language L but are not fooled by inputs not belonging to L; the well-known equivalence-test for read-once branching programs of Blum, Chandra, and Wegman [BCW80] is an example. The properties of these algorithms are captured by the following definition of the class $\mathcal{RP}$. Also the opposite behavior (which is captured by the class co-$\mathcal{RP}$) is possible for quite natural algorithms as the well-known randomized primality test by Solovay and Strassen [SS77] shows.

Definition 1.13. *The class $\mathcal{RP}$ (randomized polynomial time) consists of all languages L for which a polynomially time bounded randomized Turing machine M_L exists which for all inputs $w \in L$ accepts with probability strictly more than $\frac{1}{2}$ and for all inputs $w \notin L$ rejects with probability 1.*

Obviously, $\mathcal{RP}$-algorithms are more powerful than $\mathcal{BPP}$-algorithms: repeating such an algorithm t times results in an algorithms with error probability decreased to 2^{-t}.

The three classes $\mathcal{PP}$, $\mathcal{BPP}$, and $\mathcal{RP}$ have in common that they do not rely essentially on the possibility of halting without making a decision about the input. A randomized Turing machine M_L for a decision problem L from one of these classes can easily be modified in such a way that it always accepts or rejects whether or not the given input belongs to L. To achieve this it is sufficient to define a Turing machine M_L' almost identical to M_L with some simple modifications that only affect the states. The set of states without decisions D' is defined as the empty set. The set of rejecting states R' is defined as $R \cup D$. It is obvious that the probability of accepting remains unchanged compared to M_L and the probability of rejecting can only be increased implying that M_L' is still a valid randomized Turing machine for L. We will now introduce an even more reliable class of algorithms, the so called Las Vegas algorithms which build the class $\mathcal{ZPP}$, where the possibility of not giving an answer at all can not be removed without changing the definition of the class.

Definition 1.14. *The class $\mathcal{ZPP}$ (probabilistic polynomial time with zero error) consists of all languages L for which a polynomially time bounded randomized Turing machine M_L exists which for all inputs $w \in L$ accepts with probability strictly more than $\frac{1}{2}$ and rejects with probability 0, and for all inputs $w \notin L$ rejects with probability strictly more than $\frac{1}{2}$ and accepts with probability 0.*

Clearly, modifying M_L for a language L from $\mathcal{ZPP}$ such that halting in a state from D is interpreted as rejecting is not possible since it can no longer be guaranteed that no error is made. If one needs a $\mathcal{ZPP}$-algorithm that always yields a correct answer and never refuses to decide a small change in the definition is needed. In this case we settle with expected polynomially bounded running time instead of guaranteed polynomially bounded running time. So, if a $\mathcal{ZPP}$-algorithm in the sense of Definition 1.14 wants to stop without decision about the input instead of stopping the algorithm can be started all over again. Since the probability of not giving an answer is strictly less then $\frac{1}{2}$ the expected number of runs until receiving an answer is less than two and the expected running time is obviously polynomially time bounded.

The four different complexity classes for randomized algorithms all agree in demanding polynomial running time but differ in the probabilities of accepting and rejecting. For an overview we compare these probabilities in Table 1.1.

class	$w \in L$	$w \notin L$
$\mathcal{PP}$	Prob(accept) $> \frac{1}{2}$	Prob(reject) $> \frac{1}{2}$
$\mathcal{BPP}$	Prob(accept) $> \frac{1}{2} + \varepsilon$	Prob(reject) $> \frac{1}{2} + \varepsilon$
$\mathcal{RP}$	Prob(accept) $> \frac{1}{2}$	Prob(reject) $= 1$
$\mathcal{ZPP}$	Prob(accept) $> \frac{1}{2}$ Prob(reject) $= 0$	Prob(reject) $> \frac{1}{2}$ Prob(accept) $= 0$

Table 1.1. Overview of complexity classes for polynomially time bounded randomized algorithms.

In analogy to Definition 1.5 we define ZTIME(f) as a generalization of $\mathcal{ZPP}$.

Definition 1.15. *The class* ZTIME(f) *consists of all languages L that have a randomized Turing machine M_L which is $\mathcal{O}(f)$ time bounded and which for all inputs $w \in L$ accepts with probability strictly more than $\frac{1}{2}$ and rejects with probability 0, and for all inputs $w \notin L$ rejects with probability strictly more than $\frac{1}{2}$ and accepts with probability 0.*

We close this section with some remarks about the relations between the randomized complexity classes.

Proposition 1.16. $\mathcal{P} \subseteq \mathcal{ZPP} \subseteq \mathcal{RP} \subseteq \mathcal{BPP} \subseteq \mathcal{PP}$

Proof. All the inclusions except for $\mathcal{RP} \subseteq \mathcal{BPP}$ directly follow from the definitions. Given an $\mathcal{RP}$-algorithm A we immediately get a $\mathcal{BPP}$-algorithm for the same language by running A two times. ∎

Proposition 1.17. $\mathcal{RP} \subseteq \mathcal{NP}$

Proof. A randomized Turing machine M_L for $L \in \mathcal{RP}$ may be interpreted as a nondeterministic Turing machine that accepts if M_L does and rejects else. According to the definition of $\mathcal{RP}$ there is no accepting computation path for $w \notin L$, at least one accepting computation path for $w \in L$, and M_L is polynomially time bounded. ∎

Proposition 1.18. $\mathcal{ZPP} = \mathcal{RP} \cap \text{co-}\mathcal{RP} \subseteq \mathcal{NP} \cap \text{co-}\mathcal{NP}$

Proof. Clearly, $\mathcal{ZPP} = \text{co-}\mathcal{ZPP}$ so that $\mathcal{ZPP} \subseteq \mathcal{RP} \cap \text{co-}\mathcal{RP}$ is obvious. Given a language $L \in \mathcal{RP} \cap \text{co-}\mathcal{RP}$, one can use the algorithms for L and $\bar{L}$ to construct a $\mathcal{ZPP}$-algorithm for L by simply accepting or rejecting according to the algorithm that rejects; if both algorithms do no reject then no answer is given. Finally, $\mathcal{RP} \cap \text{co-}\mathcal{RP} \subseteq \mathcal{NP} \cap \text{co-}\mathcal{NP}$ follows immediately from Proposition 1.17. ∎

Proposition 1.19. $(\mathcal{NP} \subseteq \text{co-}\mathcal{RP}) \Rightarrow (\mathcal{NP} = \mathcal{ZPP})$

Proof. $\mathcal{NP} \subseteq \text{co-}\mathcal{RP}$ implies $\text{co-}\mathcal{NP} \subseteq \text{co-co-}\mathcal{RP} = \mathcal{RP}$, so we get $\mathcal{NP} \cap \text{co-}\mathcal{NP} \subseteq \mathcal{RP} \cap \text{co-}\mathcal{RP}$ and by Proposition 1.18 this equals $\mathcal{ZPP}$. Since by Proposition 1.17 we have $\mathcal{RP} \subseteq \mathcal{NP}$, $\mathcal{NP} \subseteq \text{co-}\mathcal{RP} \subseteq \text{co-}\mathcal{NP}$ holds and $\mathcal{NP} = \text{co-}\mathcal{NP}$ follows. So altogether we have $\mathcal{NP} \subseteq \mathcal{ZPP}$ and since $\mathcal{ZPP} \subseteq \mathcal{NP}$ is obvious we have $\mathcal{NP} = \mathcal{ZPP}$. ∎

Proposition 1.20. $\mathcal{NP} \cup \text{co-}\mathcal{NP} \subseteq \mathcal{PP}$

Proof. We first show $\mathcal{NP} \subseteq \mathcal{PP}$. Let $L \in \mathcal{NP}$ and M a polynomially time bounded nondeterministic Turing machine for L. We assume without loss of generality that in each step M has at most two possibilities to proceed and there is a polynomial p such that M finishes its computation on an input w after exactly $p(|w|)$ steps. We construct a randomized Turing machine M' for L as follows. We compute a random number r equally distributed between 0 and $2^{p(|w|)+2} - 1$ (a binary number with $p(|w|) + 2$ bits). If $r > 2^{p(|w|)+1}$ then M' accepts, in the other case it simulates M on w. Obviously the running time of M' is polynomially bounded. If $w \notin L$ is given, then M will not accept w on any computation path. So M' accepts only if $r > 2^{p(|w|)+1}$ that means with probability less than $\frac{1}{2}$. If $w \in L$ there is at least one computation path on which M accepts w, so the probability for M' to accept w is more than $\frac{1}{2}2^{-p(|w|)} + \frac{2^{p(|w|)+1}-1}{2^{p(|w|)+2}} \geq \frac{1}{2}$. Since $\text{co-}\mathcal{PP} = \mathcal{PP}$ by definition the proof is complete. ∎

We summarize the relations between the different randomized complexity classes in Figure 1.1 that shows the various inclusions we recognized.

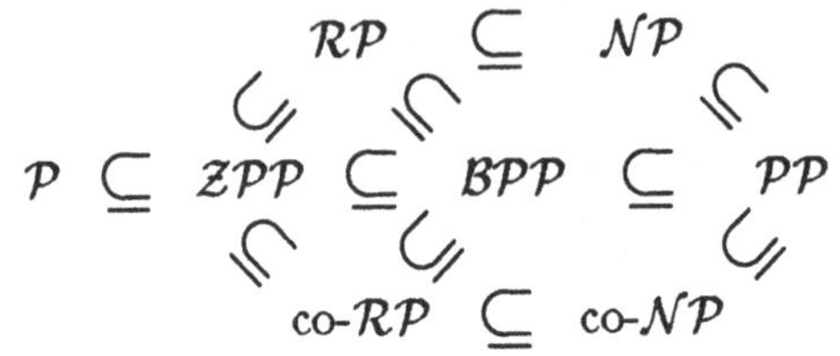

$$
\begin{array}{ccccccc}
 & \mathcal{RP} & \subseteq & \mathcal{NP} & \\
 & \cup & \cap & & \cap \\
\mathcal{P} \subseteq & \mathcal{ZPP} & \subseteq & \mathcal{BPP} & \subseteq & \mathcal{PP} \\
 & \cap & & \cup & & \cup \\
 & \text{co-}\mathcal{RP} & \subseteq & \text{co-}\mathcal{NP} &
\end{array}
$$

Fig. 1.1. Relations between complexity classes for randomized algorithms.

1.4 Optimization and Approximation

It is a natural task in practice to find a minimum or maximum of a given function since it is almost always desirable to minimize the cost and maximize the gain. The first obvious consequence of entering the grounds of optimization is that we are no longer confronted with decision problems. The problems we tackle here correlate some cost- or value-measure to possible solutions of a problem instance. Our aim is to find a solution that maximizes the value or minimizes the cost if we are asked to optimize or to find a solution that is as close as possible to the actual maximum or minimum if approximation is the task. In this section we will follow the notations of Bovet and Crescenzi [BC93] and Garey and Johnson [GJ79].

Definition 1.21. *An* optimization problem *is defined by a tuple* (I, S, v, goal) *such that* I *is the set of instances of the problem,* S *is a function that maps an instance* $w \in I$ *to all feasible solutions,* v *is a function that associates a positive integer to all* $s \in S(w)$, *and* goal *is either minimization or maximization.*

The value of an optimal solution to an instance w *is called* $\text{OPT}(w)$ *and is the minimum or maximum of* $\{v(s) \mid s \in S(w)\}$ *according to* goal.

An optimization problem can be associated with a decision problem in a natural way.

Definition 1.22. *For an optimization problem* (I, S, v, goal) *the underlying language is the set* $\{(w, k) \mid w \in S, v(w) \leqslant k\}$, *if the goal is minimization, and* $\{(w, k) \mid w \in S, v(w) \geqslant k\}$, *if the goal is maximization.*

We will define some famous optimization problems which have all in common that their underlying languages are $\mathcal{NP}$-compete.

KNAPSACK

Instance: A set V of items with weights defined by $w : V \to \mathbb{N}$ and values $v : V \to \mathbb{N}$, a maximal weight $W \in \mathbb{N}$.

Problem: Find a subset of V with weight sum not more than W such that the sum of the values is maximal.

BINPACKING

Instance: A set V of items with sizes defined by $s : V \to \mathbb{N}$ and a bin size b.

Problem: Find a partition of V that defines subsets such that each subset has size sum not more than b and has a minimal number of subsets.

We introduce two restrictions of BINPACKING, the first one is a $\mathcal{NP}$-complete decision problem and is called PARTITION.

PARTITION

Instance: A set V of items with sizes defined by $s : V \to \mathbb{N}$.

Question: Is it possible to partition V into two subsets with equal size sum?

PARTITION can be viewed as BINPACKING with the fixed bin size $\left(\sum_{v \in V} s(v) \right) / 2$ and the question whether all elements fit in just two bins. The formulation of PARTITION forces the items to be small compared to the size of the bins; otherwise PARTITION becomes simple. The second restriction of BINPACKING we define forces the items to be large compared to the size of the bins; this property does not simplify the problem significantly, though.

LARGEPACKING

Instance: A set V of items with sizes defined by $s : V \to \mathbb{N}$ and a bin size b such that $1/5b \leqslant s(v) \leqslant 1/3b$ holds for all $v \in V$.

Problem: Find a partition of V that defines subsets such that each subset has size sum not more than b and has a minimal number of subsets.

MAX3SAT

Instance: A Boolean formula Φ in conjunctive normal form such that each clauses is a disjunction of up to three literals.

Problem: Find an assignment for Φ that satisfies a maximum number of clauses simultaneously.

TRAVELINGSALESMANPROBLEM (TSP)

Instance: A graph with $G = (V, E)$ with edge weights defined by $c : E \to \mathbb{N}$.

Problem: Find a circle in G with $|V|$ edges containing all nodes that has a minimal sum of weights.

We defined the edge weights to be integers. A special case of the TSP uses points in the plane as nodes and the distances in Euclidian metric as edge weights. In this case it is more natural to allow edge weights from $\mathbb{R}$. Obtaining an optimal solution together with the length of an optimal tour may become more difficult now, since it is not clear how long a representation of the length of an optimal tour may become. However, if only approximation is the goal like in Chapter 13 this problem does not matter.

A restricted version of the TSP that is a decision problem and still $\mathcal{NP}$-complete is called HAMILTONCIRCUIT.

HAMILTONCIRCUIT
Instance: A graph G.
Question: Is there a circle in G with $|V|$ edges containing all nodes?

1.4.1 Complexity classes for optimization problems

We will concentrate on optimization problems where the main difficulty is to find an optimal solution; this rules out all problems where things like testing an input for syntactic correctness is already a hard problem.

Definition 1.23. *The class $\mathcal{NPO}$ ($\mathcal{NP}$ optimization) consists of all optimization problems (I, S, v, goal), where I is in $\mathcal{P}$, for all inputs w the set of solutions $S(w)$ is in $\mathcal{P}$, and $v(s)$ is computable in polynomial time.*

It is easy to see that for any optimization problem from $\mathcal{NPO}$ the underlying language is in $\mathcal{NP}$: to an instance $w \in I$ an appropriate solution s of polynomial length can be guessed and verified in polynomial time by computing $v(s)$ and comparing the value with k.

Definition 1.24. *The class $\mathcal{PO}$ ($\mathcal{P}$ optimization) consists of all optimization problems from $\mathcal{NPO}$ which have their underlying language in $\mathcal{P}$.*

Up to now we have defined two different classes of optimization problems without saying anything about optimization algorithms. An optimization algorithm to a given optimization problem is an algorithm that computes to any instance a solution with minimal or maximal value according to goal. For many problems it is well known that computing an optimal solution is $\mathcal{NP}$-hard, so we have to ask for less. Instead of optimization algorithms we start looking for approximation algorithms, that are algorithms that – given an instance w as input – compute a solution $\tilde{w}$ for which the value $v(\tilde{w})$ should be as close to the optimum $\text{OPT}(w)$ as possible. We will distinguish approximation algorithms by that "distance" to the optimum and start by defining what we are talking about, exactly.

Definition 1.25. *The relative error ε is the relative distance of the value of a feasible solution $\tilde{x}$ for an instance x to the optimum $\text{OPT}(x)$ and is defined as*

$$\varepsilon := \frac{|v(\tilde{x}) - \text{OPT}(x)|}{\text{OPT}(x)}.$$

The relative error is always non-negative, it becomes zero when an optimal solution is found. Notice that for minimization the relative error can be any non-negative real number while for maximization it is never greater than one.

Naturally, we are seeking for algorithms that find solutions with small relative error for all instances.

Another measure that yields numbers of the same range for both, minimization and maximization problems is the following.

Definition 1.26. *The ratio r of a solution $\tilde{x}$ to an instance x is defined as*

$$r := \max \left\{ \frac{v(\tilde{x})}{\text{OPT}(x)}, \frac{\text{OPT}(x)}{v(\tilde{x})} \right\}.$$

By definition it is clear that the ratio of a solution is never smaller than 1 and the ratio of optimal solutions is 1. There are close relations between ratio r and relative error ε for a solution $\tilde{x}$ to an instance x. For minimization problems $\varepsilon = r - 1$ holds while for maximization problems we get $\varepsilon = 1 - \frac{1}{r}$. Now we associate these measures for single solutions to approximation algorithms since our aim is to compare different approximation algorithms. As usual in complexity theory we adopt a kind of worst case perspective.

Definition 1.27. *The* performance ratio r *of an approximation algorithm A is the infimum of all r' such that for all inputs $x \in I$ the approximation algorithm A yields solutions $\tilde{x} \in S(x)$ with ratio at most r'.*

We will call an approximation algorithm with performance ratio r an r-approximation (algorithm).

One way to classify approximation problems is by the type of approximation algorithms they allow. It does not matter whether we use the relative error or performance ratio as our measure for the algorithms; we concentrate on the performance ratio here.

Definition 1.28. *We say that an optimization problem is* approximable *if there exists an approximation algorithm with polynomial time bound and performance ratio r for some constant value r. The class of all optimization problems which are approximable is called $\mathcal{APX}$. The class of all optimization problems which have polynomially time bounded approximation algorithms with performance ratio $r(|w|)$ are called $\mathcal{F}$-$\mathcal{APX}$ if $r(|w|)$ is a function in the class $\mathcal{F}$. Explicitly, we denote as $\log \mathcal{APX}$ the class of optimization problems that have polynomially time bounded approximation algorithms with performance ratio $\log(|w|)$.*

Although a specific optimization problem may be approximable in the above sense the performance ratio that can be found may be much too large for being helpful in practice. So, we are especially interested in optimization problems which allow polynomial time approximation algorithms with performance ratio r for any $r > 1$. The infimum of all achievable performance ratios is 1 for these problems.

Definition 1.29. *The class $\mathcal{PTAS}$ consists of all optimization problems L such that for any constant $r > 1$ there exists an approximation algorithm for L with running time $\mathcal{O}(p(|w|))$ and performance ratio r where p is a polynomial in the length of the input $|w|$. We call such approximation algorithms* polynomial time approximation scheme *(PTAS).*

Having a polynomial time approximation scheme means that we are able to find a solution with ratio r for any fixed r in polynomial time. However, if we are interested in very good approximations, we should be sure that the running time does not depend on the chosen performance ratio r too heavily.

Definition 1.30. *The class $\mathcal{FPTAS}$ consists of all optimization problems L such that for any $r = 1 + \varepsilon > 1$ there exists an approximation algorithm for L with running time $\mathcal{O}(p(|w|, \varepsilon^{-1}))$ and performance ratio r where p is a polynomial in both the length of the input $|w|$ and ε^{-1}. We call such approximation algorithms* fully polynomial time approximation scheme *(FPTAS).*

The classes for optimization problems we defined up to here are related in an obvious way: it is $\mathcal{PO} \subseteq \mathcal{FPTAS} \subseteq \mathcal{PTAS} \subseteq \mathcal{APX} \subseteq \mathcal{NPO}$. If $\mathcal{P} = \mathcal{NP}$, the classes are all equal, of course. On the other hand, $\mathcal{P} \neq \mathcal{NP}$ implies that the classes are all different. We will name problems that allow the separation of the classes under the assumption $\mathcal{P} \neq \mathcal{NP}$ and give an idea of how a proof can be constructed. KNAPSACK belongs to $\mathcal{FPTAS}$ as Ibarra and Kim proved [IK75] while its underlying language is $\mathcal{NP}$-complete, so we get KNAPSACK $\notin \mathcal{PO}$. The underlying language of LARGEPACKING is $\mathcal{NP}$-complete in the strong sense (it remains $\mathcal{NP}$-complete when all numbers in the input are bounded in size by a polynomial in the length of the input), so LARGEPACKING $\in \mathcal{FPTAS}$ would imply $\mathcal{P} = \mathcal{NP}$[1]. On the other hand, LARGEPACKING has a polynomial approximation time scheme presented by Karmarkar and Karp [KK82]. In Chapter 2 we will see an $\frac{8}{7}$-approximation algorithm for MAX3SAT showing that it belongs to $\mathcal{APX}$ (see Theorem 2.3). In Chapter 7 we will see that there is no better approximation algorithm possible so $\mathcal{APX} \neq \mathcal{PTAS}$ follows. Finally, a polynomially time bounded approximation algorithm for the TRAVELINGSALESMANPROBLEM with any performance ratio r can be used to construct a polynomially time bounded algorithm that solves the $\mathcal{NP}$-complete HAMILTONCIRCUIT exactly.

In order to connect the randomized algorithms from Section 1.3 with the concept of approximation algorithms we introduce the notion of expected performance ratio.

Definition 1.31. *The* expected performance ratio r *of a randomized approximation algorithm A is the infimum of all r' such that for all instances x the algorithm A yields solutions $\tilde{x}$ such that*

---

[1] This connection between very good approximations and $\mathcal{NP}$-completeness of the underlying language in the strong sense is the topic of Exercise 1.3.

$$\max \left\{ \frac{\mathbf{E}(v(\tilde{x}))}{\text{OPT}(x)}, \frac{\text{OPT}(x)}{\mathbf{E}(v(\tilde{x}))} \right\} \leqslant r'$$

where $\mathbf{E}(v(\tilde{x}))$ denotes the expected value of $\tilde{x}$.

1.4.2 Reduction concepts for optimization problems

In order to identify "hard problems" we define reduction concepts which allow us to compare different optimization problems and to find complete problems. The difficulty with reductions for optimization problems is that it is not enough to map an instance for a problem to an instance for another problem. The approximability of the given instance and the instance it is mapped to have to be related close enough to guarantee the desired ratio of approximation. Furthermore, the solutions for the second problem must admit the efficient construction of a solution of similar ratio for the first problem.

It is quite obvious that two functions and not just one like for the polynomial reductions (see Definition 1.6) are needed to construct a meaningful reduction for approximation algorithms. But it is not at all obvious how the desired properties concerning the approximation ratio can be guaranteed; so it is not too surprising that different types of reductions for approximation were defined.

A reduction that reduces an optimization problem A to an optimization problem B consists of three different parts, a function f that maps inputs x for A to inputs $y := f(x)$ for B, a function g that maps solutions $\tilde{y}$ for the input y for B to solutions $\tilde{x} := g(\tilde{y})$ for A, and some conditions that guarantee some property that is related with approximation. Given an approximation algorithm for B one can construct an approximation algorithm for A in a systematic way. An input for A is transformed into an input for B using f, then a solution for $f(x)$ is computed using the approximation algorithm for B, finally g is used to transform this solution into a valid solution for x in the domain of A. This interplay between f, g, and the approximation algorithm for B is sketched in Figure 1.2.

The idea of approximation preserving reductions is not new and was already stated in the early 80s by Paz and Moran [PM81]. The problem was that it turned out to be very difficult to prove the completeness of an optimization problem under the considered reduction type for a specific complexity class for optimization problems, e.g. for $\mathcal{APX}$ and only very artificial problems could be proven to be complete. A kind of shift happened at the beginning of the 90s when Papadimitriou and Yannakakis [PY91] defined L-reductions. The definition of this new type of reduction they gave was tailored to fit together with a new complexity class for optimization problem with a definition that does not rely on the level of approximation the problems allow (what can be said to be a computational point of view) but on a logical definition of $\mathcal{NP}$ due to Fagin

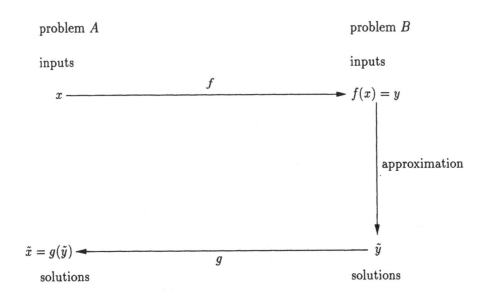

Fig. 1.2. Interplay of an (f, g)-reduction from A to B and an approximation algorithm for B.

[Fag74]. The class is called MAX-SNP and is defined in two steps. Our description here relies on the book by Papadimitriou [Pap94]. In the original definition by Papadimitriou and Yannakakis [PY91] MAX-SNP contained only maximization problems though the authors mentioned that minimization problems can be "placed" in MAX-SNP by reductions to maximization problems from MAX-SNP. Here, we begin by defining the class MAX-SNP$_0$ that can be described as search for a structure S such that a maximal number of tuples $x = (x_1, \ldots, x_k)$ fulfill a term $\varphi(S, x)$ such that φ is a quantifier-free term of first-order logic. The class MAX-SNP itself consists of all problems from $\mathcal{NPO}$ that can be L-reduced to a problem in MAX-SNP$_0$. Papadimitriou and Yannakakis proved many natural problems to belong to the class MAX-SNP, so the definition of this complexity class is not too restrictive. Using L-reductions they were able to prove that MAX3SAT is complete for MAX-SNP. The relation between MAX-SNP and computational based classes was not known for quite a time but it was well known that if any MAX-SNP-complete problem has a polynomial approximation time scheme then all MAX-SNP-problems have. So, a problem could be called hard to approximate if it was proven to be MAX-SNP-complete. In fact, many problems were shown to be MAX-SNP-complete in the following years and the notion of L-reduction became a widely accepted standard in the field. However, the use of MAX-SNP instead of $\mathcal{APX}$ was caused by the difficulties in proving natural problems to be $\mathcal{APX}$-complete. Using another type of reduction, namely AP-reductions, allows to prove MAX3SAT to be $\mathcal{APX}$-complete, so nowadays there is no need to formally define MAX-SNP and work with L-

reductions. As a consequence, we will continue by formally defining the so called
AP-reductions that were originally defined by Crescenzi, Kann, Silvestri, and
Trevisan [CKST95]. AP-reductions are not the only alternative to L-reductions,
quite a lot of different reduction concepts were introduced that all have some fa-
vors and drawbacks. An overview and comparison of different reduction concepts
for optimization problems can be found in [CKST95]. However, AP-reductions
are approximation preserving and allow the identification of complete problems
for $\mathcal{APX}$, a fact that will be important in some of the following chapters.

Definition 1.32. *Let $A := (I_A, S_A, v_A, \text{goal}_A)$ and $B := (I_B, S_B, v_B, \text{goal}_B)$
be two optimization problems from $\mathcal{NPO}$. We say that A is AP-reducible to B
($A \leqslant_{\text{AP}} B$) if there exist two functions f, g and a constant $\alpha > 0$ such that the
following conditions hold:*

1. $\forall x \in I_A, r > 1 : f(x, r) \in I_B$.

2. $\forall x \in I_A, r > 1 : S_A(x) \neq \emptyset \Rightarrow S_B(f(x, r)) \neq \emptyset$.

3. $\forall x \in I_A, r > 1, \tilde{y} \in S_B(f(x, r)) : g(x, \tilde{y}, r) \in S_A(x)$.

*4. The functions f and g can be computed in polynomial time for any fixed
$r > 1$.*

*5. $\forall x \in I_A, r > 1, \tilde{y} \in S_B(f(x, r)) : \left(\max \left\{ \frac{v_B(\tilde{y})}{\text{OPT}(f(x,r))}, \frac{\text{OPT}(f(x,r))}{v_B(\tilde{y})} \right\} \leqslant r \right) \Rightarrow$
$\left(\max \left\{ \frac{v_A(g(x,\tilde{y},r))}{\text{OPT}(x)}, \frac{\text{OPT}(x)}{v_A(g(x,\tilde{y},r))} \right\} \leqslant 1 + \alpha(r-1) \right)$.*

*The pair (f, g) is called an AP-reduction. If we specify a constant α we can call
the triple (f, g, α) an α-AP-reduction and write $A \leqslant_{\text{AP}}^{\alpha} B$ if A can be α-AP-
reduced to B.*

The conditions 1–4 formalize our understanding of a reduction for optimization
problems; the interplay of the functions f and g is stated, the second condition
ensures that no "trivial reductions" are allowed. Without this condition it would
be possible to always map an input for A to an input without solution for B so
the third and fifth conditions come never into play making the reduction mean-
ingless. The fifth condition alone formalizes our understanding how a reduction
should connect the approximability of the two problems A and B. Refining Figure 1.2 for AP-reductions leads to a visualization as shown in Figure 1.3.

We will now follow Trevisan [Tre97] and show how AP-reductions can be used
to prove MAX3SAT to be complete for $\mathcal{APX}$.

Theorem 1.33. *[Tre97] MAX3SAT is $\mathcal{APX}$-complete.*

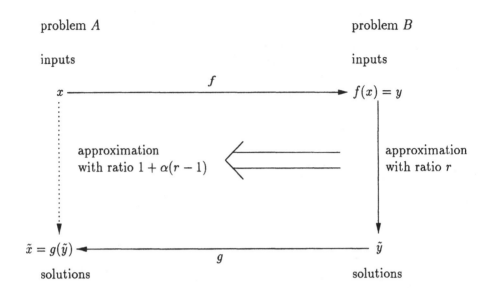

Fig. 1.3. Interplay of an α-AP-reduction from A to B and an r-approximation algorithm for B.

Proof.   Khanna, Motwani, Sudan, and Vazirani [KMSV94] proved that for any $\mathcal{APX}$-minimization problem A there exists an $\mathcal{APX}$-maximization problem B such that A can be reduced to B by a reduction for optimization problems that is called E-reduction. We adopt the method of this proof to show the same for AP-reductions.

Let $A := (I_A, S_A, v_A, \min)$ be a minimization problem from $\mathcal{APX}$. Then there is an approximation algorithm M_A for A that yields a solution $\tilde{x}$ for an instance x such that $\frac{v_A(\tilde{x})}{\mathrm{OPT}_A(x)} \leqslant c$ holds for some constant c and any input x. We define $B := (I_A, S_A, v_B, \max)$ as follows. Let $x \in I_A$ be an input for B, let $\tilde{x}$ be the solution that M_A yields for x and let $\tilde{y} \in S_A$ be any solution for x. Then we define $v_B(\tilde{y}) := \max\{\lceil (c+1)v_A(\tilde{x}) - cv_A(\tilde{y})\rceil, v_A(\tilde{x})\}$. We claim that B belongs to $\mathcal{APX}$. To proof this we show that M_A is an approximation algorithm for B with constant performance ratio. Obviously, $v_B(\tilde{x}) = v_A(\tilde{x})$ holds so we have $\mathrm{OPT}_B(x) = \lceil (c+1)v_A(\tilde{x}) - c\,\mathrm{OPT}_A(x)\rceil$. So, we can conclude for the performance ratio of M_A as approximation algorithm for B

$$\frac{\mathrm{OPT}_B(x)}{v_B(\tilde{x})} = \frac{\lceil (c+1)v_A(\tilde{x}) - c\,\mathrm{OPT}_A(x)\rceil}{v_A(\tilde{x})} \leqslant \frac{\lceil cv_A(\tilde{x})\rceil}{v_A(\tilde{x})} \leqslant c+1.$$

We claim that with $f(x, r) := x$ and

$$g(x, \tilde{y}, r) := \begin{cases} \tilde{y} & \text{if } v_A(\tilde{y}) < v_A(\tilde{x}) \\ \tilde{x} & \text{if } v_A(\tilde{y}) \geqslant v_A(\tilde{x}) \end{cases}$$

we define an AP-reduction (f, g) from A to B. Without loss of generality we may assume that c is an integer. Assume $\text{OPT}_B(x) \leqslant r v_B(\tilde{y})$ holds. By the considerations above we can deduce that $(c + 1) v_A(\tilde{x}) - c \, \text{OPT}_A(x) \leqslant r v_B(\tilde{y})$ holds. We distinguish two cases. First, assume that $v_B(\tilde{y}) = v_A(\tilde{x})$ holds. We then have $(c + 1 - r) v_A(\tilde{x}) \leqslant c \, \text{OPT}_A(x)$, so

$$\frac{v_A(g(x, \tilde{y}, r))}{\text{OPT}_A(x)} = \frac{v_A(\tilde{x})}{\text{OPT}_A(x)} \leqslant 1 + (r - 1) \frac{1}{c + 1 - r}$$

follows.

Now, assume that $v_B(\tilde{y}) = (c + 1) v_A(\tilde{x}) - c v_A(\tilde{y})$ holds. Then,

$$r c v_A(\tilde{y}) \leqslant c \, \text{OPT}_A(x) + (r - 1)(c + 1) v_A(\tilde{x})$$

follows and we have

$$\frac{v_A(\tilde{y})}{\text{OPT}_A(x)} \leqslant 1 + (r - 1)(c + 1).$$

So, we indeed defined a $(c + 1)$-AP-reduction from A to B.

From the transitivity of AP-reductions if follows that it is sufficient to show that any $\mathcal{APX}$-maximization problem A can be AP-reduced to MAX3SAT.

Since A belongs to $\mathcal{APX}$ we may assume that M_A is an approximation algorithm for A with performance ratio c.

To complete the proof we need two results as tools that will both be used but not proven in this section. The first tool follows directly from Cook's theorem [Coo71]. Given an $\mathcal{NP}$-problem Π for any input w to Π we can compute in polynomial time an instance Φ for SAT such Φ is satisfiable if and only if w is accepted by a Turing machine for Π. As we discussed in Section 1.2 (page 10) an $\mathcal{NP}$-decision problem can be decided by a nondeterministic Turing machine that guesses a proof for the membership of the input in the language and verifies it. This "proof" can be viewed as a "solution" for the decision problem Π; we will clarify this statement when we make use of this tool.

The second tool relies on a statement that can be proven to be equivalent to the PCP-Theorem (see Chapter 5). It uses a special version of SAT that is called ROB3SAT and has the following properties. For some constant $\varepsilon > 0$ either there is an assignment that satisfies all m clauses of an instance or for all assignments there are at least εm clauses that are not satisfied. The statement we need here is the following and is discussed in Chapter 4. There is a constant $\varepsilon > 0$ such that given an instance Φ for SAT it is possible to compute in polynomial time an instance Φ' for ε-ROB3SAT such that Φ' can be satisfied if and only if Φ can. Furthermore, given an assignment that satisfies more than $(1 - \varepsilon) m$ clauses of Φ' we can compute in polynomial time an assignment that satisfies Φ.

So, we now can assume that $A = (I_A, S_A, v_A, \max)$ is an $\mathcal{APX}$-maximization problem, M_A is an approximation algorithm for A with performance ratio c, and ε is the constant for ROB3SAT. We have to define f, g, and α so that (f, g, α) is an AP-reduction from A to MAX3SAT; we write B for MAX3SAT in the following. We have to show that $\frac{\text{OPT}_B(x)}{v_B(\tilde{y})} \leqslant r$ implies $\frac{\text{OPT}_A(x)}{v_A(\tilde{y})} \leqslant 1 + \alpha(r - 1)$ for a constant α that we will specify later. We distinguish two cases. First, assume that $1 + \alpha(r - 1) \geqslant c$ holds; in this case the approximation given by M_A is already good enough. So, we see it is sufficient to use the solution $\tilde{x}$ that M_A yields for the input x to fulfill the property of an AP-reduction and we define $g(x, \tilde{y}, r) := \tilde{x}$ in this case.

Let now $d := 1 + \alpha(r - 1) < c$ hold. Since we have the approximation algorithm M_A with performance ratio c that yields a solution $\tilde{x}$ for an input x we know that the value of an optimal solution $\text{OPT}_A(x)$ is in the interval $[v_A(\tilde{x}), cv_A(\tilde{x})]$. We divide this interval in $k := \lceil \log_d c \rceil$ subintervals with lower bounds $d^i v_A(\tilde{x})$ and upper bounds $d^{i+1} v_A(\tilde{x})$ for $0 \leqslant i < k$. Since $A \in \mathcal{APX} \subseteq \mathcal{NPO}$ the problem of deciding whether the value of an optimal solution is at least q belongs to $\mathcal{NP}$ for all constants q. We now have k $\mathcal{NP}$-decision problems such that the "membership proof" we mentioned in the discussion of Cook's theorem is a solution for our optimization problem A with value at least $d^i v_A(\tilde{x})$. Using the second tool mentioned above we can compute in polynomial time an instance for ε-ROB3SAT φ with m clauses that is equivalent to the i-th $\mathcal{NP}$-decision problem for all $0 \leqslant i < k$. Given an assignment τ_i that satisfies more than $(1 - \varepsilon)m$ clauses we can compute in polynomial time a satisfying assignment for φ_i and use this to compute a solution $s \in S_A(x)$ to the i-th $\mathcal{NP}$-decision problem with $v_A(s) \geqslant d^i v_A(\tilde{x})$.

We assume without loss of generality that each instance φ_i has the same number m of clauses; using an appropriate number of copies of φ_i suffices to achieve this. With i_0 we denote the maximum index i such that φ_i is satisfiable; we have $d^{i_0} v_A(\tilde{x}) \leqslant \text{OPT}_A(x) < d^{i_0+1} v_A(\tilde{x})$ then. We define $\Psi := \bigwedge_{i=0}^{k-1} \varphi_i$ as the instance for MAX3SAT. Assume that τ is an assignment for Ψ such that $\text{OPT}_B(\Psi) \leqslant rv_B(\tau)$ holds. We can use τ as an assignment of φ_i for all i by ignoring all assignments for variables that do not occur in φ_i. If τ satisfies more than $(1 - \varepsilon)m$ clauses in φ_i for some i then φ_i is satisfiable. We know that we can compute in polynomial time a satisfying assignment for φ_i; from Cook's theorem we know that this assignment can be used to compute deterministically in polynomial time a membership proof for the input to Π, in this case this proof is a solution for the problem of finding a solution to our optimization problem A with value at least $d^i v_A(\tilde{x})$. Given the assignment τ we are able to compute such a solution deterministically in polynomial time. It follows, that if we can show that τ satisfies at least $(1 - \varepsilon)m$ clauses in φ_{i_0} we can compute in polynomial time a solution with value at least $d^{i_0} v_A(\tilde{x})$ that has ratio at most $d = 1 + \alpha(r - 1)$. So, the defined reduction would indeed be an AP-reduction and the proof would be complete.

We are still free to define the constant α and will use this freedom to show that τ satisfies more than $(1 - \varepsilon)m$ clauses in φ_{i_0}. We know that $\text{OPT}_B(\Psi) \leqslant rv_B(\tau)$ holds so $\text{OPT}_B(\Psi) - v_B(\tau) \leqslant \frac{r-1}{r} \text{OPT}_B(\Psi) \leqslant \frac{r-1}{r} km$ follows. We call the number of clauses in φ_i that τ satisfies $(1 - \varepsilon_0)m$ and get $\text{OPT}_B(\Psi) - v_B(\tau) \geqslant$ $m - (1 - \varepsilon_0)m = \varepsilon_0 m$. Combining the two inequalities we can conclude that $\varepsilon_0 \leqslant \frac{r-1}{r}k$ holds. Obviously, it is sufficient to prove that $\varepsilon > k\frac{r-1}{r}$ holds for any $r > 1$. Recall that we defined k as $\lceil \log_d c \rceil$ with $d = 1 + \alpha(r - 1)$. So we want to prove

$$\varepsilon > \left\lceil \frac{\ln c}{\ln(1 + \alpha(r - 1))} \right\rceil \cdot \frac{r - 1}{r},$$

so it is sufficient to have

$$\alpha > \frac{c^{(2r-2)/(\varepsilon r)} - 1}{r - 1}$$

which is possible for a constant α since the right hand side of the inequality is monotone decreasing for all sufficiently large r and converges to a constant for $r \to 1$. ∎

As mentioned above, the class MAX-SNP and along with it L-reductions were the standard topics in the area of approximation. This implies that it is desirable to show the relations between the "syntactical class" MAX-SNP and the "computational class" $\mathcal{APX}$. But from the definition of L-reductions it took several years before the relationship between MAX-SNP and $\mathcal{APX}$ was discovered. The key idea is to find a way that allows the class MAX-SNP to be brought together with the reduction concepts defined along with $\mathcal{APX}$. This can be done by considering the closure of MAX-SNP. The closure of a class $\mathcal{C}$ under a certain type of reduction is the set of all problems $L \in \mathcal{NPO}$ such that L can be reduced to some problem in $\mathcal{C}$. By Khanna, Motwani, Sudan, and Vazirani [KMSV94] it was shown that using E-reductions the closure of MAX-SNP equals the subclass of $\mathcal{APX}$ which contains all problems from $\mathcal{APX}$ that have only solutions with values that are polynomially bounded in the length of the input. Using AP-reductions it is possible to show that the closure of MAX-SNP equals $\mathcal{APX}$ so finally the more artificial class MAX-SNP and the more natural class $\mathcal{APX}$ were unified.

Exercises

Exercise 1.1. Proof that $\mathcal{NP} \subseteq \mathcal{BPP}$ implies that $\mathcal{NP}$ equals $\mathcal{RP}$.

Exercise 1.2. Place $\mathcal{BPP}$ in the polynomial hierarchy $\mathcal{PH}$, i.e. try to find the smallest i such that $\mathcal{BPP}$ can be proven to belong to the i-th level of $\mathcal{PH}$.

Exercise 1.3. Assume that L is an $\mathcal{NPO}$-optimization problem with the following properties. For any input x the values of all solutions from $S(x)$ are positive integers, the optimal value $\text{OPT}(x)$ is bounded above by a polynomial in both the length of x and the maximal number appearing in x, and the underlying language of L is $\mathcal{NP}$-hard in the strong sense: it remains $\mathcal{NP}$-hard even if all numbers in the input are bounded above by a polynomial p in the length of the input. Show that under the assumption $\mathcal{P} \neq \mathcal{NP}$ the problem L can not belong to $\mathcal{FPTAS}$.

2. Introduction to Randomized Algorithms

Artur Andrzejak

2.1 Introduction

A *randomized algorithm* is a randomized Turing machine (see Definition 1.10). For our purposes we can consider such an algorithm as a "black box" which receives input data and a stream of random bits for the purpose of making random choices. It performs some computations depending on the input and the random bits and stops. Even for a fixed input, different runs of a randomized algorithm may give different results. For this reason, a description of the properties of a randomized algorithm will involve probabilistic statements. For example, its running time is a random variable.

Randomized algorithms may be faster, more space-efficient and/or simpler than their deterministic counterparts. But, "in almost all cases it is not at all clear whether randomization is really necessary. As far as we know, it may be possible to convert *any* randomized algorithm to a deterministic one without paying any penalty in time, space, or other resources" [Nis96].

So why does randomization make algorithms in many cases simpler and faster? A partial answer to this question gives us the following list of some paradigms for randomized algorithms, taken from [Kar91, MR95b].

Foiling the adversary. In a game theoretic view, the computational complexity of a problem is the value of the following two-person game. The first player chooses the algorithm and as answer the second player, called the *adversary*, chooses the input data to foil this particular algorithm. A randomized algorithm can be viewed as a probability distribution on a set of deterministic algorithms. While the adversary may be able to construct an input that foils one (or a small fraction) of the deterministic algorithms in the set, it is difficult to devise a single input that would foil a *randomly* chosen algorithm.

Abundance of witnesses. A randomized algorithm may be required to decide whether the input data has a certain property; for example, 'is a boolean formula satisfiable?'. Often, it does so by finding an object with the desired property. Such an object is called a *witness*. It may be hard to find a witness deterministically if it lies in a search space too large to be searched exhaustively. But if it can be shown that the search space contains a large number of witnesses, then it often suffices to choose an element at random from the space. If the input possesses the required property, then a witness is likely to be found within a few trials. On the other hand, the failure to find a witness in a large number of trials gives strong evidence (but not absolute proof) that the input does not have the required property.

Random sampling. Here the idea is used that a random sample from a population is often representative for the whole population. This paradigm is frequently used in selection algorithms, geometric algorithms, graph algorithms and approximate counting.

2.2 Las Vegas and Monte Carlo Algorithms

The popular (but imprecise) understanding of Las Vegas and Monte Carlo algorithms is the following one. A Las Vegas algorithm is a randomized algorithm which always gives a correct solution but does not always run fast. A Monte Carlo algorithm is a randomized algorithm which runs fast but sometimes gives a wrong solution. Depending on the source, the definitions vary. We allow here that an algorithm stops without giving an answer at all, as defined in [Weg93]. Below we state the exact definitions.

A *Las Vegas algorithm* is an algorithm which either stops and gives a correct solution or stops and does not answer. For each run the running time may be different, even on the same input. The question of interest is the probability distribution of the running time. We say that a Las Vegas algorithm is an *efficient Las Vegas algorithm* if on every input its worst-case running time is bounded by a polynomial function of the input size.

A *Monte Carlo algorithm* is an algorithm that either stops and does not answer or stops and gives a solution which is possibly not correct. We require that it is possible to bound the probability of the event that the algorithm errs. For decision problems there are two kinds of Monte Carlo algorithms. A Monte Carlo algorithm has *two-sided error* if it may err when it outputs either YES or NO. It has a *one-sided error* if it never errs when it outputs NO or if it never errs when it outputs YES. Note that a Monte Carlo algorithm with one-sided error is a special case of a Monte Carlo algorithm with two-sided error.

The running time of a Monte Carlo algorithm is usually deterministic (but sometimes also algorithms in which both the running time and the correctness of the output are random variables are called Monte Carlo algorithms). We say that a Monte Carlo algorithm is an *efficient Monte Carlo algorithm* if on every input its worst-case running time is bounded by a polynomial function of the input size. The nice property of a Monte Carlo algorithm is that running it repeatedly makes the failure probability arbitrarily low (at the expense of running time), see Exercises 2.1 and 2.2.

Note that a Las Vegas algorithm is a Monte Carlo algorithm with error probability 0. Every Monte Carlo algorithm can be turned into a Las Vegas algorithm, see Exercise 2.3. The efficiency of such Las Vegas algorithm depends heavily on the time needed to verify the correctness of a solution to a problem. Thus, for decision problems, a Las Vegas algorithm derived from a Monte Carlo algorithm is in general useless.

We may apply these definitions to the complexity classes introduced in the previous chapter:

- The problems in the class $\mathcal{ZPP}$ have efficient Las Vegas decision algorithms.

- The problems in the class $\mathcal{RP}$ have efficient Monte Carlo decision algorithms with one-sided error (the same applies to the class co-$\mathcal{RP}$).

- The problems in the class $\mathcal{PP}$ have efficient Monte Carlo decision algorithms with two-sided error.

In the remainder of this section we show that the class $\mathcal{ZPP}$ can be defined via the following class $\mathcal{C}$ of algorithms. The class $\mathcal{C}$ contains (randomized) decision algorithms which always output a correct answer YES or NO and have expected running time bounded by a polynomial in the size of the input. The difference between an algorithm in the class $\mathcal{C}$ and an efficient Las Vegas decision algorithm is that the earlier cannot output "don't know", but its expected running time is bounded instead of the worst-case running time.

Lemma 2.1. (i) *If L is a language in $\mathcal{ZPP}$, then there is a decision algorithm $A \in \mathcal{C}$ for L.*

(ii) *If a language L has a decision algorithm A in $\mathcal{C}$, then $L \in \mathcal{ZPP}$.*

Proof. To (i): Let A_1 be an efficient Las Vegas decision algorithm for L with worst-case running time bounded by a polynomial $p(n)$. We construct A as follows. We run A_1 at most $k_0 = p(n)$ times until it gives us an answer YES or NO. If after k_0 runs we still have no definitive answer, we apply the following deterministic algorithm. We may regard A_1 as a deterministic algorithm, once the random string has been fixed. Now we run such a deterministic version of A_1 for every possible random string which A_1 can use, i.e. for each string (of bits) of length $p(n)$. This will give us an answer YES or NO since A_1 outputs "don't know" for at most the half of the random strings by the definition of $\mathcal{ZPP}$.

Let X_n be the running time of A. To bound $\mathbf{E}[X_n]$ observe that for an integer k the probability that there is no definitive answer after running A_1 for k times is at most $(\frac{1}{2})^k$. For each of the first k_0 runs we need time at most $p(n)$. If all these runs fail, our deterministic algorithm needs time at most $2^{p(n)}p(n)$. Thus, the expected running time $\mathbf{E}[X_n]$ of A is at most

$$p(n) + \sum_{k=1}^{k_0-1} \left(\frac{1}{2}\right)^k p(n) + \left(\frac{1}{2}\right)^{p(n)} 2^{p(n)}.$$

But

$$\sum_{k=1}^{k_0-1} \left(\frac{1}{2}\right)^k < 1$$

and so $\mathbf{E}[X_n] \leqslant 3p(n)$.

To (ii): Assume that the expected running time of A is bounded by a polynomial $p(n)$, where n is the size of the input. First we show that there is an efficient Las Vegas decision algorithm A_1 for L with worst-case running time bounded by $2p(n)$. The algorithm is constructed in the following manner. We start A and allow it to run at most $2p(n)$ steps. If after this time there is no answer of A, we stop A and A_1 outputs "don't know". Clearly A_1 is an efficient Las Vegas decision algorithm.

If X_n is the running time of A, then by Markov's inequality we have

$$\mathrm{Prob}[X_n \geqslant 2p(n)] \leqslant \frac{1}{2}.$$

Therefore, the probability that A_1 outputs "don't know" is at most $\frac{1}{2}$ and so $L \in \mathcal{ZPP}$.   ■

Corollary 2.2. *The class $\mathcal{ZPP}$ is the class of all languages which have an algorithm in $\mathcal{C}$.*

In [MR95b] the above statement is used as a definition for $\mathcal{ZPP}$.

Furthermore, each efficient Las Vegas decision algorithm for a language in $\mathcal{ZPP}$ can be simulated by an algorithm in $\mathcal{C}$ and vice versa. This can be used to yet another definition of an efficient Las Vegas decision algorithm, as done by of Motwani and Raghavan [MR95b].

2.3 Examples of Randomized Algorithms

A good way to learn the principles of randomized algorithms is to study the examples. In this section we learn two randomized approximation algorithms, for the problems MAXEkSAT and MAXLINEQ3-2. Both of them will be derandomized in Chapter 3. The obtained deterministic approximation algorithms are important as their respective performance ratios are best possible, unless $\mathcal{P} \neq \mathcal{NP}$. This in turn will be shown in Chapter 7.

Recall that by Definition 1.31 the infimum of $\mathrm{OPT}(x)/\mathbf{E}[v(\tilde{x})]$ over all problem instances is the expected performance ratio of a randomized approximation algorithm for some maximization problem. Here $\mathbf{E}[v(\tilde{x})]$ is the expected value of the objective function. (To be consistent with the definitions in [MR95b, MNR96], the value r of the expected performance ratio for maximization problems has to be interchanged with $1/r$.)

2.3.1 A Randomized Approximation Algorithm for the MAXEkSAT-Problem

We will present a simple randomized algorithm for the MAXEkSAT-problem with expected performance ratio $1/(1-2^{-k})$. Especially, for $k = 3$, the expected performance ratio is $8/7$.

Let us state the problem:

MAXEkSAT

Instance: A boolean formula Φ in conjunctive normalform with exactly k different literals in each clause.
Problem: Find an assignment for Φ which satisfies the maximum number of clauses.

Let Φ be a boolean formula (instance of the problem MAXEkSAT) with m clauses over n variables. Assume that no clause contains both a literal and its complement, since then it is always satisfied. We can use then the following randomized algorithm:

RANDOM EkSAT ASSIGNMENT
Set each of the n variables of Φ independently to TRUE or FALSE with probability $1/2$.

Theorem 2.3. *The algorithm* RANDOM EkSAT ASSIGNMENT *has expected performance ratio* $1/(1-2^{-k})$.

Proof. A clause of an input formula is *not* satisfied only if each complemented variable is set to TRUE and each uncomplemented variable is set to FALSE. Since each variable is set independently, the probability of this event is $1-2^{-k}$. By the linearity of expectation the expected number of satisfied clauses is $m(1-2^{-k})$, and so $\mathrm{OPT}(\Phi)/\mathbf{E}[v(\Phi)] \leqslant 1/(1-2^{-k})$. On the other hand, for each m there is an input Φ such that $\mathrm{OPT}(\Phi)/\mathbf{E}[v(\Phi)] = 1/(1-2^{-k})$, namely a satisfiable boolean formula with m clauses. It follows that the expected performance ratio of the algorithm is $1/(1-2^{-k})$. ∎

2.3.2 A Randomized Approximation Algorithm for the MAXLINEQ3-2-Problem

The problem to be solved is the following one:

MAXLINEQ3-2

Instance: A set of linear equations modulo 2 with exactly three different variables in each equation.

Problem: Find an assignment of values for the variables which maximize the number of satisfied equations.

The following randomized algorithm has expected performance ratio 2.

RANDOM MAXLINEQ3-2 ASSIGNMENT

Given a set of m MAXLINEQ3-2-equations on n variables, set each of the n variables independently to 0 or 1 with probability $1/2$.

Theorem 2.4. *The algorithm* RANDOM MAXLINEQ3-2 ASSIGNMENT *has expected performance ratio 2.*

Proof. Whether an equation is satisfied or not depends only on the sum of the variables with coefficients different from 0. This sum is 0 or 1 modulo 2 with probability $1/2$ each since all variables are set to 0 or 1 independently. Thus, the probability that an equation is satisfied is $1/2$ and so the expected number of satisfied equations is $m/2$. In addition, for every m there is a system of equations such that all equations can be satisfied. It follows that the expected performance ratio of the considered randomized algorithm is 2.   ∎

2.4 Randomized Rounding of Linear Programs

In this section we explain a technique of solving optimization problems by randomized rounding of linear programs. The technique can be applied for solving a variety of optimization problems, for example the integer multicommodity flow problem, the set cover problem or the computation of approximate shortest paths in a graph. In this section we illustrate the technique by obtaining an optimization algorithm for the MAXSAT problem. A detailed treatment of some of these applications and a list of many others can be found in the paper of Motwani, Naor and Raghavan [MNR96].

The technique works in four steps. First, given an optimization problem, we formulate it as an integer linear program with variables assuming the values in $\{0, 1\}$, if possible. Thereafter the integer program is relaxed by allowing the variables to take the values in $[0, 1]$. (This is necessary as no good algorithms are known for solving large zero-one linear programs). In the next step the relaxed linear program is solved by one of the common methods mentioned below. Finally, *randomized rounding* is used as follows to obtain a zero-one solution vector $\bar{x}$ for the integer linear program. Let $\hat{x}$ denote the solution vector of the relaxed linear program. Then, for each i, we set $\bar{x}_i$ to 1 with probability $\hat{x}_i \in [0, 1]$. Note that the last step can be also interpreted in another way. For each i, we pick a random number y uniformly in $[0, 1]$. If $\hat{x}_i > y$, we set $\bar{x}_i$ to 1, otherwise $\bar{x}_i$ is set to 0. This view has advantages for implementing the algorithm, as now we only need to compare $\hat{x}_i$ to a randomly chosen threshold.

The advantage of randomized rounding is that the zero-one solution vector $\bar{x}$ does not violate "too much" the constraints of the linear program, which is explained in the following. Let a be a row of the coefficient matrix of the linear program. Then, by linearity of expectation we have $\mathbf{E}[a\bar{x}] = a\hat{x}$. Therefore, the expected value of the left-hand side of every constraint will not violate the bound given by the right-hand side of this constraint.

Several methods for solving rational (i.e., in our case relaxed) linear programs are known. One of the most widely used algorithms is the simplex method due to Dantzig [Dan63]. It exists in many versions. Although fast in practice, the most versions of the simplex method have exponential worst-case running time. On the other hand, both the ellipsoid method of Khachian [Kha79] and the method of Karmarkar [Kar84] have polynomially bounded running time, but in practice they cannot compete with the simplex method. The reader may find more informations about solving linear programs in [PS82].

We begin with the definition of the optimization problem.

MAXSAT

Instance: A boolean formula Φ in conjunctive normalform.
Problem: Find an assignment for Φ which satisfies maximum number of clauses.

The randomized algorithm for MAXSAT presented below has some pleasant properties:

- it gives us an upper bound on the number of clauses which can be simultaneously satisfied in Φ,

- its expected performance ratio is $1/(1 - 1/e)$ (or even $1/(1 - (1 - 1/k)^k)$ for MAXEkSAT), as we will see below,

- if the the relaxed linear program is solved by the simplex method, then the above algorithm is efficient in practice,

- once the solution of the relaxed linear program is found, we can repeat the randomized rounding and take the best solution. In this way we can improve the solution without great penalty on running time.

RANDOMIZED ROUNDING FOR MAXSAT

1. For an instance Φ of MAXSAT formulate an integer linear program L_Φ. The integer program L_Φ has following variables. For each variable x_i, $i \in \{1, \ldots, n\}$, of Φ let y_i be an indicator variable which is 1 when x_i assumes the value TRUE and 0 otherwise. Also, for each clause C_j, $j \in 1, \ldots, m$ of Φ we introduce an indicator variable z_j. z_j has the value 1 if C_j is satisfied, otherwise 0. For each clause C_j let C_j^+ be the set of indices of variables that appear uncomplemented in C_j and let C_j^- be the set of indices of variables that appear complemented in C_j. Then MAXSAT can be formulated as follows:

$$\text{maximize} \quad \sum_{j=1}^{m} z_j, \tag{2.1}$$

where $y_i, z_j \in \{0, 1\}$ for all $i \in \{1, \ldots, n\}$, $j \in \{1, \ldots, m\}$,

subject to $\displaystyle\sum_{i \in C_j^+} y_i + \sum_{i \in C_j^-} (1 - y_i) \geqslant z_j$ for all $j \in \{1, \ldots, m\}$.

Note that the right-hand side of the jth constraint for L_Φ may take the value 1 only if one of the uncomplemented variables in C_j takes the value TRUE or if one of the complemented variables in C_j takes the value FALSE. In other words, $z_j = 1$ if and only if C_j is satisfied.

2. Relax the program L_Φ, allowing the variables $y_1, \ldots, y_n$ and $z_1, \ldots, z_m$ to take the values in $[0, 1]$.

3. Solve this relaxed program. Denote the value of the variable y_i in the solution as $\hat{y}_i$ and the value of z_j in the solution as $\hat{z}_j$. Note that $\sum_{j=1}^{m} \hat{z}_j$ bounds from above the number of clauses which can be simultaneously satisfied in Φ, as the value of the objective function for the relaxed linear program bounds from above the value of the objective function for L_Φ.

4. In this step the actual randomized rounding takes place: we obtain the solution vector $\bar{y}$ for the program L_Φ by setting $\bar{y}_i$ independently to 1 with probability $\hat{y}_i$, for each $i \in \{1, \ldots, n\}$.

For the remainder of this section, we put $\beta_k = 1 - (1 - 1/k)^k$.

What is the expected number of clauses satisfied by the algorithm RANDOMIZED ROUNDING FOR MAXSAT? To calculate this value, we show in Lemma 2.6 that for a clause C_j with k literals, the probability that C_j is satisfied is at least $\beta_k \hat{z}_j$. Furthermore, for every $k \geqslant 1$ we have $1 - 1/e < \beta_k$. By linearity of expectation the expected number of satisfied clauses is at least $(1 - 1/e) \sum_{j=1}^{m} \hat{z}_j$. It follows that the expected performance ratio of the considered algorithm is at most $1/(1 - 1/e)$.

Remark 2.5. In the case of the MAXEkSAT problem, each clause C_j is satisfied with probability at least $\beta_k \hat{z}_j$. It follows that for MAXEkSAT the expected performance ratio is $1/\beta_k$.

Lemma 2.6. *The probability that a clause C_j with k literals is satisfied by the algorithm* RANDOMIZED ROUNDING *is at least $\beta_k \hat{z}_j$.*

Proof. Since we consider a clause C_j independently from other clauses, we may assume that it contains only uncomplemented variables and that it is of the form $x_1 \vee \ldots \vee x_k$. Accordingly, the jth constraint of the linear program is

$$y_1 + \ldots + y_k \geqslant z_j.$$

Only if all of the variables $y_1, \ldots, y_n$ are set to zero, the clause C_j remains unsatisfied. By Step 4 of the algorithm, this may occur with probability

$$\prod_{i=1}^{k}(1 - \hat{y}_i).$$

It remains to show that

$$1 - \prod_{i=1}^{k}(1 - \hat{y}_i) \geqslant \beta_k \hat{z}_j.$$

The left-hand side is minimized (under the condition of the equation) when each $\hat{y}_i$ assumes the value $\hat{z}_j/k$. Thus, to complete the proof we have to show that $1 - (1 - z/k)^k \geqslant \beta_k z$ for every positive integer k and each real number $z \in [0, 1]$.

Consider the functions $f(z) = 1 - (1 - z/k)^k$ and $g(z) = \beta_k z$. To show that $f(z) \geqslant g(z)$ for all $z \in [0, 1]$ notice that $f(z)$ is a concave function and $g(z)$ a linear one. Therefore we have to check the inequality $f(z) \geqslant g(z)$ only for the endpoints of the interval $[0, 1]$. Indeed, for $z = 0$ and $z = 1$ the inequality $f(z) \geqslant g(z)$ holds, and so we are done. ∎

Consequently, we obtain the following theorem:

Theorem 2.7. *For every instance of* MAXSAT *the expected number of clauses satisfied by the algorithm* RANDOMIZED ROUNDING FOR MAXSAT *is at least $(1 - 1/e)$ times the maximum number of clauses that can be satisfied. For the* MAXEkSAT-*problem, the factor $(1 - 1/e)$ can be interchanged with β_k.*

2.5 A Randomized Algorithm for MAXSAT with Expected Performance Ratio 4/3

The algorithm RANDOM EkSAT ASSIGNMENT is obviously works for the problem MAXSAT, too. Thus, together with the algorithm RANDOMIZED ROUNDING we

k	$1 - 2^{-k}$	β_k
1	0.5	1.0
2	0.75	0.75
3	0.875	0.704
4	0.938	0.684
5	0.969	0.672

Table 2.1. A comparison of probabilities of satisfying a clause C_j with k literals

have two algorithms for the MAXSAT-problem. Which one is better? Table 2.1 shows that depending on the length k of a clause C_j, one of the two algorithms may satisfy C_j with higher probability than the other algorithm.

It is sensible to combine both algorithms: for an instance of MAXSAT, we run each algorithm and choose the solution satisfying more clauses. We will show now that this procedure yields a randomized algorithm for MAXSAT with expected performance ratio 4/3. (In Chapter 11 an even better randomized algorithm for MAXSAT is presented with expected performance ratio 1.324. Some other algorithms for this problem are also discussed there.)

We use the same notations as in the previous subsection.

Theorem 2.8. *We obtain an approximation algorithm for* MAXSAT *with expected performance ratio 4/3 by taking the better solution of the algorithms* RANDOM MAXEkSAT ASSIGNMENT *and* RANDOMIZED ROUNDING FOR MAXSAT *for an instance of* MAXSAT.

Proof.   Let n_1 denote the expected number of clauses that are satisfied by the algorithm RANDOM MAXEkSAT ASSIGNMENT. Let n_2 denote the expected number of clauses satisfied by the algorithm RANDOMIZED ROUNDING FOR MAXSAT. It suffices to show that

$$\max\{n_1, n_2\} \geqslant \frac{3}{4} \sum_{j=1}^{m} \hat{z}_j.$$

To this aim we prove that

$$(n_1 + n_2)/2 \geqslant \frac{3}{4} \sum_{j=1}^{m} \hat{z}_j.$$

Put S^k to be the set of clauses of the problem instance with exactly k literals, for $k \geqslant 1$. By Lemma 2.6 we have

$$n_2 \geqslant \sum_{k \geqslant 1} \sum_{C_j \in S^k} \beta_k \hat{z}_j.$$

On the other hand, obviously

$$n_1 = \sum_{k \geqslant 1} \sum_{C_j \in S^k} (1 - 2^{-k}) \geqslant \sum_{k \geqslant 1} \sum_{C_j \in S^k} (1 - 2^{-k}) \hat{z}_j.$$

Thus,

$$(n_1 + n_2)/2 \geqslant \sum_{k \geqslant 1} \sum_{C_j \in S^k} \frac{(1 - 2^{-k} + \beta_k)}{2} \hat{z}_j.$$

It can be shown that $(1 - 2^{-k}) + \beta_k \geqslant 3/2$ for all k, so that we have

$$(n_1 + n_2)/2 \geqslant \frac{3}{4} \sum_{k \geqslant 1} \sum_{C_j \in S^k} \hat{z}_j = 3/4 \sum_{j=1}^{m} \hat{z}_j.$$

■

Exercises

Exercise 2.1. Let $0 < \varepsilon_2 < \varepsilon_1 < 1$. Consider a Monte Carlo algorithm A that gives the correct solution to a problem with probability at least $1 - \varepsilon_1$, regardless of the input. Assume that it is possible to verify a solution of the algorithm A efficiently. How can we derive an algorithm based on A which gives the correct solution with probability at least $1 - \varepsilon_2$, regardless of the input?

Exercise 2.2. Consider the same problem as in Exercise 2.1, but now $0 < \varepsilon_2 < \varepsilon_1 < 1/2$ and it is not possible to verify a solution of the Monte Carlo algorithm A efficiently (this is the case for decision problems).

Exercise 2.3. [MR95b] Let A be a Monte Carlo algorithm for a problem Π. Assume that for every problem instance of size n the algorithm A produces a correct solution with probability $\gamma(n)$ and has running time at most $T(n)$. Assume further that given a solution to Π, we can verify its correctness in time $t(n)$. Show how to derive a Las Vegas algorithm for Π that runs in expected time at most $(T(n) + t(n))/\gamma(n)$.

Exercise 2.4. (trivial) Which expected performance ratio does the algorithm RANDOM MAXLINEQ3-2 ASSIGNMENT have if the number of variables in an equation is $k \neq 3$? Which expected performance ratio do we obtain if the number of variables may be different in each equation?

3. Derandomization

Detlef Sieling[*]

3.1 Introduction

Randomization is a powerful concept in computer science. There are many problems for which randomized algorithms are more efficient than the best known deterministic algorithms. Sometimes it is easier to design randomized algorithms than deterministic algorithms. Randomization is also useful for the proof of the existence of objects with certain properties. It suffices to prove that a randomly chosen object has the desired properties with positive probability. But such a proof is not constructive.

Of course, it is much more desirable to have deterministic algorithms that obtain a required performance and a correct result in all cases rather than merely with high probability. Also proofs that construct the object with the desired properties are often more useful than proofs that merely imply the existence of such an object. Furthermore, it is often not clear whether randomization is necessary at all, i.e., there may be deterministic algorithms having the same performance as the known randomized algorithms or there may be constructive proofs. The subject of this chapter are methods that allow in some cases to transform randomized algorithms into deterministic algorithms. These deterministic algorithms work in a similar way as the given randomized algorithms. This transformation is called derandomization.

We consider two methods of derandomization.

- The first method is the method of conditional probabilities. For each random bit of a given randomized algorithm it is computed deterministically which choice of this bit is "the right one".

- The second method is the reduction of the size of the probability space to polynomial size. If this can be done, one may try all possible samples of the probability space in polynomial time (or in parallel by a polynomial number of processors). Out of this polynomial number of samples the right one can be chosen.

[*] The author was partially supported by the Deutsche Forschungsgemeinschaft as part of the Collaborative Research Center "Computational Intelligence" (SFB 531).

This chapter is organized as follows. We start with the method of conditional probabilities. First we describe the method in an abstract way. Then we apply this method to several problems. We start with the derandomization of the algorithms for MAXEkSAT, MAXSAT and MAXLINEQ3-2 given in Chapter 2. Then we derandomize the randomized rounding approach for approximating linear integer programs. In the second part of this chapter we present a method for reducing the size of the probability space. We apply this method to two examples. First we show how to derandomize the randomized algorithm for MAXE3SAT. Then we describe a randomized parallel algorithm for the problem of computing maximal independent sets. By derandomizing this algorithm we obtain an $\mathcal{NC}^2$ algorithm for this problem.

3.2 The Method of Conditional Probabilities

In this section we describe the method of conditional probabilities in a general way. Applications of this method to specific randomized algorithms can be found in the following sections. We follow the presentations of the method of conditional probabilities of Raghavan [Rag88] and Spencer [Spe94]. In this section we only consider some maximization problem. The adaptation to minimization problems is straightforward.

Let us consider some maximization problem Π for which a randomized polynomial time algorithm A is known. Let x denote the input for A. The random choices of the algorithm may be described by random bits $y_1, \ldots, y_n$ where y_i takes the value 1 with probability p_i and the value 0 with probability $1 - p_i$. We assume that the output of A is a legal solution for the maximization problem Π on input x. The value of this output is described by the random variable Z. Our goal is to construct a deterministic polynomial time algorithm which computes an output with value of at least $\mathbf{E}[Z]$.

If we fix the random bits $y_1, \ldots, y_n$ to constants $d_1, \ldots, d_n$, we may consider A as a deterministic algorithm. This algorithm performs the steps that are done by A if the random bits are equal to $d_1, \ldots, d_n$. The problem is to choose $d_1, \ldots, d_n$ deterministically.

We remark that there is always a choice for $d_1, \ldots, d_n$ so that the output of this deterministic algorithm is at least $\mathbf{E}[Z]$. This follows from a simple average argument. If for all choices $d_1, \ldots, d_n$ the output of A is smaller than some number C, then also the weighted mean of these outputs, i.e. $\mathbf{E}[Z]$, is smaller than C. Hence, it is impossible that the output of A is smaller than $\mathbf{E}[Z]$ for all $d_1, \ldots, d_n$.

The derandomized algorithm chooses the bits $d_1, \ldots, d_n$ sequentially in the order $d_1, \ldots, d_n$. The choice of d_{i+1} depends on $d_1, \ldots, d_i$ that are chosen before. We may consider a choice $d_1, \ldots, d_i$ of $y_1, \ldots, y_i$ as an intermediate state of the derandomization procedure. With each such state we associate a weight w. The weight $w(d_1, \ldots, d_i)$ is defined as the conditional expectation $\mathbf{E}[Z|y_1 = d_1, \ldots, y_i = d_i]$. In other words, we consider a modification of A where only $y_{i+1}, \ldots, y_n$ are chosen randomly with probabilities $p_{i+1}, \ldots, p_n$ and where the bits $y_1, \ldots, y_i$ have been fixed to $d_1, \ldots, d_i$. Then the weight $w(d_1, \ldots, d_i)$ is the expected output of this modified algorithm.

In particular, we have $w() = \mathbf{E}[Z]$. Furthermore, $w(d_1, \ldots, d_n) = \mathbf{E}[Z|y_1 = d_1, \ldots, y_n = d_n]$ is the output of A when all bits $y_1, \ldots, y_n$ are fixed, i.e. the output of a deterministic algorithm.

The derandomization by the method of conditional probabilities is based on the assumption that the weights $w(d_1, \ldots, d_i)$ can be computed in polynomial time. If this is not possible, one may try to estimate the weights. This method is called method of pessimistic estimators. An example of an application of this method is presented in Section 3.4.

Now we can give an outline of the derandomized algorithm.

for $i := 1$ **to** n **do**
    if $w(d_1, \ldots, d_{i-1}, 0) > w(d_1, \ldots, d_{i-1}, 1)$
        then $d_i := 0$
        else $d_i := 1$
Run the algorithm A where the random choices are replaced by $d_1, \ldots, d_n$ and output the result of A.

In order to prove that the output of A for this choice of $d_1, \ldots, d_n$ is at least $\mathbf{E}[Z]$ we consider the relation between $w(d_1, \ldots, d_i)$ and the weights $w(d_1, \ldots, d_i, 0)$ and $w(d_1, \ldots, d_i, 1)$. A simple manipulation yields

$$
\begin{aligned}
w(d_1, \ldots, d_i) &= \mathbf{E}[Z|y_1 = d_1, \ldots, y_i = d_i] \\
&= (1 - p_{i+1})\, \mathbf{E}[Z|y_1 = d_1, \ldots, y_i = d_i, y_{i+1} = 0] \\
&\quad + p_{i+1}\, \mathbf{E}[Z|y_1 = d_1, \ldots, y_i = d_i, y_{i+1} = 1] \\
&= (1 - p_{i+1})w(d_1, \ldots, d_i, 0) + p_{i+1}w(d_1, \ldots, d_i, 1). \quad (3.1)
\end{aligned}
$$

This equality should be also intuitively clear. It says that after choosing $y_1 = d_1, \ldots, y_i = d_i$ the following two procedures yield the same result:

1. We randomly choose $y_{i+1}, \ldots, y_n$ and compute the expectation of the output.

2. First we choose $y_{i+1} = 0$ and afterwards $y_{i+2}, \ldots, y_n$ randomly and compute the expectation of the output. Then we choose $y_{i+1} = 1$ and afterwards $y_{i+2}, \ldots, y_n$ randomly and compute the expectation of the output. Finally, we compute the weighted mean of both expectations.

From (3.1) we conclude

$$
w(d_1, \ldots, d_i) \leqslant \max\{w(d_1, \ldots, d_i, 0), w(d_1, \ldots, d_i, 1)\}, \quad (3.2)
$$

since the weighted mean of two numbers cannot be larger than their maximum. If $d_1, \ldots, d_n$ are the bits computed by the derandomized algorithm, we obtain

$$\mathbf{E}[Z] = w() \leqslant w(d_1) \leqslant w(d_1, d_2) \leqslant \ldots \leqslant w(d_1, \ldots, d_n).$$

Since $w(d_1, \ldots, d_n)$ is the value of the output of the derandomized algorithm, we have proved the following lemma.

Lemma 3.1. *If all weights $w(d_1, \ldots, d_i)$ can be computed in polynomial time, then the derandomized algorithm computes deterministically in polynomial time a solution of Π with a value of at least $\mathbf{E}[Z]$.*

3.3 Approximation Algorithms for MaxE*k*Sat, MaxSat and MaxLinEq3-2

We start with the problem MaxE*k*Sat. An instance of MaxE*k*Sat consists of the set $\{x_1, \ldots, x_n\}$ of variables and the set $C = \{C_1, \ldots, C_m\}$ of clauses. In the algorithm Random E*k*Sat Assignment in Chapter 2 each variable $x_1, \ldots, x_n$ is independently chosen to be 0 and 1 with probability 1/2 each. With each clause C_ℓ we associate an indicator variable I_ℓ that takes the value 1 iff the clause C_ℓ is satisfied. A clause consisting of k literals is satisfied with probability $(2^k - 1)/2^k$. Hence, $\mathbf{E}[I_\ell] = (2^k - 1)/2^k$ and the expected number of satisfied clauses is $\mathbf{E}[Z] = \mathbf{E}[I_1 + \ldots + I_m] = m \cdot (2^k - 1)/2^k$. Hence, the expected performance ratio of this algorithm is bounded by $2^k/(2^k - 1)$.

In order to derandomize this algorithm we choose an appropriate weight function and show that this weight function can be computed in polynomial time. Let Z be the random variable describing the number of satisfied clauses. As in the last section we choose

$$w(d_1, \ldots, d_i) := \mathbf{E}[Z | x_1 = d_1, \ldots, x_i = d_i].$$

We show how to compute the weight $w(d_1, \ldots, d_i)$ in polynomial time. In each clause we replace each occurrence of x_j, $j \in \{1, \ldots, i\}$, by the constant d_j and simplify the resulting clause. For each clause C_ℓ there are the following possibilities.

1. After replacing $x_1, \ldots, x_i$ and simplifying we have $C_\ell = 0$. Then $\mathbf{E}[I_\ell | x_1 = d_1, \ldots, x_i = d_i] = \text{Prob}[I_\ell = 1 | x_1 = d_1, \ldots, x_i = d_i] = 0$.

2. After replacing $x_1, \ldots, x_i$ and simplifying we have $C_\ell = 1$. Then $\mathbf{E}[I_\ell | x_1 = d_1, \ldots, x_i = d_i] = \text{Prob}[I_\ell = 1 | x_1 = d_1, \ldots, x_i = d_i] = 1$.

3. After replacing $x_1, \ldots, x_i$ and simplifying we obtain a clause containing h variables. Then $\mathbf{E}[I_\ell | x_1 = d_1, \ldots, x_i = d_i] = \text{Prob}[I_\ell = 1 | x_1 = d_1, \ldots, x_i = d_i] = 1 - 1/2^h$.

Hence, we can compute

$$w(d_1, \ldots, d_i) = \mathbf{E}[Z | x_1 = d_1, \ldots, x_i = d_i] = \sum_{\ell=1}^{m} \mathbf{E}[I_\ell | x_1 = d_1, \ldots, x_i = d_i].$$

It is clear that the replacements, the simplification of the clauses and the computation of the weights can be done in polynomial time. By combining the above definition of the weights and the algorithm from Section 3.2 we obtain a deterministic polynomial time approximation algorithm for MAXEkSAT with a performance ratio bounded by $2^k/(2^k - 1)$. In particular, we obtain a polynomial time approximation algorithm for MAXE3SAT with a performance ratio of at most 8/7. This algorithm was already presented by Johnson [Joh74].

Theorem 3.2. *There is a deterministic polynomial time approximation algorithm for MAXEkSAT with a performance ratio of at most $1 + 1/(2^k - 1)$.*

In Chapter 2 a randomized approximation algorithm for MAXSAT with an expected performance ratio of 4/3 is presented. It consists of two algorithms where the better output is chosen. The first algorithm is the algorithm RANDOM EkSAT ASSIGNMENT for which we already presented a derandomized version. The second algorithm is based on a randomized rounding approach. The instance is transformed into a linear integer program. By solving the linear relaxation of the integer program probabilities $p_1, \ldots, p_n$ are computed and for $j \in \{1, \ldots, n\}$, $x_j = 1$ is chosen with probability p_j. In Chapter 2 it is shown that the combination of both algorithms yields an expected performance ratio of at most 4/3.

Also the second algorithm can be derandomized in the same way. We choose the same weight function. For the computation of the weights we consider the three cases as above. The first two cases remain unchanged. Now consider some clause consisting of h literals after replacing $x_1, \ldots, x_i$ and simplifying. Without loss of generality these are h nonnegated literals $x_{j(1)}, \ldots, x_{j(h)}$. Then C_ℓ is satisfied with probability $1 - (1 - p_{j(1)}) \cdot \ldots \cdot (1 - p_{j(h)})$, i.e., we have $\mathbf{E}[I_\ell | x_1 = d_1, \ldots, x_i = d_i] = 1 - (1 - p_{j(1)}) \cdot \ldots \cdot (1 - p_{j(h)})$. We may combine the two derandomized algorithms and choose the better result. By the analysis of the randomized algorithm for MAXSAT the following theorem follows.

Theorem 3.3. *There is a deterministic polynomial time approximation algorithm for MAXSAT with a performance ratio of at most 4/3.*

The algorithm RANDOM MAXLINEQ3-2 ASSIGNMENT from Chapter 2 is derandomized in the same way. An instance of MAXLINEQ3-2 consists of the set $\{x_1, \ldots, x_n\}$ of variables and the set $C = \{C_1, \ldots, C_m\}$ of equations. As proved in Chapter 2 choosing the variables randomly with probability 1/2 for 0 and probability 1/2 for 1 yields a randomized approximation algorithm with an expected performance ratio of at most 2 for MAXLINEQ3-2.

Let I_ℓ be the indicator variable describing whether equation C_ℓ is satisfied. For the derandomization we choose the same weight function as for MAXE3SAT. In order to compute the weight $w(d_1, \ldots, d_i)$ we replace x_j, $j \in \{1, \ldots, i\}$, by d_j and simplify the equations. Now we may have the following cases.

1. After simplification equation C_ℓ does not contain any variable and it is not satisfied. Then $\mathbf{E}[I_\ell | x_1 = d_1, \ldots, x_i = d_i] = 0$.

2. After simplification equation C_ℓ does not contain any variable and it is satisfied. Then $\mathbf{E}[I_\ell | x_1 = d_1, \ldots, x_i = d_i] = 1$.

3. After simplification equation C_ℓ contains some variable. Then $\mathbf{E}[I_\ell | x_1 = d_1, \ldots, x_i = d_i] = 1/2$.

Again we obtain the weight $w(d_1, \ldots, d_i)$ by adding the conditional expectations of the indicator variables. Hence, derandomization yields a deterministic approximation algorithm for MAXLINEQ3-2.

Theorem 3.4. *There is a deterministic polynomial time approximation algorithm for* MAXLINEQ3-2 *with a performance ratio of at most 2.*

3.4 Approximation Algorithms for Linear Integer Programs

In this section we present the derandomization of the randomized rounding approach for approximating linear integer programs due to Raghavan [Rag88]. We start with the problem LATTICEAPPROXIMATION, which is a subproblem for approximating certain linear integer programs. First we explain how to apply the randomized rounding approach to LATTICEAPPROXIMATION. Then we argue as in the previous sections. By a probabilistic and non-constructive proof we show that there are good solutions for LATTICEAPPROXIMATION. In a second step the probabilistic proof is derandomized by the method of conditional probabilities. Afterwards we apply the techniques developed for LATTICEAPPROXIMATION to a more practical problem, namely VECTORSELECTION.

3.4.1 LATTICEAPPROXIMATION

We start with the definition of LATTICEAPPROXIMATION.

LATTICEAPPROXIMATION

Instance: An $n \times r$-matrix C, where $c_{ij} \in [0, 1]$, a vector $p \in \mathbb{R}^r$.

Problem: Find an integer vector (lattice point) $q = (q_1, \ldots, q_r)$ so that every coordinate of $C \cdot (p - q)$ has a "small" absolute value.

One may consider p as a solution of the linear relaxation of an integer program. The rows of C describe the constraints of the integer program. The goal is to find an integer vector q that approximates p "well". No constraint is violated too much since we require the absolute values of the inner products of $p - q$ and each row of C to be small.

In the following the absolute values of the inner products are called discrepancies Δ_i. This means $\Delta_i = |\sum_{j=1}^r c_{ij}(p_j - q_j)|$. Furthermore, let $s_i = \sum_{j=1}^r c_{ij}p_j$. Then the goal is to choose q so that all discrepancies Δ_i are bounded by some function in s_i. Without loss of generality the reals p_j are assumed to be in the interval $[0, 1]$. Otherwise one may replace p_j by $p_j - \lfloor p_j \rfloor$.

We consider only those solutions for q, where each q_j is chosen from $\{0, 1\}$ by randomized rounding of p_j. This means that we choose $q_j = 1$ with probability p_j and $q_j = 0$ with probability $1 - p_j$. The choice of the coordinates of q is done mutually independently. Then each q_j is a Bernoulli trial and $\mathbf{E}[q_j] = p_j$. We define the random variables Ψ_i by

$$\Psi_i = \sum_{j=1}^r c_{ij}q_j.$$

Then we have $\mathbf{E}[\Psi_i] = \sum_{j=1}^r c_{ij}\mathbf{E}[q_j] = s_i$. Since $\Delta_i = |\Psi_i - s_i|$, it is intuitive that randomized rounding may lead to small Δ_i.

Our first goal is a probabilistic proof that there is some vector q so that all discrepancies Δ_i are small. The random variables Ψ_i are weighted sums of Bernoulli trials. We shall apply bounds on the tail of the distribution of the weighted sum of Bernoulli trials, which we now derive.

Let $a_1, \ldots, a_r \in (0, 1]$ and let $X_1, \ldots, X_r$ be independent Bernoulli trials with $\mathbf{E}[X_j] = p_j$. Let $X = \sum_{i=1}^r a_i X_i$. The mean of X is $m := \mathbf{E}[X] = \sum_{j=1}^r a_j p_j$. The following theorem gives Chernoff bounds on the deviation of X above and below its mean.

Theorem 3.5. *Let $\delta > 0$ and let $m = \mathbf{E}[X] \geqslant 0$. Then*

1. $\text{Prob}[X > (1 + \delta)m] < \left(\frac{e^\delta}{(1+\delta)^{1+\delta}}\right)^m$,

2. $\text{Prob}[X < (1 - \delta)m] < \left(\frac{e^\delta}{(1+\delta)^{1+\delta}}\right)^m$.

Proof. We only prove the first statement. The second statement can be proved in a similar way. We apply Markov's inequality to the moment generating function e^{tX} and obtain for each $t > 0$:

$$\text{Prob}[X > (1 + \delta)m] = \text{Prob}[e^{tX} > e^{t(1+\delta)m}] < e^{-t(1+\delta)m}\,\mathbf{E}[e^{tX}].$$

Now we exploit the independence of the X_j.

$$
\begin{aligned}
e^{-t(1+\delta)m}\, \mathbf{E}[e^{tX}] &= e^{-t(1+\delta)m}\, \mathbf{E}[e^{ta_1 X_1} \cdot \ldots \cdot e^{ta_r X_r}] \\
&= e^{-t(1+\delta)m}\, \mathbf{E}[e^{ta_1 X_1}] \cdot \ldots \cdot \mathbf{E}[e^{ta_r X_r}] \\
&= e^{-t(1+\delta)m} \prod_{j=1}^{r} [p_j e^{ta_j} + (1-p_j)1].
\end{aligned}
\tag{3.3}
$$

Now we may choose $t = \ln(1+\delta)$ and obtain

$$
(1+\delta)^{-(1+\delta)m} \prod_{j=1}^{r} [p_j(1+\delta)^{a_j} + 1 - p_j] \leqslant (1+\delta)^{-(1+\delta)m} \prod_{j=1}^{r} \exp[p_j[(1+\delta)^{a_j} - 1]].
$$

The inequality follows from $x + 1 \leqslant e^x$. Since $a_j \in (0,1]$ we have $(1+\delta)^{a_j} - 1 \leqslant \delta a_j$. Hence, the last expression is bounded by

$$
(1+\delta)^{-(1+\delta)m} \exp\left(\sum_{j=1}^{r} \delta a_j p_j \right) = \left(\frac{e^\delta}{(1+\delta)^{1+\delta}} \right)^m.
$$

∎

Let $B(m, \delta)$ be the bound of Theorem 3.5 on the probability that X is larger than $(1+\delta)m$, i.e.

$$
B(m, \delta) = [e^\delta/(1+\delta)^{1+\delta}]^m.
$$

Due to the symmetry of Theorem 3.5 the same bound holds for deviations below the mean. We define $D(m, x)$ as the deviation δ for which the bound $B(m, \delta)$ is x, i.e., the deviation for which $B(m, D(m, x)) = x$. In [Rag88] the following estimates for $D(m, x)$ are given:

Case 1. $m > \ln 1/x$. Then

$$
D(m, x) \leqslant (e-1) \left[\frac{\ln 1/x}{m} \right]^{1/2}.
\tag{3.4}
$$

Case 2. $m \leqslant \ln 1/x$. Then

$$
D(m, x) \leqslant \frac{e \ln 1/x}{m \ln[(e \ln 1/x)/m]}.
\tag{3.5}
$$

Now we are ready to prove the existence of an approximation vector q for which the discrepancies are "small".

Theorem 3.6. *There is a 0-1-vector q so that for all $i \in \{1, \ldots, n\}$ it holds that $\Delta_i \leqslant s_i D(s_i, 1/2n)$.*

Proof. Let $q_j \in \{0,1\}$ be chosen by randomized rounding. It suffices to prove that the resulting vector satisfies the bound of the theorem with positive probability. Let β_i be the (bad) event that Δ_i exceeds the bound of the theorem. By definition of $D(m,\delta)$ we have

$$\text{Prob}[\Psi_i > s_i + s_i D(s_i, 1/2n)] < 1/2n \quad \text{and}$$
$$\text{Prob}[\Psi_i < s_i - s_i D(s_i, 1/2n)] < 1/2n.$$

Hence, the probability of each bad event is less than $1/n$ and the probability that any of the n bad events occurs is less than one. This implies that there is some vector q for which no bad event occurs. This vector q satisfies the bounds of the theorem. ∎

Now we derandomize the randomized rounding approach in order to obtain an efficient deterministic algorithm that achieves approximately the bound of Theorem 3.6. We use the same algorithm as in Section 3.2. The problem is to find a suitable weight function. The proof of Theorem 3.6 suggests to define the weight of a choice $d_1, \ldots, d_j$ of the first j bits of q as the conditional probability that a bad event occurs given $q_1 = d_1, \ldots, q_j = d_j$. (Only $q_{j+1}, \ldots, q_r$ are chosen randomly using randomized rounding.) Let us denote this probability by $P(d_1, \ldots, d_j)$. Then by Theorem 3.6 we have $P() < 1$. By the total probability theorem

$$P(d_1, \ldots, d_j) = (1 - p_{j+1})P(d_1, \ldots, d_j, 0) + p_{j+1}P(d_1, \ldots, d_j, 1).$$

This implies that for all $d_1, \ldots, d_j \in \{0,1\}$ it holds that

$$P(d_1, \ldots, d_j) \geqslant \min\{P(d_1, \ldots, d_j, 0), P(d_1, \ldots, d_j, 1)\}.$$

Hence, we may choose for q_{j+1} that d_{j+1} leading to a smaller probability of a bad event. If we choose all q_j in such a way, then we have as in Section 3.2

$$1 > P() \geqslant P(d_1) \geqslant \ldots \geqslant P(d_1, \ldots, d_r).$$

This implies $P(d_1, \ldots, d_r) = 0$, since no random choices are done. We obtain a vector $d = (d_1, \ldots, d_r)$ satisfying the bound of Theorem 3.6.

The problem is that it is not clear how to compute $P(d_1, \ldots, d_j)$ efficiently. Hence, we use a different weight function U with the following properties:

1. $U() < 1$.

2. $U(d_1, \ldots, d_j) \geqslant \min\{U(d_1, \ldots, d_j, 0), U(d_1, \ldots, d_j, 1)\}$.

3. U is an upper bound on P.

4. U can be computed in polynomial time.

The first two properties ensure that we can use U as a weight function. By the same inequality as for P we obtain $U(d_1, \ldots, d_r) < 1$. Since U is an upper bound on P, this implies $P(d_1, \ldots, d_r) = 0$. Hence, the resulting vector $(d_1, \ldots, d_r)$ satisfies the bound of Theorem 3.6. Because of the last property the derandomized algorithm runs in polynomial time. Since the probability of failure is bounded by U, this method is called the method of pessimistic estimators. It remains to find a suitable function U.

We derive the function U based on the proofs of Theorem 3.5 and Theorem 3.6. In the proof of Theorem 3.6 the bad event β_i was defined as the event that Ψ_i exceeds either of the limits of Theorem 3.6. Let us denote the limits on Ψ_i by L_{i+} and L_{i-}, i.e. $L_{i+} = s_i(1 + D(s_i, 1/2n))$ and $L_{i-} = s_i(1 - D(s_i, 1/2n))$.

By equation (3.3) for each $t_i > 0$ the probability that Ψ_i becomes larger than L_{i+} is bounded by

$$\text{Prob}[\Psi_i > L_{i+}] < e^{-t_i L_{i+}} \prod_{j=1}^{r} \mathbf{E}[e^{t_i c_{ij} q_j}] = e^{-t_i L_{i+}} \prod_{j=1}^{r} [p_j e^{t_i c_{ij}} + 1 - p_j]. \quad (3.6)$$

In a similar way it can be shown that for each $t_i > 0$ the probability that Ψ_i becomes smaller than L_{i-} is bounded by

$$\text{Prob}[\Psi_i < L_{i-}] < e^{t_i L_{i-}} \prod_{j=1}^{r} [p_j e^{-t_i c_{ij}} + 1 - p_j]. \quad (3.7)$$

We sum up the estimates of equations (3.6) and (3.7) in order to obtain a bound on the probability that any of the Ψ_i exceeds either of its limits. We define $U()$ as this sum.

$$U() := \sum_{i=1}^{n} \left[e^{-t_i L_{i+}} \prod_{j=1}^{r} [p_j e^{t_i c_{ij}} + 1 - p_j] + e^{t_i L_{i-}} \prod_{j=1}^{r} [p_j e^{-t_i c_{ij}} + 1 - p_j] \right]. \quad (3.8)$$

We may choose $t_i = \ln[1 + D(s_i, 1/2n)]$. For this choice of t_i the proofs of Theorem 3.5 and Theorem 3.6 were done. Hence, these proofs imply $U() < 1$. But we have the problem that the definition of $D(\cdot, \cdot)$ does not show how to compute $D(s_i, 1/2n)$ in polynomial time. For this reason we have to replace $D(s_i, 1/2n)$ by some upper bound $D^*(s_i, 1/2n)$. Such an upper bound may be obtained, e.g., by the estimates (3.4) and (3.5). By replacing $D(s_i, 1/2n)$ by an upper bound the allowed deviation of Ψ_i becomes larger and, therefore, the probability of a bad event cannot become larger. Hence, also after this replacement we have $U() < 1$.

Equation (3.8) only shows how to compute $U()$. Now we discuss how to compute $U(d_1, \ldots, d_l)$. First let us consider the case that only one random variable q_k was assigned the value $d_k = 1$. Then the probability that Ψ_i is larger than L_{i+} is equal to the probability that the sum of the random variables q_j, where $j \neq k$, is larger than $L_{i+} - c_{ik}$. This probability is bounded above by

$$e^{-t_i(L_{i+}-c_{ik})} \prod_{\substack{j\in\{1,\dots,r\}\\ j\neq k}} \mathbf{E}[e^{t_i c_{ij} q_j}] = e^{-t_i L_{i+}} e^{t_i c_{ik}} \prod_{\substack{j\in\{1,\dots,r\}\\ j\neq k}} [p_j e^{t_i c_{ij}} + 1 - p_j].$$

Hence, we have replaced in (3.6) the term $\mathbf{E}[e^{t_i c_{ik} q_k}] = p_k e^{t_i c_{ik}} + 1 - p_k$ by $e^{t_i c_{ik}}$. In other words, we have replaced the mean of $e^{t_i c_{ik} q_k}$ by the value that this term takes for $q_k = 1$. Similarly, we obtain a bound on the probability for $\Psi_i > L_{i+}$ given $q_k = 0$ by replacing the term $\mathbf{E}[e^{t_i c_{ik} q_k}] = p_k e^{t_i c_{ik}} + 1 - p_k$ by 1. By performing the corresponding changes for the assignments $q_1 = d_1, \dots, q_\ell = d_\ell$ on formula (3.8) we obtain a formula for $U(d_1, \dots, d_\ell)$.

In order to show that the second condition from above is fulfilled for U, Raghavan [Rag88] shows that $U(d_1, \dots, d_j)$ is a convex combination of $U(d_1, \dots, d_j, 0)$ and $U(d_1, \dots, d_j, 1)$. We omit these computations. By the derivation of U it follows that U is an upper bound on P. Altogether, we derandomized the randomized rounding approach for LATTICEAPPROXIMATION.

Theorem 3.7. *By the method of pessimistic estimators a 0-1-vector q satisfying $\forall i \in \{1, \dots, n\} : \Delta_i \leqslant s_i D^*(s_i, 1/2n)$ can be computed deterministically in polynomial time.*

3.4.2 An Approximation Algorithm for VECTORSELECTION

We apply the techniques developed for LATTICEAPPROXIMATION to the problem VECTORSELECTION. We start with the definition of the problem.

VECTORSELECTION

Instance: A collection $\Lambda = \{\lambda_1, \dots, \lambda_r\}$ of sets of vectors in $\{0,1\}^n$, where $\lambda_j = \{V_1^j, \dots, V_{k_j}^j\}$.

Problem: Select from each set λ_j exactly one vector V^j so that $\|\sum_{j=1}^r V^j\|_\infty$ is minimum.

The problem VECTORSELECTION has applications for routing problems [Rag88] and it is known to be $\mathcal{NP}$-hard. In order to apply the techniques developed above we formulate VECTORSELECTION as a linear 0-1-program. Let x_k^j ($1 \leqslant j \leqslant r, 1 \leqslant k \leqslant k_j$) be an indicator variable that describes whether V_k^j is selected. In the following 0-1-program the constraints (3.10) ensure that exactly one vector from each set is chosen, and the constraints (3.11) ensure that the ∞-norm of the sum of selected vectors, i.e. the maximum of the coordinates of this sum, is smaller than W, where the goal is to minimize W. The constraints are

$$x_k^j \in \{0,1\}, \qquad\qquad 1 \leqslant j \leqslant r, 1 \leqslant k \leqslant k_j, \qquad (3.9)$$

$$\sum_{k=1}^{k_j} x_k^j = 1, \qquad\qquad 1 \leqslant j \leqslant r, \qquad (3.10)$$

$$\sum_{j=1}^{r}\sum_{k=1}^{k_j} x_k^j V_k^j(i) \leqslant W, \qquad\qquad 1 \leqslant i \leqslant n. \qquad (3.11)$$

In order to get an approximate solution we replace the constraints (3.9) by $x_k^j \in [0,1]$ and solve this linear relaxation of the 0-1-program. Linear programs can be solved, e.g., by the simplex algorithm due to Dantzig [Dan63] or by the ellipsoid algorithm due to Khachian [Kha79]. The latter algorithm has a polynomial worst case run time. Both algorithms are also presented in [PS82]. Another method for solving linear programs with a polynomial worst-case run time is the interior point method due to Karmarkar [Kar84].

Let p_k^j denote the solutions of the linear relaxation of the 0-1-program. Let W' be the value of the objective function for this solution. Now we apply randomized rounding. In order to make sure that all constraints (3.10) are satisfied we perform the randomized rounding in such a way that from each set λ_j exactly one vector is chosen, where the vector V_k^j is chosen with probability p_k^j. This is done independently for all j. Let W^E be the maximum of the left-hand sides of constraints (3.11) for the solution obtained by randomized rounding. For the value W^E of the objective function all constraints are satisfied. We see that for this special 0-1-program randomized rounding always leads to solutions satisfying all constraints. Of course, W^E may be larger than W'. A bound on W^E in terms of W' is given by the following theorem.

Theorem 3.8. *There is a solution for the 0-1-program where the value W^E of the objective function is bounded by $W' \cdot (1 + D(W', 1/n))$.*

Proof. Let Z_i be the random variable describing the i-th coordinate of $\sum_{j=1}^r V^j$. Then we have $Z_i = \sum_{j=1}^r \sum_{k=1}^{k_j} x_k^j V_k^j(i)$. Hence, Z_i is the sum of those random variables x_k^j for which $V_k^j(i) = 1$, i.e., a sum of Bernoulli trials. The mean of Z_i is

$$\mathbf{E}[Z_i] = \sum_{j=1}^r \sum_{k=1}^{k_j} p_k^j V_k^j(i).$$

Since the p_k^j are a solution of the linear relaxation program, it follows from constraints (3.11) that $\mathbf{E}[Z_i] \leqslant W'$ for $1 \leqslant i \leqslant n$. By definition of $D(\cdot, \cdot)$ the probability of Z_i being larger than $W' \cdot (1 + D(W', 1/n))$ is less than $1/n$. Since there are n random variables Z_i the probability that any Z_i exceeds that bound is less than 1. Hence, there is some solution for which all Z_i do not exceed the bound. ∎

In order to derandomize the proof of Theorem 3.8 we need a slight generalization of the results of the last subsection. The reason is that we have the choice between k_j alternatives when performing randomized rounding, while the results for LATTICEAPPROXIMATION were derived for two alternatives. We omit this straightforward generalization. The result is the following theorem.

Theorem 3.9. *By the method of pessimistic estimators an integer solution for which the value of the objective function is bounded by $\lceil W' \cdot (1 + D^*(W', 1/n)) \rceil$ can be obtained deterministically in polynomial time.*

Since W' is an optimum solution for the linear relaxation, it is also a lower bound on the optimum solution of the 0-1-program. The performance ratio of the approximation can be estimated by applying formulae (3.4) and (3.5) to the result of Theorem 3.9. It can be shown that the performance ratio is bounded by some constant and that it converges to 1 as W' goes to infinity.

3.5 Reduction of the Size of the Probability Space

Let A be some randomized polynomial time algorithm for some optimization problem. If there is only a polynomial number of possible random choices for this algorithm, then it is easy to obtain a deterministic polynomial time algorithm from A. In polynomial time we can run A for all possible random choices and select the best result. But in general the number of possible random choices is not bounded by a polynomial. In this section we present a method that allows to reduce the size of the sample space from which the random choices are done.

We start with an outline of the method. When analyzing some randomized algorithm one usually assumes that the random choices are mutually independent. The reason is that it is often much easier to handle independent random variables rather than dependent random variables. But sometimes it is possible to analyze the randomized algorithm without the assumption of mutually independent random choices. E.g., it may suffice for the analysis to assume pairwise or d-wise independence. This weaker assumption on the independence may lead to a weaker result on the expected performance ratio of the algorithm (cf. Exercise 3.3).

We present a result which shows how to obtain a probability space (Ω, P) and d-wise independent random variables where the sample space Ω has only polynomial size. But the distribution of these random variables is merely an approximation for the distribution of the random variables used in the analysis of the randomized algorithm. Hence, one has to check whether the analysis is also valid for these random variables with approximated distributions. Possibly the analysis or even the algorithm has to be adapted. Finally, we can derandomize the algorithm by trying all samples from the polynomial size sample space.

We apply this method to two examples. First we derandomize the randomized algorithm for MAXE3SAT presented in Chapter 2. Then we show how to apply this method to a parallel randomized algorithm for the problem of computing maximal independent sets. In the derandomized version of this algorithm the randomized algorithm is run for all samples of the polynomial size sample space in parallel. This is possible by a polynomial number of processors.

3.5.1 Construction of a Polynomial Size Sample Space and d-wise Independent Random Variables

In this subsection we describe the method of Alon, Babai and Itai [ABI86] for the construction of a polynomial size sample space and d-wise independent random variables. First let us recall the definition of d-wise independence.

Definition 3.10. *The random variables $X_1, \ldots, X_n$ are called d-wise independent if each subset consisting of d of these random variables is independent, i.e., if for all $1 \leqslant i_1 < \ldots < i_d \leqslant n$ and for all $x_1, \ldots, x_d$ it holds that*

$$\text{Prob}[X_{i_1} = x_1, \ldots, X_{i_d} = x_d] = \text{Prob}[X_{i_1} = x_1] \cdot \ldots \cdot \text{Prob}[X_{i_d} = x_d].$$

Alon, Babai and Itai [ABI86] consider the general case where the random variables may take an arbitrary finite number of values. For our purposes it suffices to consider the special case where the random variables only take the two values 0 and 1. By the construction given in the proof of the following lemma, we obtain a uniform probability space (Ω, P), i.e., each elementary event in Ω has the same probability $1/|\Omega|$.

Lemma 3.11. *Let $X_1, \ldots, X_n$ be random variables where each X_i only takes the values 0 and 1. Let q be a prime power, $q \geqslant n$ and $d \geqslant 1$. Then there are a uniform probability space (Ω, P), where $|\Omega| = q^d$, and d-wise independent random variables $X_1', \ldots, X_n'$ over (Ω, P) and it holds for each $i \in \{1, \ldots, n\}$ that $|\text{Prob}[X_i' = 1] - \text{Prob}[X_i = 1]| \leqslant 1/2q$.*

By the last condition we also have $|\text{Prob}[X_i' = 0] - \text{Prob}[X_i = 0]| \leqslant 1/2q$.

Proof of Lemma 3.11. Let $p_i = \text{Prob}[X_i = 1]$ and let $p_i' = \lceil qp_i - 1/2 \rceil / q$. This means that we obtain p_i' by rounding p_i to the nearest multiple of $1/q$. In particular, we have $|p_i - p_i'| \leqslant 1/2q$.

Let $\mathbb{F}_q = \{0, \ldots, q - 1\}$, $A_{i,1} = \{0, 1 \ldots, qp_i' - 1\}$ and $A_{i,0} = \{qp_i', \ldots, q - 1\}$. Then $|A_{i,1}|/|\mathbb{F}_q| = p_i'$ and $|A_{i,0}|/|\mathbb{F}_q| = 1 - p_i'$.

The sample space Ω consists of all q^d polynomials over $\mathbb{F}_q$ of degree at most $d-1$. The probability space is uniform, i.e., each polynomial is chosen with probability $1/q^d$. The random variables $X_i' : \Omega \to \mathbb{R}$ are defined by $X_i'(p) = 0$ iff $p(i) \in A_{i,0}$ and $X_i'(p) = 1$ iff $p(i) \in A_{i,1}$.

Then $\text{Prob}[X_i' = 0] = 1 - p_i$ and $\text{Prob}[X_i' = 1] = p_i$. Hence, it holds that $|\text{Prob}[X_i' = 1] - \text{Prob}[X_i = 1]| \leqslant 1/2q$.

In order to prove that the X_i' are d-wise independent let $1 \leqslant i_1 < \ldots < i_d \leqslant n$. For $\ell \in \{1, \ldots, d\}$ let $x_\ell \in \{0, 1\}$. Then

$$\text{Prob}[X'_{i_1} = x_1, \ldots, X'_{i_d} = x_d] = \text{Prob}[p(i_1) \in A_{i_1,x_1}, \ldots, p(i_d) \in A_{i_d,x_d}].$$

There are $|A_{i_1,x_1}| \cdot \ldots \cdot |A_{i_d,x_d}|$ polynomials p fulfilling $p(i_1) \in A_{i_1,x_1}, \ldots, p(i_d) \in A_{i_d,x_d}$, since p is a polynomial of degree at most $d - 1$ and, hence, it is uniquely determined by fixing its value at d different places. Therefore,

$$
\begin{aligned}
&\text{Prob}[p(i_1) \in A_{i_1,x_1}, \ldots, p(i_d) \in A_{i_d,x_d}] \\
&= \frac{|A_{i_1,x_1}| \cdot \ldots \cdot |A_{i_d,x_d}|}{|\Omega|} \\
&= \text{Prob}[X'_{i_1} = x_1] \cdot \ldots \cdot \text{Prob}[X'_{i_d} = x_d].
\end{aligned}
$$

Hence, the X'_i are d-wise independent.   ■

We describe how to apply Lemma 3.11 for the derandomization of some randomized algorithm. The random variables $X_1, \ldots, X_n$ are the random variables describing the random choices of the algorithm. We assume that the probabilities for which $X_1, \ldots, X_n$ take the values 0 and 1 are rational numbers. Hence, we can compute the values p'_i. We replace $X_1, \ldots, X_n$ by $X'_1, \ldots, X'_n$. Since $|\Omega|$ is bounded by some polynomial in n, we can consider all $|\Omega|$ elementary events in polynomial time or in parallel by a polynomial number of processors. For each elementary event p we evaluate the random variables $X'_1, \ldots, X'_n$, i.e., we compute $X'_1(p), \ldots, X'_n(p)$ and run the given randomized algorithm for these values of the random variables. From the $|\Omega|$ results a best one is selected.

The random variable $X'_i(p)$ can be evaluated in constant time, if d is a constant. We merely have to evaluate a polynomial of constant degree and to compare the result with qp'_i. Hence, the evaluation of all $X'_i(p)$ for all $i \in \{1, \ldots, n\}$ and all $p \in \Omega$ is possible by an EREW PRAM with q^{d+1} processors in constant time.

3.6 Another Approximation Algorithm for MAXE3SAT

In this section we show how Lemma 3.11 can be applied in order to derandomize the algorithm RANDOM EkSAT ASSIGNMENT given in Chapter 2. The derandomization only works for constant k and we only consider the case $k = 3$, i.e. the algorithm for the problem MAXE3SAT. Recall that an instance of MAXE3SAT consists of the set $\{x_1, \ldots, x_n\}$ of variables and the set $C = \{C_1, \ldots, C_m\}$ of clauses. In the randomized algorithm each variable $x_1, \ldots, x_n$ is randomly and independently chosen to be 0 and 1 with probability $1/2$ each. Then the probability of a clause to be satisfied is $7/8$ and the expected number of satisfied clauses is $7/8 \cdot m$.

It is easy to see that the same analysis works if $x_1, \ldots, x_n$ are chosen to be 0 and 1 with probability $1/2$ each and if we only assume threewise independence of $x_1, \ldots, x_n$. Hence, we may apply Lemma 3.11.

We choose for q the smallest power of 2 such that $q \geq n$. Since the random variables x_i take the values 0 and 1 with probability $1/2$ each and since q is even, we may apply Lemma 3.11 and replace x_i by X_i' where $\mathrm{Prob}[x_i = 1] = \mathrm{Prob}[X_i' = 1]$. In other words, the distribution of the random variables does not change when applying Lemma 3.11.

The sample space Ω consists of all polynomials p of degree 2 over $\mathbb{F}_q$. Each such polynomial can be represented by a triple $(r_2, r_1, r_0) \in \mathbb{F}_q^3$ of its coefficients. Hence, there are q^3 such polynomials. By the proof of Lemma 3.11 we choose $X_i'(p) = 0$ iff $p(i) \in A_{i,0}$ and $X_i'(p) = 1$ iff $p(i) \in A_{i,1}$. This can be done sequentially in polynomial time for all polynomials of degree 2. Then the assignment to the x_i satisfying the largest number of clauses is chosen. Hence, we obtain the following algorithm.

DERANDOMIZED ALGORITHM FOR MAXE3SAT
Input: A set $C = \{C_1, \ldots, C_m\}$ of clauses over the variables $\{x_1, \ldots, x_n\}$ where each clause consists of exactly three literals.

    let q be the smallest power of 2 such that $q \geq n$
    for all $(r_2, r_1, r_0) \in \mathbb{F}_q^3$ **do**
        for $i := 1, \ldots, n$ **do**
            if $(r_2 i^2 + r_1 i + r_0) \bmod q \in \{0, \ldots, q/2 - 1\}$
                then $x_i := 0$
                else $x_i := 1$
        count the number of satisfied clauses
Output: An assignment to the x-variables maximizing the number of satisfied clauses.

By the analysis given in Chapter 2 and by Lemma 3.11 the algorithm is an approximation algorithm for MAXE3SAT with a performance ratio of at most $8/7$. The algorithm runs in polynomial time, since $q^3 = \Theta(n^3)$ iterations of the for-all-loop are performed. For each iteration $\Theta(m + n)$ steps are sufficient. On the other hand the derandomized algorithm described in Section 3.3 yields the same performance ratio and is faster.

3.7 A PRAM Algorithm for MAXIMALINDEPENDENTSET

In this section we present a randomized parallel algorithm for MAXIMALINDE-PENDENTSET due to Luby [Lub85, Lub86] and its analysis under the assumption of mutually independent random variables. Then we show how to adapt the analysis and the algorithm for the case of pairwise independent random variables and for the case that the distributions of the original random variables are merely approximated. In this section we follow the presentations of Alon, Babai and Itai [ABI86] and of Motwani and Raghavan [MR95b].

We do not discuss the implementation of the algorithm on an EREW PRAM. For more details on EREW PRAMs see, e.g., [KR90].

3.7.1 A Randomized Algorithm for MAXIMALINDEPENDENTSET

We start with the definition of the problem.

MAXIMALINDEPENDENTSET

Instance: An undirected graph $G = (V, E)$.
Problem: Compute a set $I \subseteq V$ that is independent and maximal. Independent means that there is no edge in E connecting nodes in I. Maximal means that I is not a proper subset of some independent set.

Note that the problem is different from the $\mathcal{NP}$-complete problem of computing an independent set of maximum size. The condition on a maximal independent set is merely that we cannot obtain a larger independent set by adding nodes.

It is easy to compute a maximal independent set for a graph sequentially. Iteratively a node v is chosen and v and all neighbors of v are removed from the graph. This is done until the graph is empty. Then the chosen nodes form a maximal independent set.

By a similar idea we can obtain a fast parallel algorithm. Now we choose some independent set S rather than a single node. Then all nodes in S and all neighbors of nodes in S are removed from the graph. This procedure is iterated until the graph is empty. Then the union of all chosen sets S is a maximal independent set.

In order to make sure that this algorithm is fast it is essential that in each iteration the number of removed nodes is large. In the following algorithm the computation of S uses randomization. The algorithm is due to Luby [Lub85, Lub86] and is also presented in [MR95b].

In the following the degree of some node v is denoted by $d(v)$. For some node set S let $\Gamma(S)$ denote the set of all neighbors of S.

A RANDOMIZED PARALLEL ALGORITHM FOR MAXIMALINDEPENDENTSET
Input: A graph $G = (V, E)$.
$I := \emptyset$
while $V \neq \emptyset$ **do**
    for all $v \in V$ **do in parallel**
        if $d(v) = 0$
            then $I := I \cup \{v\}$; delete v from G
            else mark v with probability $1/(2d(v))$
    for all $\{u, v\} \in E$ **do in parallel**
        if u and v are marked
            then unmark the node with lower degree (if $d(u) = d(v)$, then unmark the node with the smaller number)
    let S be the set of marked nodes
    $I := I \cup S$
    delete all nodes in $S \cup \Gamma(S)$ from G
Output: The maximal independent set I.

It is easy to check that at the end of each iteration of the while-loop the set S is an independent set and that the algorithm eventually computes a maximal independent set. By choosing $1/(2d(v))$ as probability for marking v, nodes with smaller degree are marked with higher probability. This decreases the probability of marking both endpoints of an edge. The analysis consists of the proof that the expected number of edges deleted in some iteration (of the while-loop) is $\Omega(|E^*|)$ where E^* denotes the set of edges at the beginning of this iteration. Then the expected number of iterations is $\mathcal{O}(\log|E|) = \mathcal{O}(\log|V|)$. One may easily check that each iteration can be performed by an EREW PRAM with $\mathcal{O}(|V| + |E|)$ processors in time $\mathcal{O}(\log|E|)$. Altogether, we shall obtain the following theorem.

Theorem 3.12. *A maximal independent set of a graph $G = (V, E)$ can be computed by a randomized algorithm on an EREW PRAM with $\mathcal{O}(|V| + |E|)$ processors where the expected run time is $\mathcal{O}(\log^2 |V|)$.*

For the analysis we assume that the random markings are mutually independent for all nodes. In the following we consider only one iteration of the while-loop. We call a node v *bad* if (at the beginning of the iteration) the degree of at least $2/3$ of its neighbors is larger than the degree of v. Otherwise it is called *good*. An edge $\{u, v\}$ is *bad* if u and v are bad. If at least one endpoint of an edge is good, then the edge is called *good*. We show that the probability for each good edge to be deleted in this iteration is at least some positive constant. Hence, the expected number of good edges deleted in each iteration is at least a constant fraction of the number of good edges at the beginning of the iteration. Together with the fact that at least half of the edges of each graph are good it follows that in each iteration $\Omega(|E^*|)$ edges are deleted.

Lemma 3.13. *Let v be a good node with a degree larger than 0. Then during an iteration with probability of at least $1 - e^{-1/6}$ some node $w \in \Gamma(v)$ is marked.*

Proof. There are at least $d(v)/3$ neighbors of v with a degree at most $d(v)$ since v is good. In order to estimate the probability that some neighbor of v is marked, we consider only these nodes. Each node w among these neighbors of v is marked with probability $1/(2d(w)) \geqslant 1/(2d(v))$. Then the probability that at least one of these nodes is marked is at least

$$1 - \left(1 - \frac{1}{2d(v)}\right)^{d(v)/3} \geqslant 1 - e^{-1/6}.$$

■

Lemma 3.14. *The probability that during an iteration some marked node w is not unmarked is at least $1/2$.*

Proof. A marked node w may get unmarked only if one of its neighbors with a degree of at least $d(w)$ is also a marked node. Let u be such a neighbor. Then u is marked with probability $1/(2d(u)) \leqslant 1/(2d(w))$. The number of such neighbors u is bounded by $d(w)$. Then the probability of w not to become unmarked is at least

$$1 - \mathrm{Prob}[\exists u \in \Gamma(w) : u \text{ is marked} \wedge d(u) \geqslant d(w)] \geqslant 1 - d(w) \cdot \frac{1}{2d(w)} = \frac{1}{2}.$$

∎

Since in each iteration exactly those nodes are deleted that are marked and not unmarked, Lemma 3.13 and Lemma 3.14 imply the following lemma.

Lemma 3.15. *During an iteration the probability of a good node to become deleted is at least* $(1 - e^{-1/6})/2$.

In order to complete the estimate of the expected number of iterations we apply the following lemma. We omit the proof, which can be found in [ABI86] and in [MR95b].

Lemma 3.16. *In each graph $G = (V, E)$ the number of good edges is at least* $|E|/2$.

By definition a good edge is incident to at least one good node. By Lemma 3.15 in each iteration each good node is deleted with probability at least $(1 - e^{-1/6})/2$. Hence, the same holds for each good edge. Since at least half of the edges are good, the expected number of deleted edges is at least $|E^*|(1 - e^{-1/6})/4$. This completes the proof of Theorem 3.12.

3.7.2 Derandomization of the Randomized Algorithm for MAXIMALINDEPENDENTSET

The derandomization of the algorithm of the last section is based on the observation that mutual independence of the random markings is not necessary for the analysis of the algorithm. It suffices to assume pairwise independence. For this case we obtain a slightly weaker version of Lemma 3.13. As in the analysis of the randomized algorithm we consider only one iteration of the while-loop. Let n be the number of nodes in the graph at the beginning of the iteration and let E_i for $i \in \{1, \ldots, n\}$ be the event that the i-th node is marked.

Lemma 3.17. *Assume that the events $E_1, \ldots, E_n$ are pairwise independent. Let v be a good node with degree larger than 0. Then during an iteration with probability of at least $1/12$ some node $w \in \Gamma(v)$ is marked.*

Proof. Let k be the number of neighbors of v with degree at most $d(v)$. Since v is good, we have $k \geqslant d(v)/3$. After renumbering the nodes we may assume that the neighbors of v with degree at most $d(v)$ are the first k nodes $v_1, \ldots, v_k$. Then $E_1, \ldots, E_k$ denote the events that these nodes are marked. We have $\text{Prob}[E_i] = 1/(2d(v_i)) \geqslant 1/(2d(v))$. Define $sum := \sum_{i=1}^{k} \text{Prob}[E_i]$. Then $sum \geqslant k/(2d(v)) \geqslant 1/6$. The event that at least one node in $\Gamma(v)$ is marked is $\bigcup_{i=1}^{k} E_i$. The claim of the lemma follows directly from the following lemma due to Luby [Lub85]. ∎

Lemma 3.18. *Let $E_1, \ldots, E_k$ be pairwise independent events and let $sum :=$ $\sum_{i=1}^{k} \text{Prob}[E_i]$. Then $\text{Prob}[\bigcup_{i=1}^{k} E_i] \geqslant \min(sum, 1)/2$.*

A proof of this lemma can be found in [Lub85].

A careful inspection of the proof of Lemma 3.14 shows that this proof remains valid even when replacing the mutually independent marking by pairwise independent marking. Hence, the expected number of iterations of the algorithm is also $\mathcal{O}(\log |V|)$ when using pairwise independent marking. Let X_i be the random variable describing the marking of the i-th node. We can apply Lemma 3.11 and replace X_i by X_i'. For q we may choose the smallest prime larger than n. We obtain a sample space of size $q^2 = \mathcal{O}(|V|^2)$. The iterations of the original algorithm are derandomized separately. Each iteration is executed for all q^2 polynomials in Ω and the corresponding values of the X_i' in parallel. A largest set S is selected and all nodes from $S \cup \Gamma(S)$ are removed from the graph.

The only problem is that the probability of marking a node was slightly changed by applying Lemma 3.11. The probability of a node to become marked may change by $1/2q$. Luby [Lub86] shows how to overcome this problem. Before starting an iteration of the algorithm we search for a node v where $d(v) \geqslant n/16$. If such a node v is found, v is included into I and v and all neighbors of v are deleted. In this case at least $1/16$ of the nodes of the graph are removed. This implies that this situation may occur at most 16 times.

If no such node is found, for all nodes v we have $d(v) < n/16 \leqslant q/16$ or $q \geqslant 16d(v)$. Let v be the i-th node. Then $|\text{Prob}[X_i = 1] - \text{Prob}[X_i' = 1]| \leqslant 1/2q \leqslant 1/(32d(v))$. The probability that v is marked is at least $15/16 \cdot 1/(2d(v))$ and at most $17/16 \cdot 1/(2d(v))$. Now we have to check, whether the estimates in Lemma 3.14 and Lemma 3.17 change.

In Lemma 3.14 the probability that some neighbor of u is marked is now bounded by $17/16 \cdot 1/(2d(w))$. Hence, we obtain a slightly weaker version of this lemma.

Lemma 3.19. *If the events $E_1, \ldots, E_n$ are pairwise independent, the probability that during an iteration some marked node w is not unmarked is at least $15/32$.*

In Lemma 3.17 the bound on the probability of E_i changes to $15/16 \cdot 1/(2d(w))$. Again we get a slightly weaker version.

Lemma 3.20. *Assume that the events $E_1, \ldots, E_n$ are pairwise independent. Let v be a good node with degree larger than 0. Then during an iteration with probability of at least $5/64$ some node $w \in \Gamma(v)$ is marked.*

But the probability that a good node belongs to $S \cup \Gamma(S)$ is still a positive constant and, therefore, the number of iterations is $\mathcal{O}(\log |V|)$. As described above all polynomials in Ω can be tried in parallel. We have proved the following theorem.

Theorem 3.21. *The problem of computing a maximal independent set can be solved by an EREW PRAM with $\mathcal{O}(|V|^4)$ processors in time $\mathcal{O}(\log^2 |V|)$. Hence, MAXIMALINDEPENDENTSET $\in \mathcal{NC}^2$.*

Exercises

For the first two exercises we present a randomized algorithm for the problem MAXCUT from [MR95b]. MAXCUT is considered in detail in Chapter 11. We start with the definition of MAXCUT. An instance of MAXCUT is an undirected graph $G = (V, E)$. The task is to find a partition of the node set V into two sets V_1 and V_2 so that the number of edges with one endpoint in V_1 and the other one in V_2 is maximized.

A simple randomized approximation algorithm for MAXCUT works as follows. Each node is put randomly and mutually independent in V_1 or V_2 with probability $1/2$ each. For each edge the probability that one endpoint is in V_1 and the other one in V_2 is $1/2$. By linearity of the mean it follows that the expected number of edges with endpoints in different sets is $|E|/2$. Since the size of a cut cannot be larger than $|E|$, we obtain a randomized polynomial time approximation algorithm for MAXCUT with an expected performance ratio of at most 2.

Exercise 3.1. [MR95b] Derandomize this algorithm by the method of conditional probabilities.

Exercise 3.2. Apply Lemma 3.11 for $d = 2$ in order to derandomize this algorithm.

Exercise 3.3. We consider the algorithm RANDOM EkSAT ASSIGNMENT from Chapter 2 for the case $k = 3$. Estimate the expected performance ratio of this algorithm for MAXE3SAT for the case that the random choices are pairwise independent rather than threewise or mutually independent. Derandomize the algorithm by applying Lemma 3.11 for $d = 2$.

4. Proof Checking and Non-Approximability

Stefan Hougardy

4.1 Introduction

In this chapter we will present the PCP-Theorem and show how it can be used to prove that there exists no PTAS for $\mathcal{APX}$-complete problems unless $\mathcal{P} = \mathcal{NP}$. Moreover we will show how the PCP-Theorem implies that MAXCLIQUE cannot be approximated up to a factor of n^ε in an n-vertex graph and how this factor can be improved to $n^{1-\varepsilon}$ by making use of Håstad's [Hås97a] result showing that δ amortized free bits suffice for a $(\log n, 1)$-verifier to recognize any $\mathcal{NP}$ language.

4.2 Probabilistically Checkable Proofs

The class PCP (standing for **P**robabilistically **C**heckable **P**roofs) is defined as a common generalization of the two classes co-$\mathcal{RP}$ and $\mathcal{NP}$. To see this let us briefly recall the definitions of these two classes as given in Chapter 1. A language L belongs to co-$\mathcal{RP}$ if there exists a randomized polynomial time Turing machine M such that

$x \in L \quad \Rightarrow \mathrm{Prob}[M \text{ accepts } x] = 1.$

$x \notin L \quad \Rightarrow \mathrm{Prob}[M \text{ accepts } x] < 1/2.$

Using a similar notation, the class $\mathcal{NP}$ can be defined to consist of all languages L for which there exists a polynomial time Turing machine M such that

$x \in L \quad \Rightarrow \exists$ certificate c such that M accepts (x, c).

$x \notin L \quad \Rightarrow \forall$ certificates c M rejects (x, c).

It is easily seen that this definition is equivalent to the one given in Chapter 1 as the certificate c used in the above definition simply corresponds to an accepting computation path of the non-deterministic Turing machine used in Chapter 1 to define the class $\mathcal{NP}$.

The idea of the class PCP now is to allow simultaneously the use of randomness and non-determinism (or equivalently use of a certificate). The notions here are slightly different: the Turing machine will be called *verifier* and the certificates are called *proofs*.

A *verifier* V is a (probabilistic) polynomial time Turing machine with access to an input x and a string τ of random bits. Furthermore the verifier has access to a proof π via an oracle, which takes as input the position of the proof the verifier wants to query and outputs the corresponding bit of the proof π. Depending on the input x, the random string τ and the proof π the verifier V will either accept or reject the input x. We require that the verifier is *non-adaptive*, i.e., it first reads the input x and the random bits τ, and then decides which positions in the proof π it wants to query. Especially this means that the positions it queries do not depend on the answers the verifier got from previous queries. We will denote the result of V's computation on x, τ and π as $V(x, \tau, \pi)$.

As we will see later, verifiers are very powerful if we allow them to use polynomially many random bits and to make polynomially many queries to a proof. This is the reason why we introduce two parameters that restrict the amount of randomness and queries allowed to the verifier. An $(r(n), q(n))$-*restricted verifier* is a verifier that for inputs of length n uses at most $\hat{r}(n) = \mathcal{O}(r(n))$ random bits and queries at most $\hat{q}(n) = \mathcal{O}(q(n))$ bits from the proof π.

The class $\text{PCP}(r(n), q(n))$ consists of all languages L for which there exists an $(r(n), q(n))$-restricted verifier V such that

$$x \in L \quad \Rightarrow \quad \exists \pi \text{ s. t. } \text{Prob}_\tau[V(x, \tau, \pi) = ACCEPT] = 1. \qquad (completeness)$$

$$x \notin L \quad \Rightarrow \quad \forall \pi \ \text{Prob}_\tau[V(x, \tau, \pi) = ACCEPT] < 1/2. \qquad (soundness)$$

Here the notation $\text{Prob}_\tau[\ldots]$ means that the probability is taken over all random strings τ the verifier may read, i.e., over all 0-1-strings of length $\hat{r}(n) = \mathcal{O}(r(n))$.

We extend the definition of PCP to sets R and Q of functions in the natural way: $\text{PCP}(R, Q) = \cup_{r \in R, q \in Q} \text{PCP}(r(\cdot), q(\cdot))$.

Obviously, the class PCP is a common generalization of the classes co-$\mathcal{RP}$ and $\mathcal{NP}$, since we have:

$$\text{PCP}(\text{poly}, 0) = \text{co-}\mathcal{RP}$$

$$\text{PCP}(0, \text{poly}) = \mathcal{NP}$$

Here we denote by poly the set of all polynomials; similarly we use polylog for the set of all poly-logarithmic functions etc.

The natural question now is: How powerful is a (poly, poly)-restricted verifier ? As was shown by Babai, Fortnow and Lund in 1990 [BFL91] this verifier is *very* powerful; namely they proved that

$PCP(\text{poly}, \text{poly}) = \mathcal{NEXP}$

This result was "scaled down" almost at the same time independently by Babai, Fortnow, Levin and Szegedy [BFLS91] who proved

$\mathcal{NP} \subseteq PCP(\text{poly} \log n, \text{poly} \log n)$

and by Feige, Goldwasser, Lovász, Safra and Szegedy [FGL$^+$91] who proved that

$\mathcal{NP} \subseteq PCP(\log n \cdot \log \log n, \log n \cdot \log \log n)$.

The hunt for the smallest parameters for the class PCP that were still able to capture all of $\mathcal{NP}$ was opened. A natural conjecture was that $\mathcal{NP}$ equals $PCP(\log n, \log n)$ and indeed, shortly after Arora and Safra [AS92] proved an even stronger result that broke the $\log n$ barrier in the number of query bits:

$\mathcal{NP} = PCP(\log n, \text{poly} \log \log n)$.

Just some weeks later the hunt was brought to an end by Arora, Lund, Motwani, Sudan and Szegedy [ALM$^+$92] who obtained the ultimate answer in the number of queries needed to capture $\mathcal{NP}$:

Theorem 4.1 (The PCP-Theorem). $\mathcal{NP} = PCP(\log n, 1)$

At first sight this is a very surprising result as the number of queries needed by the verifier is a *constant*, independent of the input size. Let us take as an example the problem 3SAT and consider the usual way to prove that a 3SAT instance x is satisfiable, namely a truth assignment to the variables. If we are allowed to query only a constant number of the values assigned to the variables, then it is impossible to decide with constant error probability whether the given 3SAT instance is satisfiable. The reason why this does not contradict the PCP-Theorem is the following. The proofs used by the verifiers are not the kind of certificates that are usually used for problems in $\mathcal{NP}$. Rather as proofs there will be used special error correcting encodings of such certificates.

The proof of the PCP-Theorem will be given in Chapter 5. The main part in the proof is to show that the inclusion $\mathcal{NP} \subseteq PCP(\log n, 1)$ holds. The other inclusion can easily be derived from the following slightly more general statement.

Proposition 4.2. $PCP(r(n), q(n)) \subseteq NTIME(2^{\mathcal{O}(r(n))} \cdot \text{poly})$ *whenever* $q(n) = \mathcal{O}(\text{poly})$.

Proof. Simply observe that we can guess the answers to the $\mathcal{O}(q(n))$ queries to the proof π and simulate the PCP-verifier for all $2^{\mathcal{O}(r(n))}$ possible random strings to decide whether we should accept an input x or not. Each of the simulations can be carried out in polynomial time.   ∎

Interestingly, there can be proved a trade off between the randomness and the number of queries needed by the PCP-verifier such that its power is always exactly suited to capture the class $\mathcal{NP}$[Gol95].

Proposition 4.3. *There exist constants $\alpha, \beta > 0$ such that for every integer function $l(\cdot)$, so that $0 \leqslant l(n) \leqslant \alpha \log n$*

$$\mathcal{NP} = \text{PCP}(r(\cdot), q(\cdot))$$

where $r(n) = \alpha \log n - l(n)$ and $q(n) = \beta 2^{l(n)}$.

As the extreme cases this proposition shows $\mathcal{NP} = \text{PCP}(\log n, 1)$ and $\mathcal{NP} = \text{PCP}(0, \text{poly})$.

4.3 PCP and Non-Approximability

The reason why the PCP-Theorem caused a sensation is due to its close connection to the seemingly unrelated area of approximation algorithms. This connection was first discovered by Feige, Goldwasser, Lovász, Safra and Szegedy [FGL+91] who observed that if the problem MaxClique can be approximated up to a constant factor then $\text{PCP}(r(n), q(n)) \subseteq \text{DTIME}(2^{\mathcal{O}(r(n)+q(n))} \cdot \text{poly})$. This observation was the main motivation for proving $\mathcal{NP} \subseteq \text{PCP}(\log n, \log n)$ and thus showing that no constant factor approximation algorithm for Max-Clique can exist, unless $\mathcal{P}=\mathcal{NP}$.

Later on, the connection between PCPs and non-approximability has been extended to a large number of other optimization problems. The results obtained in this area are always of the type: No "good" polynomial time approximation algorithm can exist for solving the optimization problem, unless something very unlikely happens. Here the term "very unlikely" means that the existence of such an approximation algorithm would imply $\mathcal{P}=\mathcal{NP}$ or $\mathcal{NP}=\mathcal{ZPP}$ or $\mathcal{NP} \subseteq \text{DTIME}(2^{\text{poly} \log n})$ or similar statements. The precise definition of a "good" approximation algorithm heavily depends on the underlying optimization problem. Roughly, optimization problems can be divided into three classes: (1) Problems for which polynomial time constant factor approximation algorithms are known, but no PTAS can exist. As an example we will show in Section 4.4 that Max3Sat has no PTAS and therefore no problem in $\mathcal{APX}$ has a PTAS unless $\mathcal{P}=\mathcal{NP}$. (2) Problems that can be approximated up to a factor of $c_1 \cdot \log n$ but no polynomial time algorithm can achieve an approximation ratio of

$c_2 \cdot \log n$, for certain constants $c_1 > c_2$. In Chapter 10 we will see that the problem SETCOVER is an example for such an optimization problem. (3) Problems whose solution is of size $\mathcal{O}(n)$ but that cannot be approximated up to a factor of n^ε for certain constants ε. Famous examples in this class are MAXCLIQUE and CHROMATICNUMBER. We will prove in Section 4.6 that no polynomial time approximation algorithm for MAXCLIQUE can have a performance guarantee of n^ε for a certain ε, unless $\mathcal{P}=\mathcal{NP}$.

For more than twenty years no non-trivial lower bounds for the approximation guarantee of any of the above mentioned problems was known. The PCP-Theorem now gave such results for a whole bunch of optimization problems. But this is not the only consequence set off by the PCP result. Soon after getting the first non-trivial non-approximability results people started to tighten the gap between the best known approximation factors achievable in polynomial time and the lower bounds obtained by using the PCP-Theorem. Surprisingly, this challenge not only led to improvements on the just obtained new lower bounds, but also for several classical optimization problems, better polynomial time approximation algorithms have been found. As an example Goemans and Williamson [GW94a] improved the long known trivial 2-approximation algorithm for MAXCUT to an 1.139-approximation algorithm for this problem. Interestingly, the new approximation algorithms do not rely on the PCP result. The PCP-Theorem and its consequences for non-approximability results rather functioned as a new strong motivation to try to improve the best known approximation algorithms known so far.

For improving the lower bounds for non-approximability results by using the PCP-Theorem it turns out that one has to reduce the constants involved in the $\mathcal{O}$-terms of the number of random bits and query bits the verifier makes use of, while at the same time the probability of accepting wrong inputs has to be lowered.

In the definition of the class PCP we required the verifier to accept wrong inputs with error probability of at most $1/2$. However, this number $1/2$ is chosen arbitrarily: by repeating the verification process a constant number of times, say k times, the error probability can easily be decreased to $(1/2)^k$ while changing the number of random bits and queries made by the verifier by a constant factor only. Thus if we define $\mathrm{PCP}_\varepsilon(\cdot, \cdot)$ as the class of languages that can be recognized by $(\cdot, \cdot)$-restricted verifiers that have an error probability of at most ε, we obtain

$$\mathrm{PCP}(\log n, 1) = \mathrm{PCP}_{1/2}(\log n, 1) = \mathrm{PCP}_\varepsilon(\log n, 1) \quad \forall \text{ constants } \varepsilon > 0.$$

While the error probability can be reduced to an arbitrarily small constant, the number of queries made by the $(\log n, 1)$-verifier must be at least 3, as the following result shows. (Here we use the notation $\mathrm{PCP}(., queries = .)$ to express that the constant in the $\mathcal{O}$-term for the number of queries is 1). For the adaptive version of this result see Exercise 4.1.)

Proposition 4.4. $\forall \varepsilon > 0 \quad \text{PCP}_\varepsilon(\log n, queries = 2) = \mathcal{P}$

Proof. Clearly $\mathcal{P} \subseteq \text{PCP}_\varepsilon(\log n, queries = 2)$ thus we only have to prove the other inclusion.

Let L be any language in $\text{PCP}_\varepsilon(\log n, queries = 2)$ and let x be any input for which we want to decide in polynomial time whether $x \in L$ or not.

For each of the $2^{O(\log n)}$ random strings we now can simulate the $(\log n, queries = 2)$-restricted verifier to see which two positions it would have queried from the proof and for what values of the queried bits the verifier would have accepted the input. For the ith random string τ_i let b_{i1} and b_{i2} be the two bits queried by the verifier on input x and random string τ_i. Now we can express by a 2SAT-formula in the two variables b_{i1} and b_{i2} whether the verifier would accept the input x. The verifier will accept the input x if and only if all the 2SAT-formulae obtained in this way from all the random strings τ_i can be satisfied simultaneously. Thus we have reduced the problem of deciding whether $x \in L$ to a 2SAT-problem which can be solved in polynomial time. ∎

The number of queries needed by the verifier in the original proof of the PCP-Theorem of Arora et al. is about 10^4. This number has been reduced in a sequence of papers up to the current record due to Bellare, Goldreich and Sudan [BGS95] who proved that 11 queries to the proof suffice for the verifier to achieve an error probability of $1/2$:

$$\mathcal{NP} = \text{PCP}(\log n, queries = 11)$$

However, as we shall see later, to obtain tight non-approximability results 11 queries are still too much. Even if one could proof that three queries suffice, this would not be strong enough to yield the desired non-approximability results. It turned out that instead of counting the number of queries needed by the verifier the right way of measuring the query complexity is expressed in so called *amortized free bits*. We will give the precise definition of this notion in Section 4.7 where it will become clear why amortized free bits are the right way to measure the query complexity of a verifier when one is interested in getting tight non-approximability results. While Proposition 4.4 shows that at least 3 bits have to be queried, unless $\mathcal{P}=\mathcal{NP}$, it was shown in a sequence of papers that the number of amortized free bits queried by the verifier can not only be smaller than 2, but even arbitrarily small. This is roughly the result of Håstad [Hås97a] that we mentioned in the introduction and allows to prove that MAXCLIQUE cannot be approximated up to a factor of $n^{1-\varepsilon}$ for arbitrarily small $\varepsilon > 0$.

4.4 Non-Approximability of $\mathcal{APX}$-Complete Problems

Arora, Lund, Motwani, Sudan and Szegedy [ALM+92] not only proved the PCP-Theorem but at the same time they also proved as a consequence that no $\mathcal{APX}$-complete problem has a PTAS.

Theorem 4.5. *Unless* $\mathcal{P} = \mathcal{NP}$, *no* $\mathcal{APX}$-*complete problem has a PTAS.*

Proof. We will show that the existence of a PTAS for MAX3SAT implies $\mathcal{P} = \mathcal{NP}$. Since MAX3SAT is $\mathcal{APX}$-complete (see Thereom 1.33) this proves the theorem.

Let L be an arbitrary language from $\mathcal{NP}$ and let V be the $(\log n, 1)$-restricted verifier for L whose existence is guaranteed by the PCP-Theorem.

For any input x we will use the verifier V to construct a 3SAT instance S_x such that S_x is satisfiable if and only if x is an element of L. Moreover, if x does not belong to L then at most some constant fraction of the clauses in S_x can simultaneously be satisfied. Therefore a PTAS for MAX3SAT could be used to distinguish between these two cases and the language L could be recognized in polynomial time, i.e., we would have $\mathcal{P} = \mathcal{NP}$.

We now describe how to construct the 3SAT instance S_x for a given input x using the verifier V. We interpret the proof queried by V as a sequence $x_1, x_2, x_3, \ldots$ of bits where the ith bit of the proof is represented by the variable x_i. For a given random string τ the verifier V will query $q = \mathcal{O}(1)$ bits from the proof which we will denote as $b_{\tau 1}, b_{\tau 2}, \ldots, b_{\tau q}$. It will accept the input x for the random string τ, if the bits $b_{\tau 1}, b_{\tau 2}, \ldots, b_{\tau q}$ have the correct values. Therefore we can construct a q-SAT formula F_τ that contains at most 2^q clauses such that F_τ is satisfiable if and only if there exists a proof π such that the verifier V accepts input x on random string τ. This q-SAT formula F_τ can be transformed into an equivalent 3SAT formula F_τ' containing at most $q \cdot 2^q$ clauses and possibly some new variables. We define S_x to be the conjunction of all formulae F_τ' for all possible random strings τ.

If x is an element of L then by definition of a restricted verifier there exists a proof π such that V accepts x for every random string τ. Thus the formula S_x is satisfiable.

If x is not an element of L then for every proof π the verifier V accepts x for at most $1/2$ of all possible random strings τ. This means that at most $1/2$ of the formulae F_τ' are simultaneously satisfiable. Since every F_τ' consists of at most $q \cdot 2^q$ clauses we get that at least a $\frac{1}{2 \cdot q \cdot 2^q}$ fraction of the clauses of S_x are not satisfiable.

The existence of a PTAS for MAX3SAT therefore would allow to distinguish between these two cases and thus it would be possible to recognize every language in $\mathcal{NP}$ in polynomial time. ∎

From this proof and the $\mathcal{NP} = \text{PCP}(\log n, queries = 11)$-result stated in Section 4.3, we immediately get the following corollary.

Corollary 4.6. *Unless* $\mathcal{P} = \mathcal{NP}$ *no polynomial time approximation algorithm for* MAX3SAT *can have a performance guarantee of* $45056/45055 = 1.000022\ldots$

Proof. The proof of Theorem 4.5 has shown that there exists no polynomial time approximation algorithm for MAX3SAT with a performance guarantee of $1/(1 - \frac{1}{2 \cdot q \cdot 2^q})$, unless $\mathcal{P} = \mathcal{NP}$. Setting $q = 11$ we get the claimed result. ∎

The constant we achieved in this corollary of course is far from being optimal. The first reasonable constant for the non-approximability of MAX3SAT was obtained by Bellare, Goldwasser, Lund and Russell [BGLR93]. They obtained $94/93 = 1.0107\ldots$ which was improved by Bellare and Sudan [BS94] to $66/65 = 1.0153\ldots$ and by Bellare, Goldreich and Sudan [BGS95] to $27/26 = 1.0384\ldots$ until very recently Håstad[Hås97b] improved this to the best possible result of $8/7 - \varepsilon = 1.142\ldots$ (see Chapter 7).

The proof of Theorem 4.5 shows that there exists a constant $\varepsilon > 0$ such that the following promise problem, called ROBE3SAT, is $\mathcal{NP}$-hard:

ROBE3SAT

Instance: A 3SAT formula Φ such that Φ is either satisfiable or at least an ε-fraction of the clauses of Φ is not satisfiable

Question: Is Φ satisfiable ?

If we use 3SAT as the language L in the proof of Theorem 4.5 we see that instances of the ordinary 3SAT problem can be transformed in polynomial time into instances of ROBE3SAT such that satisfiable instances are transformed into satisfiable instances and unsatisfiable instances are transformed into instances where at least an ε-fraction of the clauses is unsatisfiable.

In Chapter 5 it will be shown (see Theorem 5.54) that for the verifier constructed in the proof of the PCP-Theorem we have the following property: given a proof π that is accepted with probability larger than the soundness probability for an input x, one can construct in polynomial time a new proof π' that is accepted with probability 1 for input x.

If we apply this result to the instances of ROBE3SAT that are constructed from ordinary 3SAT instances as described in the proof of Theorem 4.5 we get the following result.

Corollary 4.7. *There exists a polynomial time computable function g that maps* 3SAT *instances to* ROBE3SAT *instances and has the following properties:*

$-$ *if x is a satisfiable* 3SAT *instance then $g(x)$ is.*

– *if x is an unsatisfiable* 3Sat *instance then at most an* $1 - \varepsilon$ *fraction of the clauses of* $g(x)$ *are simultaneously satisfiable.*

– *given an assignment to* $g(x)$ *that satisfies more than an* $1 - \varepsilon$ *fraction of the clauses of* $g(x)$ *one can construct in polynomial time a satisfying assignment for* x.

4.5 Expanders and the Hardness of Approximating MaxE3Sat-*b*

In this section we will show that the $\mathcal{NP}$-hardness of RobE3Sat also holds for the special case of RobE3Sat-*b*, i.e., for 3Sat-formulae where each clause contains exactly three literals and each variable appears at most *b* times. This result will be used in Chapter 7 and Chapter 10.

For a constant k, a k-regular (multi-)graph $G = (V, E)$ is called an *expander*, if for all sets $S \subset V$ with $|S| \leqslant |V|/2$ there are at least $|S|$ edges connecting S and $V - S$. The next lemma shows that for every sufficiently large n, there exist sparse expanders.

Lemma 4.8. *For every sufficiently large n and any constant* $k \geqslant 5$ *there exist* $4k$*-regular expanders.*

Proof.   We construct a bipartite graph on sets $A = B = \{1, \ldots, n\}$ by choosing k random permutations $\pi_1, \ldots, \pi_k$ and connecting vertex i in A with vertices $\pi_1(i), \ldots, \pi_k(i)$ in B. This way we get a k-regular bipartite graph on n vertices. We claim that there exist permutations $\pi_1, \ldots, \pi_k$ such that whenever we choose a set $S \subset A$ with $|S| < n/2$ then there are at least $3|S|/2$ vertices in B that are adjacent to some vertex in S.

Let S be a subset of A with $t := |S| \leqslant n/2$ and T be the set containing all vertices in B that are adjacent to some vertex in S, such that $m := |T| = \lfloor (3|S| - 1)/2 \rfloor$. We will call such pairs (S, T) *bad*. The probability that for randomly chosen permutations $\pi_1, \ldots, \pi_k$ and fixed sets S and T these sets form a bad pair equals

$$\left(\binom{m}{t} \frac{t!(n-t)!}{n!} \right)^k .$$

If we sum this up over all possible choices for S and T we see that the probability that there exists a bad pair is bounded by

$$\sum_{t \leqslant \frac{n}{2}} \binom{n}{m} \left(\binom{n}{t} \frac{m!(n-t)!}{n!(m-t)!} \right)^k .$$

For $1 \leqslant t < n/3$ it is easily verified that the function

$$f(t) := \binom{n}{t}\binom{n}{m}\frac{m!(n-t)!}{n!(m-t)!}$$

satisfies $f(t) > f(t+1)$, so the maximum of f in this range is attained at $t = 1$. For $n/3 \leqslant t \leqslant n/2$ the expression $\frac{m!(n-t)!}{(m-t)!}$ reaches its maximum for $t = n/3$ or $t = n/2$. Now a simple computation using Stirling's formula shows that $\frac{n}{2}(f(1) + f(n/3) + f(n/2))$ tends to zero for $k \geqslant 5$ and n going to infinity.

Therefore there exist k-regular bipartite graphs with n vertices on each side such that every set S from the left side with $|S| \leqslant n/2$ has at least $3/2|S|$ neighbors on the right side. Now, if we identify vertices with the same number from both sides and duplicate each edge, we obtain a $4k$-regular (multi-)graph on n vertices such that from every set S of vertices of size at most $n/2$ there are at least $|S|$ edges leaving the set S. ∎

While the above lemma shows the *existence* of expanders only, it can be shown that expanders can be *constructed* explicitly in polynomial time. See for example [Mar75] or [GG81] for such constructions.

With the help of expanders we are now able to prove the desired hardness result for RoBE3SAT-b.

Lemma 4.9. *There exists a constant b such that* RoBE3SAT-b *is $\mathcal{NP}$-hard.*

Proof. We prove this by reducing from RoBE3SAT. Given a 3SAT formula F with clauses $C_1, \ldots, C_m$ and variables $x_1, \ldots, x_n$, we replace each of the say k_i occurrences of the variable x_i by new variables $y_{i,1}, \ldots, y_{i,k_i}$ and choose an expander G_i with $y_{i,1}, \ldots, y_{i,k_i}$ as its vertices. Now we add the clauses $(y_{i,a} \vee \overline{y}_{i,b})$ and $(\overline{y}_{i,a} \vee y_{i,b})$ whenever $y_{i,a}$ and $y_{i,b}$ are connected by an edge in G_i. We call this new 3SAT formula F'. Since the expanders G_i have constant degree, each variable in F' appears only a constant number of times. Moreover, F' still has $\mathcal{O}(m)$ many clauses. Whenever we have an assignment to F' we may assume that for all i the variables $y_{i,1}, \ldots, y_{i,k_i}$ have the same value. If this was not the case, we simply could change the values of $y_{i,1}, \ldots, y_{i,k_i}$ to the value of the majority of these variables. This way, we may loose up to $k_i/2$ satisfied clauses, but at the same time we gain at least $k_i/2$. This follows from the fact that G_i is an expander and therefore every set S of vertices of size at most $k_i/2$ has at least $|S|$ edges leaving it. Each of these edges yields an unsatisfied clause before changing the values of $y_{i,1}, \ldots, y_{i,k_i}$ to the value of their majority. Therefore a solution to F' can be used to define a solution to F and both formulae have the same number of unsatisfiable clauses. ∎

Corollary 4.10. *There exist constants $\delta > 0$ and b such that no polynomial time algorithm can approximate* MaxE3SAT-b *up to a factor of $1 + \delta$, unless $\mathcal{P}=\mathcal{NP}$.*

Proof. Suppose there exists a polynomial time $1+\delta$ approximation algorithm for MAXE3SAT-b for all $\delta > 0$. Given an instance of ROBE3SAT-b such that either all clauses are satisfiable or at least an ε-fraction of the clauses is not satisfiable, this algorithm with $\delta < \varepsilon/(1-\varepsilon)$ could be used to distinguish between these two cases in polynomial time. This implies $\mathcal{P}=\mathcal{NP}$ because of Lemma 4.9. ∎

4.6 Non-Approximability of MAXCLIQUE

A *clique* in a graph is a set of pairwise adjacent vertices. The problem CLIQUE is defined as follows.

CLIQUE
Instance: Given a graph G and an integer k
Question: Is there a clique of size $\geq k$ in G ?

The corresponding optimization problem is called MAXCLIQUE.

MAXCLIQUE
Instance: Given a graph G
Problem: What is the size of a largest clique in G ?

While it is a classical result due to Karp [Kar72] that CLIQUE is $\mathcal{NP}$-complete there was not any non-approximability result known for the problem MAX-CLIQUE up to the year 1991 when Feige, Goldwasser, Lovász, Safra and Szegedy [FGL+91] observed a connection between PCPs and MAXCLIQUE. The only result known long before this is the fact that MAXCLIQUE is a *self-improving* problem:

Proposition 4.11. *If for any constant c there is a c-approximation-algorithm for* MAXCLIQUE, *then there also exists a $\sqrt{c}$-approximation algorithm for* MAX-CLIQUE.

Proof. This result follows immediately from the fact that the product of a graph G with itself (replace each vertex of G by a copy of G and join two such copies completely if the corresponding vertices in G are adjacent) yields a new graph G' whose maximum clique size is the square of the size of a maximum clique in G. Thus a c-approximation algorithm for G' can be used to obtain a $\sqrt{c}$-approximation algorithm for G. ∎

This self-improving property implies that if there exists *any* constant factor approximation algorithm for MAXCLIQUE then there even exists a PTAS for this problem. As the best known approximation algorithm for MAXCLIQUE due to Boppana and Halldórsson [BH92] has a performance guarantee of $\mathcal{O}(n/\log^2 n)$, the existence of a PTAS for MAXCLIQUE was assumed to be extremely unlikely but could not be ruled out before 1991.

In this section we will see how the PCP-Theorem implies the nonexistence of a polynomial time n^ε-approximation algorithm for MAXCLIQUE for some $\varepsilon > 0$, while in the next section we will show that even an $n^{1-\delta}$ approximation algorithm does not exist for this problem for arbitrarily small δ, unless $\mathcal{NP}=\mathcal{ZPP}$. We start by showing that no polynomial time constant factor approximation algorithm for MAXCLIQUE can exist.

Proposition 4.12. *Unless $\mathcal{P}=\mathcal{NP}$, no polynomial time constant factor approximation algorithm for* MAXCLIQUE *can exist.*

Proof.   We use the standard reduction from 3SAT to CLIQUE to prove this result.

For a given 3SAT-formula F with clauses $C_1,\ldots,C_m$ and variables $x_1,\ldots,x_n$ we construct a graph G on $3m$ vertices $(i,j), i = 1,\ldots,m; j = 1,2,3$ as follows. The vertices (i,j) and (i',j') are connected by an edge if and only if $i \neq i'$ and the jth literal in clause i is not the negation of the j'th literal in clause i'.

If there exists a clique in G of size k then it contains at most one literal from each clause, and it contains no two literals that are the negation of each other. Therefore, by setting all literals corresponding to vertices of this clique to *true* one gets a truth assignment for F that satisfies at least k of its clauses. On the other hand, given a truth assignment for F that satisfies k of the clauses, one gets a clique of size k in G by simply selecting from each satisfied clause one literal that evaluates to *true* in this assignment.

Thus we have shown that the graph G has a clique of size k if and only if there exists a truth assignment for F that satisfies k of its clauses. This shows that a PTAS for MAXCLIQUE cannot exist as otherwise we would also get a PTAS for MAX3SAT which is ruled out by Theorem 4.5. Proposition 4.11 now implies that for no constant c a c-approximation algorithm for MAXCLIQUE can exist. ∎

To prove better non-approximability results for MAXCLIQUE, especially for proving the n^ε non-approximability we have to make a more direct use of the PCP-Theorem. To start with we first present a reduction from 3SAT to CLIQUE that is slightly different from the one used in the proof of Proposition 4.12 and was used by Papadimitriou and Steiglitz [PS82] to prove the $\mathcal{NP}$-completeness of CLIQUE.

Proof.   (Second proof for Proposition 4.12)

Again we are given a 3SAT-formula F with clauses $C_1,\ldots,C_m$ and variables $x_1,\ldots,x_n$. The idea this time is that for each clause we want to list all truth assignments that make this clause true. Instead of listing this exponential number of assignments, we only list 7 *partial* truth assignments for each clause.

A partial truth assignment assigns the values *true* and *false* to certain variables only; the rest of the variables has the value '·', meaning that the value is undefined. As an example, a partial truth assignment for variables $x_1, \ldots, x_9$ might look like ·10·0··01. We say that two different truth assignments t and t' are *compatible*, if for all variables x for which $t(x) \neq \cdot$ and $t'(x) \neq \cdot$ we have $t(x) = t'(x)$. For each clause C_i there are exactly 7 satisfying partial truth assignments with values defined only on the three variables appearing in C_i (we assume here without loss of generality that every clause contains exactly three different variables). We construct for each of the m clauses of F these 7 partial truth assignments and take them as the vertices of our graph G. Two vertices in G are connected if the corresponding truth assignments are compatible.

First note that no two partial truth assignments corresponding to the same clause of F can be compatible and therefore G is an m-partite graph. Now if there exists a clique of size k in G then this means that there is a set of k pairwise compatible partial truth assignments for k different clauses of F. Thus there exists one truth assignment that satisfies all these k clauses simultaneously. On the other hand, if there is a truth assignment for F that satisfies k of its clauses, then there is one partial truth assignment for each of these clauses that is compatible to this truth assignment and therefore these k partial truth assignments are pairwise compatible yielding a clique of size k in G.   ∎

We will now see – as was discovered by Feige, Goldwasser, Lovász, Safra and Szegedy [FGL+91] – that the reduction of Papadimitriou and Steiglitz applied to the PCP-result will achieve the n^ε non-approximability result for CLIQUE. As a first step we will prove once more Proposition 4.12.

Proof. (Third proof for Proposition 4.12)

Let L be an $\mathcal{NP}$-complete language and V be its $(\log n, 1)$-restricted verifier whose existence is guaranteed by the PCP-Theorem. Let $r(n) = \mathcal{O}(\log n)$ and $q(n) = \mathcal{O}(1)$ be the number of random bits respectively query bits used by the verifier V. Now for an input x we construct a graph G_x in an analogous way as Papadimitriou and Steiglitz did in their reduction from 3SAT to CLIQUE as described in the second proof of Proposition 4.12. The role of a clause appearing in the 3SAT formula is now taken by a random string read by the verifier and the 3 variables appearing in a given clause correspond to the $q(n)$ positions queried from the proof by the verifier.

For each of the possible $2^{r(n)}$ random strings we list all of the at most $2^{q(n)}$ partial proofs (i.e., assignments of 0 and 1 to the positions queried by the verifier for the given random string, and assignment of '·' to all other positions) that will make the verifier V accept the input x. All these partial proofs are vertices in our graph G_x and we connect two such vertices if they are compatible (as defined above). The graph G_x has at most $2^{r(n)+q(n)}$ vertices and since for two given partial proofs it can be decided in polynomial time whether they are compatible, the graph can be constructed in polynomial time.

For a fixed proof π any two vertices of G_x that are compatible with π are adjacent. Therefore, if there exists a proof π such that the verifier V accepts the input x for k different random strings, then the graph G_x contains a clique of size k.

If on the other hand, the graph G_x contains a clique of size k, then the k partial proofs corresponding to the vertices of the clique are pairwise compatible and as no two partial proofs that correspond to the same random string are compatible with each other, there must exist a proof π such that the verifier accepts the input x for k different random strings.

Thus we have shown that the size of a maximum clique in G_x equals the maximum number of random strings for which the verifier accepts a proof π, where the maximum is taken over all proofs π.

Now if $x \in L$ then by the definition of PCP there exists a proof π such that the verifier accepts for all possible random strings. Thus in this case $\omega(G_x) = 2^{r(n)}$.

If $x \notin L$ then by the definition of PCP for each proof π the verifier accepts x for at most $1/2$ of the random strings. Therefore we have $\omega(G_x) \leqslant \frac{1}{2} 2^{r(n)}$ in this case.

Now a 2-approximation algorithm for MaxClique could be used to recognize the $\mathcal{NP}$-complete language L in polynomial time.   ∎

For the reduction we used in the above proof, the non-approximability factor we obtain for MaxClique solely depends on the error probability of the verifier. We have already seen, that this error probability can be made arbitrarily, but constantly small as we know that $\mathcal{NP} = PCP_\varepsilon(\log n, 1)$ for all $\varepsilon > 0$. To obtain an n^ε non-approximability result for MaxClique we had to reduce the error probability of the verifier to $n^{-\varepsilon}$. Clearly this can be done by running the $(\log n, 1)$-restricted verifier for $\mathcal{O}(\log n)$ independent random strings; however this would result in a total number of $\mathcal{O}(\log^2 n)$ random bits needed by the verifier which results in a graph that can no longer be constructed in polynomial time.

The idea here now is that instead of using truly random bits one can make use of so called *pseudo random bits* that can be generated by performing a random walk on an expander graph. It can be shown that by this method one can generate $\alpha \log n$ pseudo random strings of length $\mathcal{O}(\log n)$ by using only $c \cdot \alpha \log n$ truly random bits (for more details on this see for example [HPS94]). Thus, starting with an $(\log n, 1)$-verifier V that has error probability of $1/2$ we can construct a new verifier V' that simulates the verifier V $\alpha \log n$ times. If q is the number of bits queried by the verifier V, then we get for the new verifier V' :

error probability	:	$n^{-\alpha}$
# random bits	:	$c \cdot \alpha \log n$
# query bits	:	$q \cdot \alpha \log n$

Now if we use this verifier to construct for an input x a graph G_x as described above, we get that the clique number of G_x cannot be approximated up to n^α for arbitrarily large but constant α.

The graph G_x has

$$N := 2^{c \cdot \alpha \log n + q \cdot \alpha \log n} = n^{c \cdot \alpha + q \cdot \alpha}$$

vertices. Thus we get:

$$n^{-\alpha} = N^{-\frac{\alpha}{c\alpha + q\alpha}} = N^{-\frac{1}{c+q}}$$

As c and q are constants, we have shown:

Theorem 4.13. *Unless $\mathcal{P}=\mathcal{NP}$, there exists a constant $\varepsilon > 0$ such that no n^ε approximation algorithm for MAXCLIQUE can exist.*

4.7 Improved Non-Approximability Results for MAXCLIQUE

In the last section we have seen, that MAXCLIQUE cannot be approximated up to n^ε for some constant ε. Here we now want to see how large this ε can be.

The value of ε was $\varepsilon = 1/(c + q)$ where c was a constant that came in from the generation of pseudo random bits and q is the number of queries made by the $(\log n, 1)$-restricted verifier. Thus to achieve a small value for ε we have to try to minimize these two constants. It can be shown that by using Ramanujan-expanders due to Lubotzky, Phillips and Sarnak [LPS86] for the generation of pseudo random bits, the constant c can almost achieve the value 2. As we already know that 11 queries are enough for an $(\log n, 1)$-restricted verifier, this shows that we can choose $\varepsilon = 0.076$.

From this value for ε up to the ultimate result due to Håstad [Hås97a] showing that ε can be chosen arbitrarily close to 1, there was a long sequence of improvements which is surveyed in Table 4.1.

Friedman [Fri91] has shown that the Ramanujan-expanders of Lubotzky, Phillips and Sarnak are best possible, meaning that the constant c must have at least the value 2. On the other hand we know from Proposition 4.4 that q must be at least 3. Thus to get values of ε that are larger than $1/5$ we need some new ideas.

First we note, that in the construction of the graph G_x in the third proof of Proposition 4.12 we listed for each of the $2^{r(n)}$ random bits all partial proofs that made the verifier V accept the input x. As V queries at most $q(n)$ bits there can be at most $2^{q(n)}$ partial proofs for which V accepts x. However, a close look at the proof of the PCP-Theorem reveals, that for a fixed random string there are usually much less than $2^{q(n)}$ accepted partial proofs. The reason for

Authors	Factor	Assumption
Feige, Goldwasser, Lovász, Safra, Szegedy 91	$\exists \varepsilon > 0, 2^{\log^{1-\varepsilon} n}$	$\mathcal{N}\tilde{\mathcal{P}} \neq \tilde{\mathcal{P}}$
Arora, Safra 92	$\exists \varepsilon > 0, 2^{\log^{1-\varepsilon} n}$	$\mathcal{N}\mathcal{P} \neq \mathcal{P}$
Arora, Lund, Motwani, Sudan, Szegedy 92	$n^{1/10000}$	$\mathcal{N}\mathcal{P} \neq \mathcal{P}$
Bellare, Goldwasser, Lund, Russell 93	$n^{1/30}$ $n^{1/25}$	$\mathcal{N}\mathcal{P} \neq \text{co-}\mathcal{R}\mathcal{P}$ $\mathcal{N}\tilde{\mathcal{P}} \neq \text{co-}\mathcal{R}\tilde{\mathcal{P}}$
Feige, Kilian 94	$n^{1/15}$	$\mathcal{N}\mathcal{P} \neq \text{co-}\mathcal{R}\mathcal{P}$
Bellare, Sudan 94	$\forall \varepsilon, n^{1/6-\varepsilon}$ $\forall \varepsilon, n^{1/5-\varepsilon}$ $\forall \varepsilon, n^{1/4-\varepsilon}$	$\mathcal{N}\mathcal{P} \neq \mathcal{P}$ $\mathcal{N}\mathcal{P} \neq \text{co-}\mathcal{R}\mathcal{P}$ $\mathcal{N}\tilde{\mathcal{P}} \neq \text{co-}\mathcal{R}\tilde{\mathcal{P}}$
Bellare, Goldreich, Sudan 95	$\forall \varepsilon, n^{1/4-\varepsilon}$ $\forall \varepsilon, n^{1/3-\varepsilon}$	$\mathcal{N}\mathcal{P} \neq \mathcal{P}$ $\mathcal{N}\mathcal{P} \neq \text{co-}\mathcal{R}\mathcal{P}$
Håstad 96	$\forall \varepsilon, n^{1/2-\varepsilon}$	$\mathcal{N}\mathcal{P} \neq \text{co-}\mathcal{R}\mathcal{P}$
Håstad 96	$\forall \varepsilon, n^{1-\varepsilon}$	$\mathcal{N}\mathcal{P} \neq \text{co-}\mathcal{R}\mathcal{P}$

Table 4.1. Non-approximability results for MaxClique.

this is, that the verifier will often query some bits, say b_1, b_2, b_3 and then it will test whether these queried bits satisfy a certain relation, say $g(b_1) + g(b_2) = g(b_3)$ for some function g. If this relation is not satisfied, then V will not accept the input x. For this example it follows, that instead of 8 possible answers for the bits b_1, b_2, b_3 there can be at most 4 answers for which V accepts the input x. Roughly speaking, instead of counting the number of bits queried by the verifier from the proof, it is only of interest, how many of these queried bits have no predetermined value. These bits are called *free bits* and denoted by f.

More precisely the number f of *free bits* queried by a verifier is defined as

$$f := \log(\max_{\tau, x} \# \text{ partial proofs for which } V \text{ accepts } x).$$

Following [BS94] we define the class FPCP as the free bit variant of PCP, i.e., the class of languages where we measure the number of free query bits instead of query bits.

The idea of free bits appears for the first time in the paper of Feige and Kilian [FK94] who proved that MAXCLIQUE cannot be approximated up to $n^{1/15}$ unless $\mathcal{NP}$=co-$\mathcal{RP}$. The name 'free bits' was invented by Bellare and Sudan [BS94].

Thus, to improve the non-approximability factor for MAXCLIQUE, which we now know is $\varepsilon = 1/(c+f)$ one carefully has to look at the proof of the PCP-Theorem to see how many free bits are needed. Bellare, Goldreich and Sudan [BGS95] have shown in a result similar to Proposition 4.4 that at least 2 free bits are needed.

Proposition 4.14. $\forall \varepsilon > 0$ $\text{FPCP}_\varepsilon(\log n,\ \text{free bits} = 1) = \mathcal{P}$

On the other hand they also showed that 2 free bits suffice for a $(\log n, 1)$-restricted verifier to recognize any $\mathcal{NP}$-language.

Theorem 4.15. $\mathcal{NP} \subseteq \text{FPCP}_{0.794}(\log n,\ \text{free bits} = 2)$

They also proved the best known result for error probability $1/2$.

Theorem 4.16. $\mathcal{NP} \subseteq \text{FPCP}_{1/2}(\log n,\ \text{free bits} = 7)$

Still these results only yield that we cannot get a polynomial time $n^{1/4}$-approximation algorithm for MAXCLIQUE. Before further improving on the query complexity we will see how the constant c in the expression for ε can be decreased to 1 by using a more efficient method of generating pseudo random bits due to Zuckerman [Zuc93].

An (m, n, d)-*amplification scheme* is a bipartite graph $G = (A \cup B, E)$ with $|A| = m$ and $|B| = n$ such that every vertex in A has degree d. We construct (m, n, d)-amplification schemes uniformly at random by choosing for each vertex in A uniformly at random d elements of B as neighbors. We are interested in amplification schemes that satisfy a certain expansion property.

An (m, n, d, a, b)-*disperser* is a bipartite graph $G = (A \cup B, E)$ with m vertices on the left and n vertices on the right such that every vertex on the left has degree d and each subset of size a on the left has at least b neighbors. The following result shows that for certain parameter sets (m, n, d, a, b)-dispersers can randomly be constructed in a very simple way.

Lemma 4.17. *The probability that a uniformly at random chosen $(2^R, 2^r, R+2)$-amplification scheme is a $(2^R, 2^r, R+2, 2^r, 2^{r-1})$-disperser is at least $1 - 2^{-2^r}$.*

Proof. For $S \subseteq 2^R$ and $T \subseteq 2^r$ let $A_{S,T}$ be the event that all neighbors of S are in T. Then the probability that the randomly chosen $(2^R, 2^r, R+2)$-amplification scheme is not the desired disperser equals

$$\text{Prob}[\bigcup_{\substack{|S|=2^r \\ |T|=2^{r-1}-1}} A_{S,T}] \leqslant \sum_{\substack{|S|=2^r \\ |T|=2^{r-1}-1}} \text{Prob}[A_{S,T}]$$

$$= \binom{2^R}{2^r}\binom{2^r}{2^{r-1}-1}\left(\frac{2^{r-1}-1}{2^r}\right)^{(R+2)2^r}$$

$$< 2^{R2^r} 2^{2^r} 2^{-(R+2)2^r}$$

$$= 2^{-2^r}$$

∎

We will use these $(2^R, 2^r, R+2, 2^r, 2^{r-1})$-dispersers to generate $R+2$ pseudo random strings of length r in a very simple way: We simply choose a vertex from the left side and take all its $R+2$ neighbors as pseudo random strings. The following result shows that by doing so we can reduce the constant c to 1.

Theorem 4.18. *Unless $\mathcal{NP}=\mathcal{ZPP}$ no polynomial time algorithm can achieve an approximation factor of $n^{\frac{1}{1+f}-\epsilon}$ for* MaxClique *for arbitrarily small ϵ.*

Proof. Let V be a verifier for recognizing a language L that uses $r(n)$ random bits, queries f free bits and achieves an error probability of $1/2$. We will construct a verifier V' now as follows.

The verifier V' first uniformly at random chooses a $(2^R, 2^{r(n)}, R+2)$-amplification scheme which by Lemma 4.17 is with very high probability a $(2^R, 2^{r(n)}, R+2, 2^{r(n)}, 2^{r(n)-1})$-disperser. Now V' randomly selects a vertex from the left side of the disperser and uses its $R+2$ neighbors as random strings of length $r(n)$. For each of these $R+2$ random strings the verifier V' simulates the verifier V for input x. It accepts x, if and only if V accepts x for all $R+2$ runs.

The verifier V' uses R random bits and its free bit complexity is $(R+2)f$. Thus the graph G_x constructed for input x by the same construction as described in the third proof of Proposition 4.12 has $N := 2^{R+(R+2)f}$ vertices.

If $x \in L$ then there exists a proof π such that V accepts input x for all random strings and therefore V' accepts for all 2^R random strings. Thus the graph G_x has a clique of size 2^R.

If $x \notin L$ then we claim that G_x has a clique of size at most $2^{r(n)}$. Assume this is not the case, i.e., there exist $p > 2^{r(n)}$ random strings for which V' accepts input x. Since V accepts input x for less than $\frac{1}{2}2^{r(n)}$ random strings, this means that there are $p > 2^{r(n)}$ vertices on the left of the disperser whose neighborhood contains at most $2^{r(n)-1} - 1$ vertices. This contradicts the definition of a $(2^R, 2^{r(n)}, R+2, 2^{r(n)}, 2^{r(n)-1})$-disperser.

Thus we cannot distinguish in polynomial time whether there is a clique of size 2^R or whether every clique has size at most $2^{r(n)}$, i.e., MaxClique cannot be approximated up to a factor of $2^{R-r(n)}$.

Now we have:

$$N = 2^{R+(R+2)f}$$

$$= 2^{R(1+f)} \cdot 2^{2f}$$

$$\Rightarrow \quad 2^R = \left(\frac{N}{2^{2f}}\right)^{\frac{1}{1+f}}$$

If we choose $R = \alpha \log n$ we get for $\alpha \to \infty$ that MaxClique cannot be approximated up to $N^{\frac{1}{1+f}-\varepsilon}$ for arbitrarily small ε, unless $\mathcal{NP}$=co-$\mathcal{RP}$ (we used a randomized construction for the disperser). ∎

Since by Theorem 4.15 we know that we can set $f = 2$ we get that no $n^{1/3-\varepsilon}$ approximation algorithm can exist for MaxClique. On the other hand we know that f must be larger than 1. Thus we need to refine the notion of query complexity once more to arrive at the final tight non-approximability result for MaxClique.

The observation here now is, that for our non-approximability results we only made use of the fact that there is a certain gap between the completeness and soundness probability of the verifier. So far we have assumed that the completeness probability is always 1; in this case the verifier is said to have *perfect completeness*. All the non-approximability results for MaxClique we have seen so far do not need perfect completeness. We already have observed the trade off between the error gap and the number of queries: we can square the error gap by just doubling the number of queries, i.e., if we want to enlarge the logarithm of the error gap by a factor of k, then we have to allow the verifier to use k times as many query bits. A careful analysis of the proof of Theorem 4.18 reveals that in fact the non-approximability-factor for MaxClique does not just depend on f but on the ratio of f and the logarithm of the error gap.

This motivates the definition of so called *amortized free bit complexity* . If a verifier has completeness probability c and soundness probability s and has free bit complexity f, then its amortized free bit complexity $\bar{f}$ is defined as

$$\bar{f} := f/\log(c/s).$$

We define the class $\overline{\text{FPCP}}$ as the amortized free bit complexity variant of PCP. In the proof of Theorem 4.18 we used an error gap of 2. Bellare and Sudan

[BS94] have shown that the same result as Theorem 4.18 can be proved in terms of amortized free bit complexity.

Theorem 4.19. *Unless* $\mathcal{NP}=\mathcal{ZPP}$ *no polynomial time algorithm can achieve an approximation factor of* $n^{\frac{1}{1+7}-\varepsilon}$ *for* MAXCLIQUE *for arbitrarily small* ε.

Håstad [Hås97a] has shown (see Chapter 9) that $\mathcal{NP} \subseteq \overline{\text{FPCP}}(\log n,$ *amortized free bits* $= 0$). Together with Theorem 4.19 this yields the desired non-approximability result for MAXCLIQUE.

Theorem 4.20. *Unless* $\mathcal{NP}=\mathcal{ZPP}$ *no polynomial time algorithm can achieve an approximation factor of* $n^{1-\varepsilon}$ *for* MAXCLIQUE *for arbitrarily small* ε.

Exercises

Exercise 4.1. Prove Proposition 4.4 for the adaptive case.

Exercise 4.2. $\text{PCP}(1, \log n) = ?$

Exercise 4.3. What non-approximability result for MAXCLIQUE can be obtained by rerunning the $(\log n, 1)$-restricted verifier $\mathcal{O}(\log n)$ times, i.e., without making use of pseudo random bits ?

Exercise 4.4. $\text{FPCP}(\log n,$ *free bits* $= 1) = \mathcal{P}$

Exercise 4.5. Why did we not have to take into account in the proof of Theorem 4.18 the random bits needed to create the disperser ?

Exercise 4.6. Show that the PCP-Theorem implies that the average number of queries needed by a $(\log n, 1)$-restricted verifier can be made arbitrarily small, while increasing the error probability.

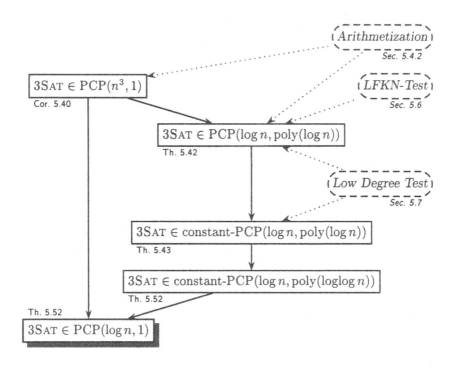

Fig. 5.1. Structure of the proof of the PCP-Theorem
(The small numbers refer to the corresponding theorems and sections)

all values of the considered linear function using some fixed ordering on the arguments.

The redundancy of such a proof has the following three important properties. It is possible to check whether a given proof is nearly a linear encoding of an assignment for 3SAT by reading only a constant number of bits of the proof. Moreover, the linear encoding is fault-tolerant, i.e., whenever only a small fraction of the proof differs from a linear code, we are able to reconstruct the correct values using only a constant number of bits of the proof. Finally, a constant number of bits of the proof will suffice to decide whether the linear encoding corresponds to a satisfying assignment. Since the dimension of the considered vector space is polynomial in the size of the Boolean formula, the length of the corresponding proof is exponential in the size of the given Boolean formula and hence a polynomial number of random bits is required to get random access to the proof. This establishes that $3\text{S}\text{AT} \in \text{PCP}(n^3, 1)$. The details of this construction will be explained in Section 5.4 summarized in Corollary 5.40.

The key idea for reducing the number of random bits, which also implies a reduction of the length of the proof, is a more concise encoding of an assignment. Instead of linear functions we use multivariate polynomials of low degree as an

5. Proving the PCP-Theorem

Volker Heun, Wolfgang Merkle, Ulrich Weigand

5.1 Introduction and Overview

The PCP-Theorem states that $\mathcal{NP}$ is equal to the class $\mathrm{PCP}(\log n, 1)$ of languages which are recognized by verifiers which on an input of length n use at most $\mathcal{O}(\log n)$ random bits and query only $\mathcal{O}(1)$ positions of their proofs. In the following we give an essentially self-contained account of the PCP-Theorem and its proof which is based on the original work due to Arora [Aro94], Arora, Lund, Motwani, Sudan, and Szegedy [ALM+92], and Arora and Safra [AS92]. We also use ideas and concepts from the presentation given by Hougardy, Prömel, and Steger [HPS94].

In this section, we give an outline of the proof of the PCP-Theorem, which states that $\mathcal{NP} = \mathrm{PCP}(\log n, 1)$. As an implication of Proposition 4.2, we obtain $\mathrm{PCP}(\log n, 1) \subseteq \mathcal{NP}$. Hence it remains to show the reverse containment $\mathcal{NP} \subseteq \mathrm{PCP}(\log n, 1)$. Clearly, it is sufficient to show that an $\mathcal{NP}$-complete problem is contained in $\mathrm{PCP}(\log n, 1)$. In the remainder of this chapter, we will prove that $3\textsc{Sat} \in \mathrm{PCP}(\log n, 1)$, which completes the proof of the PCP-Theorem.

The structure of the proof that $3\textsc{Sat} \in \mathrm{PCP}(\log n, 1)$ is illustrated in Figure 5.1. As introduced in Chapter 4, $\mathrm{PCP}(r, q)$ denotes the class of languages with probabilistically checkable proofs such that a verifier requires $\mathcal{O}(r)$ random bits and queries $\mathcal{O}(q)$ bits. In this figure only, constant-$\mathrm{PCP}(r, q)$ denotes the restricted class of languages with probabilistically checkable proofs, where the verifier is restricted to query the $\mathcal{O}(q)$ bits from a constant number of segments. Here a segment is a contiguous block of $\mathcal{O}(q)$ bit positions.

Obviously, a satisfying assignment of a given Boolean formula is a proof of its satisfiability. But clearly it is not sufficient to read only a constant number of bits of this proof for a verification. A key idea is to extend such a proof using linear functions from $\mathbb{F}_2^m$ into $\mathbb{F}_2$, where the images of basis vectors depend on the given assignment in a more or less natural way. Due to the linearity the images of the basis vectors of $\mathbb{F}_2^m$ determine the images of all other vectors and m will be a polynomial in the length of the given Boolean formula. The details will be explained in Section 5.3. We will refer to this encoding as *linear code* and in general we call the conversion of an assignment into a function an *arithmetization*. A proof will be constructed from the linear code by listing encoding. We will refer to such an encoding based on multivariate polynomials as the *polynomial code*. In this case the length of such an encoding will be polynomially related to the size of the given Boolean formula as we will see in Section 5.3. The polynomial code has similar properties as the linear code. It is possible to verify whether a given proof is a polynomial encoding of an assignment and the polynomial code is fault-tolerant. Unfortunately, for the verification and the reconstruction a polylogarithmic number of bits of the proof have to be inspected. Nevertheless, this is the second major step in proving the PCP-Theorem and will be stated in Theorem 5.42. The details will be given in Section 5.4. The proof of these details is rather complicated and requires mathematical tools like the so-called LFKN-Test (cf. Section 5.6) and the Low Degree Test (cf. Section 5.7). As shown in Section 5.4 the result can be improved in the following sense. There exists a probabilistically checkable proof for 3SAT such that only a constant number of 'contiguous blocks' of polylogarithmic length have to be inspected by the verifier instead of polylogarithmic bits at arbitrary positions, which will be shown in Theorem 5.43.

As we will see in Section 5.5 it is possible to compose two verifiers. This is the main idea for a further reduction of the number of queried bits while the length of the proof does not increase significantly. First, the proof system based on the polynomial code is composed with itself obtaining a probabilistically checkable proof of polynomial length, where only a constant number of blocks of size $\mathcal{O}(\text{poly}(\log\log n))$ have to be inspected by the verifier. Then these blocks of length $\mathcal{O}(\text{poly}(\log\log n))$ in the composed proof system will be encoded using the proof system based on linear codes. So we obtain a proof system of polynomial length where only a constant number of bits have to be inspected by the verifier. This will be discussed in detail in Section 5.5 and summarized in Theorem 5.52. As a consequence we obtain $\mathcal{NP} = \text{PCP}(\log n, 1)$ as stated in Theorem 5.53.

5.2 Extended and Constant Verifiers

Extended Verifiers. We formulate the proof of the PCP-Theorem in terms of extended verifiers as introduced in Definition 5.1. Extended verifiers resemble rather closely usual verifiers as introduced in Chapter 4, however, there are two main differences. Firstly, the input of an extended verifier is not a single argument x, but a pair (x, π_0) of arguments where π_0 is a designated part of the proof. Secondly, extended verifiers are not exclusively designed to accept a subset of all possible inputs in the sense that for every input (x, π_0) the extended verifier either accepts or rejects (x, π_0) with high probability: we will consider extended verifiers where for example there are such pairs which are accepted and rejected with equal probability of $1/2$.

Definition 5.1. *An* extended verifier V *is a polynomial time bounded Turing machine which, besides its input tape and its work tapes, has access to a binary random string τ and to a proof $\pi = (\pi_0, \pi_1)$ where we assume that*

- *V has random access to the binary strings π_0 and π_1, that is, V can access single bits of these strings by writing corresponding queries on a special oracle tape,*

- *V queries π non-adaptively, that is, the queries to the proof might depend on the input and the random string, but not on the answers to previous queries to the proof.*

Here we assume, firstly, that the random string τ is obtained by independent tosses of a fair coin and, secondly, that the length of τ can be computed from the length of the input x in polynomial time. (Thus in particular we cannot hide information in the length of the supplied random string.)

An extended verifier V accepts $(x, (\pi_0, \pi_1))$ on random string τ iff V eventually halts in an accepting state while working on x, π_0, π_1, and τ, and otherwise V rejects $(x, (\pi_0, \pi_1))$ on random string τ.

An extended verifier V accepts $(x, (\pi_0, \pi_1))$ iff V accepts this pair on all random strings, and V rejects $(x, (\pi_0, \pi_1))$ iff V accepts this pair on at most 1/4 of all random strings.

An extended verifier V accepts the input (x, π_0) iff there is some binary string π_1 such that V accepts $(x, (\pi_0, \pi_1))$ and V rejects the input (x, π_0) iff V rejects $(x, (\pi_0, \pi_1))$ for all binary strings π_1.

Given an extended verifier V, we denote by $A[V]$ the set of all pairs (x, π_0) which V accepts, and by $R[V]$ the set of all such pairs which V rejects.

Observe that in connection with extended verifiers we use several variants of accepting and rejecting and that we assume that it will always be clear from the context which variant is meant.

An extended verifier queries its proofs non-adaptively and thus it is convenient to assume that the answers to its queries are given simultaneously as a finite binary string written on some designated tape. In addition we assume that there is some appropriate mechanism to indicate queries beyond the end of the strings π_0 or π_1. Observe that for resource bounded extended verifiers as considered in the following it does not really matter whether we assume that the proofs π_0 and π_1 are given as binary strings or as an infinite set oracle because anyway such an extended verifier can access just a finite part of its proof.

Resource Bounds for Extended Verifiers. We are mainly interested in extended verifiers with additional bounds on the length of the supplied random string and on the number of bits the verifier reads from the proof. Observe in connection with Definition 5.2 that we use for example the notation r, as well as $r(n)$, in order to refer to the resource bound obtained by applying the function r to the length n of the input string. Observe further that for technical reasons we consider only resource bounds which are non-decreasing and can be computed from the argument n in time polynomial in n.

Definition 5.2. *An extended verifier is (r, q)-restricted iff there is some constant c such that on every input x of length n, the random tape holds a random string τ of length $c \cdot r(n)$ and for each given random string, V queries at most $c \cdot q(n)$ bits of the proof π.*

For classes $\mathcal{F}_0$ and $\mathcal{F}_1$ of functions from naturals to naturals, an extended verifier is $(\mathcal{F}_0, \mathcal{F}_1)$-restricted if it is (r, q)-restricted for functions r in $\mathcal{F}_0$ and q in $\mathcal{F}_1$.

Remark 5.3. In Definition 5.1 the constant $1/4$ can be replaced by an arbitrary value in the open interval between 0 and 1 without changing the introduced concepts. This can be shown by applying the well-known technique of probability amplification. More precisely, we run a verifier repeatedly for a constant number of times while using independent parts of the random strings for each iteration, and we accept iff all the iterations result in acceptance. Observe that this remark also applies to (r, q)-restricted extended verifiers, because iterating the computation a constant number of times will increase the amount of resources used only by a constant factor.

Verifiers and Extended Verifiers. Remark 5.4 shows that extended verifiers are a generalization of the usual concept of verifier.

Remark 5.4. Given a verifier V_0, we can easily construct an extended verifier V_1 which on input x and proof (π_0, π_1) simply behaves as V_0 does on input x and proof π_1. Both verifiers are equivalent in the sense that for all inputs x and for all proofs π_0, the verifier V_0 accepts x iff the extended verifier V_1 accepts the pair (x, π_0), and likewise for rejectance.

Conversely consider an extended verifier V_1 with the special property that for each input x, either there is some proof π_0 where V_1 accepts the input (x, π_0) or for all proofs π_0, the input (x, π_0) is rejected. Then it is straightforward to construct a verifier V_0 which accepts exactly the strings x of the former type, that is, the strings in the set

$$\{x \mid \text{ there is some proof } \pi_0 \text{ where } V_1 \text{ accepts the input } (x, \pi_0)\}.$$

More precisely, the verifier V_0 on input x and proof π interprets π as a pair (π_0, π_1) and then works as the extended verifier V_1.

For example consider solution verifiers as introduced in Definition 5.33 below. A solution verifier is a special type of extended verifier which accepts the pair (x, π_0) iff π_0 is a code for a satisfying assignment of the 3-CNF-formula x and which rejects the pair if π_0 does not agree with such an encoding on most places. As a consequence solution verifiers indeed have the special property that for each input x either there is some proof π_0 where the input (x, π_0) is accepted or for all proofs π_0, the input (x, π_0) is rejected. Here the two cases correspond to x being satisfiable, respectively being not satisfiable. By the discussion in the preceding paragraph solution verifiers can thus be viewed as verifiers which accept 3SAT, the set of satisfiable 3-CNF-formulae.

Recall that in contrast to the case of verifiers with extended verifiers there might be inputs (x, π_0) which are neither accepted nor rejected. Example 5.5 gives some indication why this relaxation is reasonable.

Example 5.5. In the sequel we will construct extended verifiers which accept a pair $(0^n, \pi_0)$ in case π_0 is a codeword in some given code C_n, and which reject the pair in case π_0 does not agree with a codeword in C_n on at least a δ_0-fraction of its places where δ_0 is close to, but strictly below 1. Moreover while working on a single random string these extended verifiers read only a small part of the alleged codeword π_0. In this situation we should expect that pairs $(0^n, \pi_0)$ where π_0 is not in C_n, but is close enough to a codeword in C_n, are accepted on more than one quarter of all random strings, but not for all random strings, that is, there are inputs which are neither accepted nor rejected.

Constant Extended Verifiers. In Section 5.4 we will introduce solution verifiers which are a special type of extended verifiers. The introduction of such solution verifiers is mainly motivated by the fact that under certain restrictions a verifier and a solution verifiers can be composed in order to yield a more efficient verifier, see Lemma 5.51. In connection with such compositions we are led to consider verifiers which query the proof not at arbitrarily scattered places, but access only a constant number of contiguous subwords of the proof. Observe that this might amount to a restriction only in the presence of additional bounds on the total number of queries to the proof, because otherwise the verifier could just query all places.

Definition 5.6. *A* segment *is a subset of the natural numbers of the form* $\{i \mid n \leqslant i \leqslant m\}$ *where n and m are arbitrary natural numbers. A verifier V is* constant *iff there is some constant c such that for all inputs x, the sets of positions at which V queries the proofs π_0 and π_1, respectively, are both equal to the union of at most c segments.*

Observe that a constant (r, q)-restricted verifier queries constantly many segments each of length at most $\mathcal{O}(q)$, and that an $(r, 1)$-restricted verifier is always constant.

5.3 Encodings

5.3.1 Codes

In Section 5.4 we will introduce solution verifiers as a special form of extended verifiers. A solution verifier interprets its input x as a Boolean formula and expects the input π_0 to be an encoded assignment for this formula. Before we consider solution verifiers in detail, we compile some material on codes.

Definition 5.7. *An* alphabet *is a finite set of symbols. A* string *over an alphabet Σ is a finite sequence of symbols from Σ.*

For a rational δ in the open interval between 0 and 1, two strings are δ-close iff they have the same length and disagree on at most a δ-fraction of their places. A string is δ-close to a set of strings iff it is δ-close to some string in the set.

A code *is a set C of strings over some fixed alphabet Σ such that all strings in C have the same length and for all $\delta < 1/2$ no pair of distinct strings in C is δ-close. The elements of a code C are denoted as* codewords *and their length is the* codeword length *of C.*

Let C be a code and let n be an arbitrary natural number. Then every function dec from C onto $\{0,1\}^n$ is called a decoding function *for C. In this situation a codeword y is a* code *for the binary string x iff $\mathrm{dec}(y)$ is equal to x.*

A coding scheme *is a sequence $(C_0, \mathrm{dec}_0), (C_1, \mathrm{dec}_1), \ldots$ where for every natural number n, the function dec_n is a decoding function for C_n with range $\{0,1\}^n$. Given such a coding scheme we write Σ_n, s_n, and k_n for the alphabet, the alphabet size, and the codeword length, respectively, of the code C_n.*

Observe that given a decoding function from some code to $\{0,1\}^n$, for every binary string of length n, there is some associated codeword, but in general this codeword is not unique. A coding scheme can be viewed as a rule which yields for every natural number n a method to encode strings of length n. The concept coding scheme is not designed to distinguish encodings of strings of mixed length and correspondingly the codes $C_0, C_1, \ldots$ of a coding scheme are not required to be disjoint and might be over arbitrary alphabets $\Sigma_0, \Sigma_1, \ldots$.

Remark 5.8. In the sequel we will consider extended verifiers which expect parts of their proof to be encoded according to some given coding scheme. Now the tape alphabet of an extended verifier contains only the symbols 0 and 1, and accordingly we have to represent codewords over the alphabets $\Sigma_0, \Sigma_1, \ldots$ as binary strings. Given an alphabet Σ of size s this is easily done by first representing the symbols in Σ in some canonical fashion by binary strings of length $\lceil \log s \rceil$, then extending this representation to a homomorphism from strings over Σ_n to binary strings. We denote the image of a word over Σ under such a homomorphism as *associated binary codeword* or string, and accordingly we speak of the length of the binary codewords. In case we use strings over some arbitrary alphabet Σ as part of the input of an extended verifier, it is to be understood that the tapes of the verifier actually hold the associated binary string. In order to avoid confusion with the symbols of the alphabet Σ we refer to the symbols of an associated binary codeword by the term bits. Observe that we have to scan $\lceil \log s \rceil$ bits of the associated binary string in order to read a single symbol of a string over an alphabet of size s.

5.3.2 Robust Coding Schemes

Efficient Operations on Codewords. The motivation for using codes in connection with extended verifiers comes from the fact that certain tests on strings can be performed more efficiently on the associated codewords than on the strings themselves. For an example see Example 5.9 where it is shown that for every $\delta < 1/8$ and for all pairs of strings which are both known to be δ-close to codewords in some code, in order to check probabilistically whether both strings are in fact close to the same codeword it suffices to scan $\mathcal{O}(1)$ randomly selected symbols.

Example 5.9. We let $\delta < 1/8$ be some arbitrary constant and consider strings v and w which are both δ-close to a code C. Then for both strings there is a unique nearest codeword in C. If we assume that the nearest codewords of v and w are the same then v and w must be 2δ-close. On the other hand, if the nearest codewords are different, then they can agree on at most $1/2$ of their places, and consequently v and w are not even δ_0-close for $\delta_0 := 1/2 - 2\delta$, where by assumption on δ we have $1/2 - 2\delta > 1/4$.

So trying to find a disagreement between v and w by comparing the two strings at some place chosen uniformly at random amounts to a Bernoulli trial with success probability $2\delta < 1/4$ in case the two strings are close to the same codeword and with success probability $\delta_0 > 1/4$, otherwise. As a consequence we can distinguish the two cases probabilistically. More precisely probability theory tells us that for every $\varepsilon > 0$ there is a constant c such that if we repeat the Bernoulli trial $4c$ times then with probability $1 - \varepsilon$ the accumulated number of successes is strictly below c in the first case and is strictly above c in the second.

Now consider a coding scheme $(C_0, \mathrm{dec}_0), (C_1, \mathrm{dec}_1), \ldots$, let s_n be the size of the alphabet Σ_n, let k_n be the codeword length of C_n and assume that s_n and k_n can be computed from n in time polynomial in n. Then by the preceding remarks for every $\delta < 1/8$ there is a $(\lceil \log k_n \rceil, \lceil \log s_n \rceil)$-restricted extended verifier which for all strings v and w over Σ_n which are δ-close to C_n, accepts the input $(0^n, vw)$ in case v is equal to w, but rejects it in case v and w are not δ-close to the same codeword.

Robust Coding Schemes. The probabilistic test for being close to the same codeword given in Example 5.9 is more efficient than comparing the corresponding non-encoded strings bit by bit. In the verification of the test however we have to assume that the input indeed contains strings which are close to codewords. We will show next that this is not a real restriction in case there is an appropriately restricted extended verifier which simply rejects all inputs where the strings under consideration are not close to codewords. Likewise we will consider probabilistic tests where during the verification we not just assume that the input contains strings close to a certain code, but also that at all positions scanned these strings indeed agree with the nearest codeword. Again this is not a real restriction in case, firstly, for each fixed random string, the probabilistic test under consideration scans the strings under consideration only at constantly many symbol and, secondly, for each such symbol, we can probabilistically test for agreement with the nearest codeword within the given resource bounds.

Unlike for the probabilistic test for being close to the same codeword given in Example 5.9 which works for arbitrary coding schemes in the case of tests for closeness to a codeword and for agreement with the nearest codeword at some given place we should not expect that there is a general procedure which works for coding schemes of different types. There are however such procedures for specific coding schemes, and in particular so for two coding schemes based on linear functions and on polynomials, respectively, which we will introduce in Sections 5.3.3 and 5.3.4. Using the notation introduced in Definition 5.10 these coding schemes will be shown to be $(n, 1)$- and $(\log n, \text{poly}(\log n))$-robust, respectively.

Definition 5.10. *Let C be a coding scheme $(C_0, \text{dec}_0), (C_1, \text{dec}_1), \ldots$.*

The coding scheme C is (r, q)-checkable iff there is a positive rational δ_0 where for every $0 < \delta \leqslant \delta_0$ there is a constant (r, q)-restricted extended verifier V such that, firstly, V accepts every pair $(0^n, \pi_0)$ where π_0 is in C_n and, secondly, V rejects every pair $(0^n, \pi_0)$ where π_0 is not δ-close to a codeword in C_n.

The coding scheme C is (r, q)-correctable iff there is a constant (r, q)-restricted extended verifier V and a positive rational $\delta_1 < 1/4$ such that, firstly, V accepts every pair $((0^n, i), \pi_0)$ where π_0 is a codeword z in C_n and, secondly, V rejects every pair $((0^n, i), \pi_0)$ where π_0 is δ_1-close to a codeword z in C_n but π_0 and z disagree at the i-th symbol. (Observe that the codeword z is uniquely determined because δ_1 is less than $1/4$.)

The coding scheme C is (r, q)-robust iff it is (r, q)-checkable and (r, q)-correctable.

A positive rational δ witnesses that C is (r, q)-checkable iff C satisfy the definition of (r, q)-checkable with $\delta_0 = \delta$, and likewise we define a concept of witness for (r, q)-correctable coding schemes. A positive rational δ witnesses that C is (r, q)-robust iff δ witnesses both that C is (r, q)-checkable and that C is (r, q)-correctable.

Observe that for all three properties introduced in Definition 5.10 the class of witnessing rationals is closed downwards in the sense that if a positive rational δ witnesses one of these properties then so does every positive rational smaller than δ.

We will show in the remainder of this section that, intuitively speaking, in case an extended verifier V expects that its proof contains at some specified places codewords according to some (r, q)-robust coding scheme then under certain conditions we can assume during the verification of V's behavior that the proof contains at these places indeed codewords and not just arbitrary binary strings. Before we give a formal statement of this remark in Proposition 5.14 we demonstrate the techniques used in Examples 5.11 and 5.12.

Example 5.11. Let C be a coding scheme with codes $C_0, C_1, \ldots$. Let r and q be resource bounds such that C_n has codeword length in $2^{O(r(n))}$ and is over an alphabet Σ_n of size in $2^{O(q(n))}$. According to Example 5.9 there is an (r, q)-restricted extended verifier V_0 and a positive rational $\delta < 1/8$ such that for all strings v and w over Σ_n which are δ-close to C_n, V_0 accepts the input $(0^n, vw)$ in case v is equal to w, but rejects it in case v and w are not δ-close to the same codeword.

Now in case the coding scheme C is in addition (r, q)-checkable then we can drop the condition that v and w are δ-close to C_n, that is, there is an (r, q)-restricted extended verifier V_1 which for all strings v and w over Σ_n, accepts the input $(0^n, vw)$ in case v and w are identical codewords in C_n and rejects in case v and w are not δ-close to the same codeword in C_n.

On an input $(0^n, vw)$ the extended verifier V_1 runs a subverifier in order to check probabilistically whether both of v and w are δ-close to a codeword in C_n. In case this test fails then V_1 rejects immediately and otherwise, V_1 just simulates V_0 on the given input.

Now consider V_1's behavior on input $(0^n, vw)$. In case v and w are identical codewords in C_n then by construction V_1 accepts. On the other hand in case v or w is not δ-close to C_n then V_1 rejects due to the initial test, and in case both are δ-close to C_n but their respective nearest codewords differ then V_1 rejects by assumption on V_0.

Example 5.12 shows a more involved application of subverifiers where the input for the subverifier depends on the random string. While the argument in Example 5.12 is basically the same as in the proof of Lemma 5.51 about composing verifiers, we refrain from stating the underlying technique in full generality.

Example 5.12. Let C be a coding scheme with codes $C_0, C_1, \ldots$. Consider the following simple test for identity of two codewords from C_n: select uniformly at random a position between 1 and the codeword length of C_n and compare the codewords at this place; accept in case the symbols read are the same and reject, otherwise. Observe that in case both codewords are the same the test accepts with probability 1, but otherwise it rejects with probability at least $1/2$. It is now straightforward to construct an extended verifier V_0 which for every input $(0^n, vw)$ where v and w are codewords in C_n, accepts in case v and w are equal and rejects, otherwise. The extended verifier V_0 simply runs the above test k times while using independent parts of its random string for each repetition and V_0 rejects in case one of the repetitions results in rejection. Thus V_0 rejects with probability of at least $1 - 1/2^k$ in case v and w are different codewords in C_n.

The extended verifier V_0 requires only $k \lceil \log k_n \rceil$ random bits and reads only $k \lceil \log s_n \rceil$ bits of its proof where s_n and k_n are the alphabet size and the codeword length of code C_n. Thus in general V_0 will be more efficient than the extended verifier constructed in Example 5.9. However the latter verifier just assumes that v and w are close to C_n while V_0 requires that both are indeed codewords. Under the assumption that the coding scheme C is (r, q)-robust where $\log k_n$ is in $\mathcal{O}(r(n))$ and $\log s_n$ is in $\mathcal{O}(q(n))$ we will now show that we can turn the extended verifier V_0 into an (r, q)-restricted extended verifier V_1 which works for all inputs. Note that we construct V_1 just in order to show an application of robust coding schemes and that in particular verifier V_1 is not necessarily more efficient than the extended verifier constructed in Example 5.9.

So let $\delta < 1/8$ be a positive rational which witnesses that the coding scheme C is (r, q)-robust and let V_a and V_b be corresponding verifiers which witness that C is (r, q)-checkable and (r, q)-correctable, respectively.

On an input $(0^n, vw)$ the new verifier V_1 first checks probabilistically whether v and w are both δ-close to C_n by simulating V_a on inputs $(0^n, v)$ and $(0^n, w)$. In case V_a rejects for at least one of these inputs, V_1 rejects immediately. Otherwise V_1 starts simulating V_0 on input $(0^n, vw)$ and random string τ. While simulating V_0 for a single random string τ the extended verifier V_1 checks probabilistically whether the constantly many symbols read by V_0 while working on random string τ agree with the corresponding nearest codeword by simulating for each such symbol the extended verifier V_b with an appropriate input. In case V_b rejects during some of the latter simulations, then so does V_1 for the random string τ of V_0 under consideration, but otherwise V_1 just simulates V_0 and accepts or rejects according to the latter simulation.

By construction V_1 accepts an input $(0^n, vw)$ in case v and w are identical codewords in C_n. We show that V_1 rejects such an input in case either, firstly, the strings v or w are not δ-close to C_n or, secondly, both are δ-close to C_n but the corresponding nearest codewords are different. In case v is not δ-close to C_n then V_a on input $(0^n, v)$ rejects for at least $3/4$ of its random strings, and then so does V_1 due its simulation of V_a. By symmetry the same holds for w. So we can restrict the remainder of the verification to the case where both strings are δ-close to C_n and we let y and z be the nearest codewords of v and w, respectively.

Now consider the simulating computation on a single random string τ of V_0. In case all symbols read by V_0 while working on this random string agree with y and z, respectively, then V will behave like V_0 would behave on input $(0^n, yz)$. On the other hand, in case some of the symbols read do not agree with y or z, respectively, then the simulation of V_b results in rejecting with constant probability which can be chosen as close to 1 as desired, say with probability $9/10$. But we can assume that also V_0 in case of rejection rejects its input for a fraction of $9/10$ of all random strings, and consequently V_1 rejects the input $(0^n, wv)$ with probability at least $81/100$.

In the proof of Proposition 5.14 we apply robust coding schemes in essentially the same way as in Example 5.12. The former application is however more involved because the alleged codewords might belong to various codes in some given

coding scheme and might occur at arbitrary positions of the proofs π_0 and π_1 of
an extended verifier. Before we state the proposition we introduce some related
notation.

Definition 5.13. *A* pattern *of length c is a finite sequence $(n_1, i_1) \ldots, (n_c, i_c)$
of pairs of natural numbers. We denote patterns by small Greek letters α, β, $\ldots$.
A* pattern function *is a function from binary strings to patterns of some fixed
length c.*

With the coding scheme C understood a binary string z obeys *a pattern α equal to
$(n_1, i_1) \ldots, (n_c, i_c)$ iff the string z contains non-overlapping codewords $w_1, \ldots, w_c$
such that w_j is in the n_j-th code C_{n_j} of C and starts at position i_j of z. If we
require the w_j only to be δ-close to C_{n_j} instead of being elements of C_{n_j} the
string z is said to be δ-*close to the pattern α.*

For a string z which is $1/4$-close to some pattern of length c the nearest obeying *string of z is the string obtained by replacing the subwords $w_1, \ldots, w_c$ of z
specified by the pattern with their respective nearest codewords.*

An extended verifier V is symbol constant *on π_0 w.r.t. a pattern function $x \mapsto
\alpha(x)$ in case for all $(x, (\pi_0, \pi_1))$ the extended verifier reads at most constantly
many symbols of the subwords of π_0 which are specified by $\alpha(x)$ (Note that V
might read further symbols not belonging to these subwords.) Likewise, we define
the concept of an extended verifier being symbol constant on π_1.*

Observe that in Example 5.12 we have implicitly used a pattern function which
maps the string 0^n to the sequence $(n, 1), (n, k_n + 1)$ where k_n is the codeword
length of the n-th code of the considered coding scheme.

While the formulation of Proposition 5.14 is rather technical, observe again that
its content is basically the same as in Example 5.12: given a resource bounded
extended verifier V_0 where firstly, V_0 works as intended for all inputs where
all designated alleged codewords are indeed codewords and secondly, for each
fixed random string, V_0 reads only constantly many symbols of these alleged
codewords, then there is a suitably resource bounded extended verifier V_1 which,
intuitively speaking, works for all inputs.

Proposition 5.14. *Let $x \mapsto \alpha_l(x)$, $l=0,1$, be polynomial time computable pattern functions, let C be a coding scheme and for all x, let k_x be the maximal
codeword length of the codes specified by the first components of the pairs in
$\alpha_0(x)$ and $\alpha_1(x)$ w.r.t. C. Let V_0 be an (r,q)-restricted extended verifier which is
symbol constant on π_0 and on π_1 w.r.t. α_0 and α_1, respectively. Let*

$$O := \{(x, (\pi_0, \pi_1)) \mid \pi_0 \text{ obeys } \alpha_0 \text{ and } \pi_1 \text{ obeys } \alpha_1\},$$

*and for all x let A_x and R_x contain exactly the pairs in O which V accepts and
rejects, respectively.*

Assume further that the coding scheme C is (r', q')-robust and that there is some constant c where for all x, we have $r'(k_x) \leqslant cr(|x|)$ and $q'(k_x) \leqslant cq(|x|)$.

Then there is an (r, q)-restricted extended verifier V_1 and a positive rational $\delta < 1/4$ such that on the one hand V_1 accepts all pairs in A_x, and on the other hand

- *V_1 rejects all pairs $(x, (\pi_0, \pi_1))$ where π_0 is not δ-close to $\alpha_0(x)$ or π_1 is not δ-close to $\alpha_1(x)$,*

- *V_1 rejects all pairs $(x, (\pi_0, \pi_1))$ where π_0 is δ-close to $\alpha_0(x)$, π_1 is δ-close to $\alpha_1(x)$, and where for the corresponding nearest obeying strings π'_0 and π'_1 the pair $(x, (\pi'_0, \pi'_1))$ is in R_x.*

In addition if the extended verifier V_0 is constant then we can choose V_0 to be constant, too.

Proof. The extended verifier V_1 essentially works like the verifier we have constructed in Example 5.12. Let the positive rational δ witness that the coding scheme C is (r', q')-robust.

While working on $(x, (\pi_0, \pi_1))$ where x has length n the extended verifier V_1 first tests probabilistically whether π_0 and π_1 are δ-close to the pattern $\alpha_0(x)$ and $\alpha_1(x)$, respectively, and V_1 rejects immediately in case at least one of these tests fails. In doing so we exploit that the coding scheme C is (r', q')-checkable in order to test for each of the constantly many pairs (n_j, i_j) specified by the pattern $\alpha_0(x)$ whether proof π_0 contains at position i_j a string which is δ-close to the n_j-th code of C, and likewise for α_1 and π_1. Thus testing for closeness to the patterns $\alpha_0(x)$ and $\alpha_1(x)$ uses at most $\mathcal{O}(r(n))$ random bits and reads at most $\mathcal{O}(q(n))$ bits of the proof due to the assumption on r' and q'.

In case the test for closeness does not fail V_1 proceeds by simulating V_0, however for each random string of V_0, the extended verifier V_1 checks first whether the constantly many symbols read by V_0 from the segments specified by the patterns $\alpha_0(x)$ and $\alpha_1(x)$ agree with the nearest obeying proof. By definition of nearest obeying proof this amounts to check whether the constantly many symbols read from the alleged codewords specified by the patterns $\alpha_0(x)$ and $\alpha_1(x)$ agree with the respective nearest codewords in the corresponding code. Again this test uses at most $\mathcal{O}(r(n))$ random bits and reads at most $\mathcal{O}(q(n))$ bits of the proof due to the fact that the coding scheme C is (r', q')-correctable and by assumption on r' and q'. The extended verifier V_1 rejects in case this test fails for one of the symbols and otherwise, it accepts or rejects according to the simulation of V_0.

While verifying the behavior of V_1 we proceed essentially as for the extended verifier constructed in Example 5.12 and we leave the easy modifications to the interested reader. Finally observe that by definition of robustness the subverifiers used by V_1 in order to test for δ-closeness and for agreement with the nearest codeword are constant and that during the run of V_1 for each random string of V_0 there are only constantly many invocations of these subverifiers. Thus in case V_0 is constant then so is V_1. ∎

The linear function and the polynomial coding scheme introduced next are both based on functions over finite fields. For a prime p, we denote the field of p elements by $\mathbb{F}_p$. The elements of $\mathbb{F}_p$ can be identified with the natural numbers $0, \ldots, p-1$ in the natural way, and by this identification we extend the usual ordering on the natural numbers to the field $\mathbb{F}_p$.

5.3.3 The Linear Function Coding Scheme

We introduce a coding scheme based on linear functions over $\mathbb{F}_2$. Observe in connection with Definition 5.15 that we identify the symbols 0 and 1 with the zero and the unit of $\mathbb{F}_2$ and consequently the usual operations over $\mathbb{F}_2$ extend to binary strings. Observe further that we identify a function from $\{0,1\}^n$ to $\{0,1\}$ with a binary string z of length 2^n where the i-th position of z corresponds to the value of the function at the i-th string in $\{0,1\}^n$ w.r.t. the lexicographical ordering.

Definition 5.15. – *Given a binary string $x = x_1 \ldots x_n$ of length n, let l_x denote the function from $\{0,1\}^n$ to $\{0,1\}$ defined by*

$$l_x(y) \quad = \quad \sum_{i=1}^{n} x_i y_i.$$

– For every natural number n, the linear function code C_n^{lin} *is the set*

$$C_n^{\mathrm{lin}} \quad := \quad \{l_x \mid x \in \{0,1\}^n\}.$$

The linear function coding scheme C^{lin} *is given by the codes $C_0^{\mathrm{lin}}, C_1^{\mathrm{lin}}, \ldots$ and decoding functions $\mathrm{dec}_0^{\mathrm{lin}}, \mathrm{dec}_1^{\mathrm{lin}}, \ldots$ where $\mathrm{dec}_{|x|}^{\mathrm{lin}}$ maps l_x to x.*

We show first that the codewords in every linear function code C_i^{lin} are indeed at most $1/2$-close.

Lemma 5.16. *For every pair x and y of distinct binary strings of equal length, the functions l_x and l_y agree on exactly $1/2$ of the positions in their domain.*

Proof. We let n be the length of x and y, and for ease of notation we assume without loss of generality that x and y differ on their last bit. Further we note that by symmetry it suffices to consider the case where the last bit of x is 0. Thus for each binary string w of length $n-1$, we have $l_x(w0) = l_x(w1)$ and

$l_y(w0) \neq l_y(w1)$ and hence l_x and l_y agree on exactly one of the strings $w0$ and $w1$, and disagree on the other. The assertion of the lemma then is immediate. ∎

In order to prove that the linear code is checkable the following technical lemma is required. This lemma intuitively states that whenever a function g is not close to a linear function, this function g violates the linearity on a large fraction of pairs of its arguments.

Lemma 5.17. *Let $\delta < 1/3$ be a constant and let $\tilde{g} : \mathbb{F}_2^n \to \mathbb{F}_2$ be a function which is not δ-close to a linear function. Then we have*

$$\mathrm{Prob}_{x,y}[\tilde{g}(x) \neq \tilde{g}(x+y) - \tilde{g}(y)] > \frac{\delta}{2}. \tag{5.1}$$

Proof. We show the lemma by contraposition: in case $\tilde{g}$ does not satisfy inequality (5.1), then $\tilde{g}$ is δ-close to some linear function g. The function g will be constructed as follows:

$$g(x) := \mathrm{majority}_y\{\tilde{g}(x+y) - \tilde{g}(y)\}, \tag{5.2}$$

where majority denotes the value occurring most often over all choices of y; breaking ties arbitrarily.

It remains to show that g and $\tilde{g}$ are δ-close and that g is a linear function. We first prove that g and $\tilde{g}$ are δ-close. By definition of g (cf. equation (5.2)), we have for each $x \in \mathbb{F}_2^n$

$$\mathrm{Prob}_y[g(x) = \tilde{g}(x+y) - \tilde{g}(y)] \geq \frac{1}{2}. \tag{5.3}$$

Starting with the negation of inequality (5.1) we obtain the following chain of inequalities:

$$\frac{\delta}{2} \geq \mathrm{Prob}_{x,y}[\tilde{g}(x) \neq \tilde{g}(x+y) - \tilde{g}(y)]$$

$$\geq \mathrm{Prob}_{x,y}[\tilde{g}(x) \neq g(x) \wedge g(x) = \tilde{g}(x+y) - \tilde{g}(y)]$$

$$= \mathrm{Prob}_x[\tilde{g}(x) \neq g(x)] \cdot \mathrm{Prob}_{x,y}[g(x) = \tilde{g}(x+y) - \tilde{g}(y) \mid \tilde{g}(x) \neq g(x)]$$

Since inequality (5.3) implies a lower bound of the second probability, we get:

$$\geq \frac{1}{2} \cdot \mathrm{Prob}_x[\tilde{g}(x) \neq g(x)].$$

Hence we obtain $\mathrm{Prob}_x[\tilde{g}(x) \neq g(x)] \leq \delta$, i.e., g and $\tilde{g}$ are δ-close.

To prove that g is a linear function, we first prove the following property of g:

$$p_a := \mathrm{Prob}_x[g(a) \neq \tilde{g}(a+x) - \tilde{g}(x)] \leq \delta. \tag{5.4}$$

By inequality (5.3) we know that $p_a \leqslant 1/2$. Note that for some fixed $a \in \mathbb{F}_2^n$ each $x + a \in \mathbb{F}_2^n$ is equally likely if $x \in \mathbb{F}_2^n$ is chosen uniformly at random. Using this fact and the negation of inequality (5.1), we obtain:

$$\text{Prob}_{x,y}[\tilde{g}(x + a) + \tilde{g}(y) \neq \tilde{g}(x) + \tilde{g}(y + a)]$$
$$\leqslant \quad \text{Prob}_{x,y}[\tilde{g}(x + a) + \tilde{g}(y) \neq \tilde{g}(x + a + y)]$$
$$+ \text{Prob}_{x,y}[\tilde{g}(x) + \tilde{g}(y + a) \neq \tilde{g}(x + y + a)] \quad \leqslant \quad \delta.$$

Starting with the last inequality, we obtain:

$$
\begin{aligned}
\delta \quad &\geqslant \quad \text{Prob}_{x,y}[\tilde{g}(x + a) - \tilde{g}(x) \neq \tilde{g}(y + a) - \tilde{g}(y)] \\
&= \quad \sum_{z \in \mathbb{F}_2} \text{Prob}_{x,y}[\tilde{g}(x + a) - \tilde{g}(x) = z \wedge \tilde{g}(y + a) - \tilde{g}(y) \neq z] \\
&= \quad \sum_{z \in \mathbb{F}_2} \text{Prob}_x[\tilde{g}(x + a) - \tilde{g}(x) = z] \cdot \text{Prob}_y[\tilde{g}(y + a) - \tilde{g}(y) \neq z]
\end{aligned}
$$

Using the fact that $\mathbb{F}_2 = \{g(a), 1 - g(a)\}$
and the definition of p_a in equation (5.4).

$$= \quad (1 - p_a)p_a + p_a(1 - p_a)$$

Recall that $p_a \leqslant 1/2$, which implies $2(1 - p_a) \geqslant 1$.

$$\geqslant \quad p_a$$

Now let $a, b \in \mathbb{F}_2^n$ be fixed. Using inequality (5.4), we derive:

$$
\begin{aligned}
\text{Prob}_x[g(a) + g(b) \neq g(a + b)] \quad &= \quad \text{Prob}_x[g(a) + g(b) + \tilde{g}(x) \neq g(a + b) + \tilde{g}(x)] \\
&\leqslant \quad \text{Prob}_x[g(a) + g(b) + \tilde{g}(x) \neq \tilde{g}(a + x) + g(b)] \\
&\quad + \text{Prob}_x[g(b) + \tilde{g}(a + x) \neq \tilde{g}(b + a + x)] \\
&\quad + \text{Prob}_x[\tilde{g}(a + b + x) \neq g(a + b) + \tilde{g}(x)] \\
&= \quad 3 \cdot p_a \leqslant 3 \cdot \delta < 1.
\end{aligned}
$$

Since the first term is independent of x, the probability must be either 0 or 1. Hence, with probability 0 we get $g(a) + g(b) \neq g(a+b)$, i.e., g is a linear function. ∎

Now we are able to prove the checkability of linear function coding schemes. The previous Lemma 5.17 suggests the following linearity test for the linear function coding scheme.

<u>LINEARITY TEST</u>
Assume that input is 0^n.
Interpret π_0 as $\tilde{g} \in C_n^{\text{lin}}$.
Choose $x, y \in \mathbb{F}_2^n$ uniformly at random.
if $(\tilde{g}(x + y) \neq \tilde{g}(x) + \tilde{g}(y))$
      then reject
      else accept

We simply check the linearity condition at randomly selected positions. Whenever the considered function is not linear, we find with a positive probability a pair of arguments on which the function will not behave as a linear function. Of course this probability can easily be amplified by iterating the procedure LINEARITY TEST.

Proposition 5.18. *The linear function coding scheme is* $(n, 1)$*-checkable.*

Proof. Observe that the LINEARITY TEST will never reject a function $\tilde{g}$ which is indeed linear. Observe further that selecting the pair x and y at random requires $2n$ random bits and and that the LINEARITY TEST evaluates the function $\tilde{g}$ only at three positions. Thus in order to show that the linear function coding scheme is $(n, 1)$-checkable, it suffices to show that for every positive rational $\delta < 1/6$ there is a constant k such that in case $\tilde{g}$ is not δ-close to some linear function, then iterating the LINEARITY TEST k times results in rejecting with probability at least $3/4$. But this follows from the previous Lemma 5.17. ∎

Proposition 5.19. *The linear function coding scheme is* $(n, 1)$*-correctable.*

Proof. Let $\tilde{g}$ be a function in $\mathbb{F}_2^n$ which is $1/8$-close to a linear function g. Assume that we want to check whether $\tilde{g}$ and g agree at some given position x. Since $\tilde{g}$ and g are $1/8$-close, they disagree on at most $1/8$ of all positions y, and likewise, as $y \mapsto x + y$ is a bijection on $\mathbb{F}_2^n$, $\tilde{g}(x + y)$ differs from $g(x + y)$ for at most $1/8$ of all strings y. So we find that for at least $6/8$ of all strings y we have

$$g(x) \;=\; g(x + y) - g(y) \;=\; \tilde{g}(x + y) - \tilde{g}(y) \,,$$

where the first equation follows by linearity of g and the latter by the preceding discussion.

The desired test for agreement between $\tilde{g}$ and g at position x now is straightforward: choose a random y and reject if $\tilde{g}(x)$ does not agree with $\tilde{g}(x + y) - \tilde{g}(y)$. Observe that this test will never reject a linear function $\tilde{g}$. Whenever this test is applied to a function which is $1/8$-close to a linear function g but does not agree with g at position x then by the preceding discussion the test rejects with probability at least $6/8$.

Finally, we obtain a verifier V as required in the definition of $(n, 1)$-correctable which works as follows: on input $((0^n, x), \pi_0)$, the verifier first runs LINEARITY TEST and in case π_0 does not represent some linear function in $\mathbb{F}_2^n$ rejects immediately with probability at least $3/4$, and otherwise, it runs the test for agreement at x described above. Observe that each of the two tests requires $\mathcal{O}(n)$ random bits and evaluates π_0 at $\mathcal{O}(1)$ positions. ∎

As a consequence of the previous propositions, we obtain the following corollary.

Corollary 5.20. *The linear function coding scheme is* $(n, 1)$*-robust.*

5.3.4 The Polynomial Coding Scheme

We introduce a coding scheme which is based on m-variate polynomials over $\mathbb{F}_p$ where the parameters p and m depend on the length of the strings we want to encode. Recall that an m-variate polynomial of degree d over $\mathbb{F}_p$ is a sum of terms of the form $a x_1^{i_1} \ldots x_k^{i_k}$ where a is in $\mathbb{F}_p$ and where the maximum of $i_1 + \ldots + i_k$ taken over all these terms is equal to d. Observe that such a polynomial defines canonically a function from $\mathbb{F}_p^m$ to $\mathbb{F}_p$.

We denote by $<_{\mathrm{lex}}$ the strict lexicographic ordering on $\mathbb{F}_p^m$ which we obtain in the usual way by extending the canonical ordering on $\mathbb{F}_p$ to sequences of m elements of $\mathbb{F}_p$. With the ordering $<_{\mathrm{lex}}$ on $\mathbb{F}_p^m$ fixed, we can identify functions from $\mathbb{F}_p^m$ to $\mathbb{F}_p$ with strings of length p^m over the alphabet $\mathbb{F}_p$. More precisely, each such function g corresponds to the string $g(x_1)g(x_2)\ldots g(x_{p^m})$ where $x_1 <_{\mathrm{lex}} x_2 \ldots <_{\mathrm{lex}} x_{p^m}$ are the elements of $\mathbb{F}_p^m$.

Definition 5.21. *Let p, m, and d be natural numbers where p is a prime and d is less than $p/2$. Then we denote by $C(p, m, d)$ the code which contains exactly the strings which correspond to m-variate polynomials over $\mathbb{F}_p$ of degree at most d.*

Observe that the codewords in $C(p, m, d)$ are strings over $\mathbb{F}_p$ of length p^m. By the theorem of Schwartz the assumption $d < p/2$ implies that distinct codewords in a code $C(p, m, d)$ are indeed not $1/2$-close.

Theorem 5.22 (Schwartz). *Two distinct m-variate polynomials of degree at most d over some finite field $\mathbb{F}$ agree on at most a $d/|\mathbb{F}|$ fraction of places in their domain $\mathbb{F}^m$.*

Proof. Given two different m-variate polynomials f and g over $\mathbb{F}$ of degree at most d, their difference h is a m-variate polynomial that is not the zero polynomial (i.e. the polynomial that has all coefficients equal to zero), since for at least one tuple of exponents $(i_1, \ldots, i_m)$ the coefficients of $x_1^{i_1} \cdots x_m^{i_m}$ in f and g disagree, and thus the coefficient of $x_1^{i_1} \cdots x_m^{i_m}$ in h is not zero. (Note that it is nevertheless very well possible that $h(x) = 0$ for all $x \in \mathbb{F}^m$, as the example $f(x) = x^p$ and $g(x) = x$ in the case $\mathbb{F} = \mathbb{F}_p$ and $m = 1$ shows. That example does not contradict the theorem though, since we have $d/|\mathbb{F}| = p/p = 1$ in this case.) Thus it suffices to prove that any degree d polynomial f over $\mathbb{F}$ that is not the zero polynomial takes the value zero at no more than a $d/|\mathbb{F}|$ fraction of points of $\mathbb{F}^m$. We prove this by induction on m. The case $m = 1$ is clear, since a non-zero univariate polynomial of degree at most d has at most d roots.

So we assume $m > 1$ and note that f can be written as

$$f(x_1, \ldots, x_m) = \sum_{i=0}^{d} x_1^i \cdot f_i(x_2, \ldots, x_m)$$

where each f_i is a $(m-1)$-variate polynomial of degree at most $d-i$. Note that since f is not the zero polynomial, at least one of the f_i is non-zero. Let k be the maximal index where this is the case. Thus we know by the inductive hypothesis that for at least $1 - (d-k)/|\mathbb{F}|$ of all values of $(x_2, \ldots, x_m)$, $f_k(x_2, \ldots, x_m) \neq 0$. For any such value $(a_2, \ldots, a_m)$, by choice of k the above representation yields that $f(x_1, a_2, \ldots, a_m)$ is a non-zero univariate polynomial of degree at most k, and thus is zero for at most k values of x_1.

Therefore the fraction of non-zeroes of f is at least

$$(1 - \frac{k}{|\mathbb{F}|})(1 - \frac{d-k}{|\mathbb{F}|}) \geqslant 1 - \frac{d-k}{|\mathbb{F}|} - \frac{k}{|\mathbb{F}|} = 1 - \frac{d}{|\mathbb{F}|},$$

which concludes the proof. ∎

While defining decoding functions for the codes of the form $C(p, m, d)$, first we fix a strict ordering $<$ on $\mathbb{F}_p^m$. Then in order to decode a codeword q in $C(p, m, d)$ to a binary string u of length n, we consider the first n elements $x_1 < \ldots < x_n$ of $\mathbb{F}_p^m$ and interpret the value $q(x_j)$ as the j-th bit of u. Observe that by means of the ordering $<$ we are not only able to specify the places x_j to be used in the decoding, but we can also fix an ordering on the decoded bits. The latter possibility will come in handy in connection with the splitting of codewords as discussed in Section 5.5.1.

Definition 5.23. *Let $<$ be a strict ordering on a finite set D with $|D| \geqslant 2$ and let $x_1 < \ldots < x_{|D|}$ be the elements of D. For a function g from some superset of D to a finite field $\mathbb{F}$ we let $\sigma(g, <)$ be the binary string $u_1 \ldots u_{|D|}$ of length $|D|$ defined by*

$$u_j = \begin{cases} 0 & \text{in case } g(x_j) \text{ is equal to the zero of } \mathbb{F} \\ 1 & \text{otherwise} \end{cases} . \tag{5.5}$$

For a natural $n \leqslant |D|$ we let $\sigma(g, <, n)$ be the length n prefix of $\sigma(g, <)$.

Observe in connection with the definition of $\sigma(g, <)$ that the ordering $<$ is total and hence for each pair of different elements in its domain, the relation $<$ either contains (x_i, x_j) or contains (x_j, x_i). Thus in case the domain D of $<$ contains at least two elements then D is determined by $<$.

Now we can rephrase the remark preceding Definition 5.23: we will use decoding functions of the form

$$\begin{aligned} \text{dec}: \quad C(p, m, d) &\longrightarrow \{0, 1\}^n \\ q &\longmapsto \sigma(q, <, n) \end{aligned} \tag{5.6}$$

where $<$ is a strict ordering on some appropriate subset of $\mathbb{F}_p^m$.

However we have to take care that indeed for every binary string of the considered length n there is an associated codeword. Remark 5.24 shows that this can be

achieved while working with polynomials of moderate degree in case we base the
decoding on the values of the codeword q on a set of the form H^m where H is
a subset of $\mathbb{F}_p$. In the setting of decoding functions of the form $\sigma(q, <, n)$ this
means that we specify a subset H of $\mathbb{F}_p$ with $|H^m| \geqslant n$ and a strict ordering $<$
on H^m and then decode q to $\sigma(q, <, n)$.

Remark 5.24. Let H be a subset of $\mathbb{F}_p$ of cardinality h. Given some arbitrary
function g from H^m to $\mathbb{F}$, the polynomial

$$\sum_{(h_1, \ldots, h_m) \in H^m} g(h_1, \ldots, h_m) \prod_{i=1}^m \prod_{y \in H - \{h_i\}} \frac{x_i - y}{h_i - y} \tag{5.7}$$

has degree at most $m(h - 1)$ and agrees by construction with g on H^m. Now
assume that we are given a strict ordering $<$ on H_p^m. Then for every binary
string x of length n less or equal to $|H|^m$, there is an m-variate polynomial q
over $\mathbb{F}_p$ of degree at most mh, that is, a polynomial q in $C(p, m, mh)$, such that
x is equal to $\sigma(q, <, n)$.

In case the parameters p, m and h are understood, we will denote a polynomial
of the form (5.7) as *low degree extension* of the function g.

In order to define the polynomial coding scheme, for each n, we choose parame-
ters p_n, m_n, d_n which then specify a code $C_n^{\text{poly}} = C(p_n, m_n, d_n)$. In addition we
choose a subset H_n of $\mathbb{F}_{p_n}$ and an appropriate strict ordering $<_n$ on $H_n^{m_n}$ which
then yields a decoding function which maps a codeword q to the length n binary
string $\sigma(q, <_n, n)$. More precisely we let for every natural number $n \geqslant 16$,

$$h_n := \lceil \log n \rceil, \qquad m_n := \left\lceil \frac{\log n}{\log \log n} \right\rceil + 1; \tag{5.8}$$

further we let

$$p_n \text{ be the smallest prime greater than } 36 \lceil \log n \rceil^2 \tag{5.9}$$

(which is less than $72 \lceil \log n \rceil^2$ according to Bertrand's Postulate and which can
be found in time polynomial in n) and we let H_n contain exactly the first h_n
elements in $\mathbb{F}_{p_n}$ w.r.t. the standard ordering on $\mathbb{F}_{p_n}$. For $n < 16$, we choose the
parameters as in the case $n = 16$.

We call a sequence $<_0, <_1, \ldots$ *admissible* iff firstly, for all n, the relation $<_n$ is
a strict ordering on $H_n^{m_n}$ and secondly, in time polynomial in n we can go from
n and i to the i-th element of $H_n^{m_n}$ and, conversely, from n and an element of
$H_n^{m_n}$ to its index within H_n^m. We fix an admissible sequence $<_0, <_1, \ldots$. For the
moment, the orderings $<_n$ need not be specified to greater detail and in particular
up to Section 5.5.1 all results related to the polynomial coding scheme will not
depend on our choice of an admissible sequence. In Section 5.5.1 we will then add
further restrictions which are related to the checking of split representations.

We let d_n be equal to $m_n h_n$, we let

$$C_n^{\text{poly}} := C(p_n, m_n, d_n),$$

and we define corresponding decoding functions by

$$\text{dec}_n^{\text{poly}}(q) := \sigma(q, <_n, n).$$

Definition 5.25. *The* polynomial coding scheme C^{poly} *is*

$$(C_1^{\text{poly}}, \text{dec}_1^{\text{poly}}), (C_2^{\text{poly}}, \text{dec}_2^{\text{poly}}), \ldots.$$

In connection with the definition of the polynomial coding scheme the cardinality of $H_n^{m_n}$ must be at least n because we want to decode strings of length n from the restriction of some polynomial to $H_n^{m_n}$; in fact, related to the checking of split representations as considered in Section 5.5.1 we want to ensure that even the cardinality of $H_n^{m_n-1}$ is at least n. Indeed we have

$$\left| H_n^{m_n-1} \right| = h_n^{m_n-1} \geqslant \left[(\log n)^{1/\log \log n} \right]^{\log n} = 2^{\log n} = n.$$

Thus in particular, by choice of the d_n and by Remark 5.24, each function $\text{dec}_n^{\text{poly}}$ is onto $\{0,1\}^n$ and hence is indeed a decoding function for C_n^{poly}. For further use observe that for all $n \geqslant 16$, we have $\log n \geqslant 4$ and $\log \log n \geqslant 2$ which then yields easily

$$h_n = \lceil \log n \rceil \geqslant \log n \geqslant \frac{\log n}{2} + 2 \geqslant \frac{\log n}{\log \log n} + 2 \geqslant \left\lceil \frac{\log n}{\log \log n} \right\rceil + 1 = m_n. \quad (5.10)$$

Thus the codes of the polynomial coding schemes are all defined w.r.t. parameters where h_n is greater or equal to m_n. As a consequence we have always

$$p_n \geqslant 36 \lceil \log n \rceil^2 = 36 h_n^2 \geqslant 36 h_n m_n \geqslant 36 d_n. \quad (5.11)$$

In Section 5.4.5 we will use a more general form of the polynomial coding scheme where we add a natural number $e \geqslant 2$ as an additional parameter. In order to define this coding scheme for a given value of e we let for every natural number n,

$$p_{e,n} \text{ be the smallest prime greater than } 36 \lceil \log n \rceil^e. \quad (5.12)$$

Then we define the parameters m_n, h_n, and d_n as above, and we define H_n and $<_n$ as above with p_n replaced by $p_{e,n}$. Observe that while H_n is now a subset of $\mathbb{F}_{p_{e,n}}$ and not of $\mathbb{F}_{p_n}$ the old and the new version of the set H_n and the ordering $<_n$ on $H_n^{m_n}$ are essentially the same and thus we will identify them with each other, respectively. Finally we let

$$C_{e,n}^{\text{poly}} := C(p_{e,n}, m_n, d_n),$$

and we define corresponding decoding functions by

$$\text{dec}_{e,n}^{\text{poly}}(q) := \sigma(q, <_n, n).$$

Definition 5.26. *Let e be a natural number greater or equal to 2. The* poly*nomial coding scheme with parameter e, denoted by C_e^{poly}, is*

$$(C_{e,1}^{\text{poly}}, \text{dec}_{e,1}^{\text{poly}}), (C_{e,2}^{\text{poly}}, \text{dec}_{e,2}^{\text{poly}}), \dots.$$

Observe that by definition C^{poly} is equal to C_2^{poly} and note again that in the sequel we will always work with this coding scheme, except for Section 5.4.5 where we consider coding schemes C_e^{poly} with $e \geqslant 2$.

In the sequel we will construct resource bounded extended verifiers which expect parts of their proof to be encoded according to the polynomial coding scheme. Among other requirements, the parameters p_n, m_n, etc. in the definition of the polynomial coding scheme have to be chosen such that these extended verifiers can apply certain procedures and probabilistic tests to the alleged codewords while staying within their respective resource bounds. Remark 5.27 contains some of these procedures and tests together with some corresponding restrictions on the parameters of the polynomial coding scheme.

Remark 5.27. We list some of the requirements which motivate the choice of the parameters in the definition of the polynomial coding scheme. For ease of notation we fix a natural number n and we let $p = p_n$, $m = m_n$, and so on.

First, we want $mh \log p$ to be in $\mathcal{O}(\text{poly}(\log n))$ because during the procedure EXTENDED LFKN-TEST we want to read $\text{poly}(mh)$ values in $\mathbb{F}_p$ while scanning only $\mathcal{O}(\text{poly}(\log n))$ bits in total.

Second, for the sake of the LOW DEGREE TEST and the EXTENDED LFKN-TEST we want p greater or equal to $36d$ (see equation (5.11)).

Third, we want $m \log p$ to be in $\mathcal{O}(\log n)$ because in the procedures LOW DEGREE TEST and EXTENDED LFKN-TEST we want to choose m elements of $\mathbb{F}$ at random while using only $\mathcal{O}(\log n)$ random bits.

Proposition 5.28. *For every parameter $e \geqslant 2$, the polynomial coding scheme C_e^{poly} is $(\log n, \text{poly}(\log n))$-checkable.*

Proof. We have to construct a constant $(\log n, \text{poly}(\log n))$-restricted extended verifier that checks, given a pair $(0^n, \pi_0)$, whether π_0 is δ-close to a codeword of $C_{e,n}^{\text{poly}}$. From the argument 0^n the verifier is able to compute in polynomial time the parameters defining the code $C_{e,n}^{\text{poly}}$: $p := p_{e,n}$, $h := h_n$, $m := m_n$, and $d := d_n$. If we now interpret the string π_0 as $\sigma(\tilde{g}, <_{\text{lex}})$ for some function $\tilde{g} \colon \mathbb{F}_p^m \to \mathbb{F}_p$, then by definition of $C_{e,n}^{\text{poly}}$ the string π_0 is a codeword if and only if $\tilde{g}$ can be expressed as a polynomial of degree at most d, and furthermore π_0 is δ-close to a codeword if and only if $\tilde{g}$ is δ-close to some polynomial $g \colon \mathbb{F}_p^m \to \mathbb{F}_p$ of total degree at most d.

Thus we see that the main part of the problem consists of checking whether a given m-ary function over a finite field $\mathbb{F}_p$ is δ-close to some polynomial of degree d. Since d is typically small compared to p, this problem is usually called the "Low Degree Test". (Note that using Lagrange interpolation, every m-ary function over $\mathbb{F}_p$ can be written as polynomial of total degree mp; thus the problem is only interesting for $d < mp$.) Note that the problem is made especially difficult by the constraints we have to obey: We are only allowed to use $\mathcal{O}(\log n)$ random bits, and to check only a constant number of segments of π_0 of length $\mathcal{O}(\mathrm{poly}(\log n))$. Observe that such a segment might contain up to $\mathcal{O}(\mathrm{poly}(\log n))$ symbols of a codeword in $C_{e,n}^{\mathrm{poly}}$, because coding a single symbol from the corresponding alphabet $\mathbb{F}_p$ requires $\lceil \log p_{e,n} \rceil$ bits, that is $\mathcal{O}(\log \log n)$ bits.

On the other hand, the verifier is allowed to use an additional string π_1 containing information that could help with the decision. How could this information be used? The basic idea of the low degree test is to note that if $\tilde{g}$ is indeed a m-variate polynomial of degree d over $\mathbb{F}_p$, then for each $x, h \in \mathbb{F}_p^m$ the function $t \mapsto \tilde{g}(x+th)$ is a univariate polynomial of degree d. Since the set $\{x+th \mid t \in \mathbb{F}_p\}$ can be geometrically interpreted as the line $l_{x,h}$ with slope h through the point x, we will call this function the restriction of $\tilde{g}$ to the line $l_{x,h}$. In fact, it can be shown that the converse statement holds as well: If the restrictions of $\tilde{g}$ to all lines can be written as polynomials with degree at most d, then $\tilde{g}$ has total degree at most d as well (see Lemma 5.67).

We will now use this fact to fix our interpretation of π_1. If we suppose that $\tilde{g}$ as encoded by π_0 is indeed a degree d polynomial, then all restrictions of $\tilde{g}$ to some line $l_{x,h}$ are polynomials of degree d which can be represented by their coefficients: $d + 1$ elements of $\mathbb{F}_p$. The verifier shall now expect π_1 to contain a table T that has one element $T(x, h)$ for each $x, h \in \mathbb{F}_p^m$ such that $T(x, h)$ consists of the coefficients $T(x, h)_i \in \mathbb{F}_p$ for $1 \leqslant i \leqslant d + 1$ of the polynomials

$$\tilde{P}_{x,h}(t) := \sum_{i=0}^{d} T(x, h)_{i+1} \cdot t^i.$$

The verifier will now try to ensure that those polynomials are indeed the restrictions of $\tilde{g}$ to the lines $l_{x,h}$. Note that the table T will be coded into the string π_1 in such a way that the coefficients $T(x, h)_i$, $1 \leqslant i \leqslant d + 1$, will immediately follow each other. Thus the lookup of one polynomial $T(x, h)$ can be done by reading one segment of length $\mathcal{O}(d \log p) = \mathcal{O}(\mathrm{poly}(\log n))$ of π_1, which means our verifier is allowed to use a constant number of such table entries.

The basic operation of the verifier will now be to check whether the table T does indeed contain "correct" polynomials $\tilde{P}_{x,h}$ that agree with the function $\tilde{g}$. This is done by choosing random $x, h \in \mathbb{F}_p^m$ and $t \in \mathbb{F}_p$, and verifying that

$$\tilde{P}_{x,h}(t) = \tilde{g}(x + th).$$

One such check obviously uses $\mathcal{O}(m \log p) = \mathcal{O}(\log n)$ random bits, and queries one value of $\tilde{g}$ and one entry of T. As noted above, this does not violate our resource constraints. Thus the verifier is allowed to repeat this basic step a constant number of times, reject if any test fails, and accept otherwise.

From the arguments above it is obvious that if $\tilde{g}$ is indeed a polynomial of degree d, we can construct a table $T_{\tilde{g}}$ such that the verifier will never reject given $\tilde{g}$ and $T_{\tilde{g}}$. On the other hand, if $\tilde{g}$ cannot be written as low degree polynomial, then for each table T we can of course find some x, h, and t such that $\tilde{P}_{x,h}(t) \neq \tilde{g}(x+th)$. This does not help us very much though, since our verifier can check only a constant number of possible combinations x, h, and t, and we cannot guarantee that it will find a rejecting combination among those that are tried.

But it can be shown that — under certain conditions — if $\tilde{g}$ is not δ-close to any degree d polynomial, then each test of our verifier will reject with probability at least $\delta/8$. An exact statement of this property will be given as

Theorem 5.69. *Let $0 < \delta < 1/6066$ and $d, m \in \mathbb{N}$ be constants, and let F be a finite field with $|F| \geqslant 8d$ and $|F| \geqslant 1/\delta$. Let $\tilde{g}: F^m \to F$ be a function, and let $T: F^{2m} \to F^{d+1}$ be a function such that the degree d polynomials $\tilde{P}_{x,h}$ over F given by $\tilde{P}_{x,h}(t) = \sum_{i=0}^d T(x,h)_{i+1} \cdot t^i$ satisfy*

$$\mathrm{Prob}_{x,h,t}[\tilde{P}_{x,h}(t) = \tilde{g}(x+th)] \geqslant 1 - \frac{\delta}{8}.$$

Then there exists a (unique) polynomial $g: F^m \to F$ of total degree d so that

$$\mathrm{Prob}_x[g(x) = \tilde{g}(x)] \geqslant 1 - \delta.$$

The proof of this theorem is postponed to Section 5.7, since it is technically rather difficult (in fact, as [Aro94] notes, for some time this was the last missing part of the proof of the PCP-Theorem). What remains to be checked is that the additional conditions of the theorem are satisfied in our situation. If we choose a small enough δ_0, we see that this is the case, since by construction of the polynomial codes, we have $p \geqslant 8d$. But, observe that the requirement $p \geqslant 1/\delta$ will be false for small values of n. In order to handle this problem recall that resource bounds for extended verifiers are formulated in terms of the $\mathcal{O}$-notation and so we can simply choose the corresponding constants so large that for these small values of n the verifier can read the course-of-values of the whole function and can check by brute force whether it is close to a polynomial.

We finally note that if each test of the verifier rejects with probability $\delta/8$, by performing $\lceil 2/\log(1+\delta/8) \rceil$ such tests, we get a total rejection rate of

$$1 - (1 - \frac{\delta}{8})^{2/\log(1+\delta/8)} = 1 - 4^{\frac{\log(1-\delta/8)}{\log(1+\delta/8)}} \geqslant 1 - 4^{-1} = \frac{3}{4}.$$

We conclude the proof by explicitly stating the test procedure:

<u>Low Degree Test</u>
Read input n and compute $p := p_{e,n}$, $m := m_n$, and $d := d_n$.
Assume π_0 and π_1 encode $\tilde{g}$ and T as described above.
if $p < 1/\delta$ **then**
> Perform brute-force check.

else
> **repeat** $\lceil 2/\log(1 + \delta/8) \rceil$ times
>> Choose $x, h \in \mathbb{F}_p^m$ and $t \in \mathbb{F}_p$ uniformly at random.
>> **if** $\sum_{i=0}^d T(x,h)_{i+1} \cdot t^i \neq \tilde{g}(x + th)$ **then reject**
>
> **accept**

As the above arguments have shown, the algorithm Low Degree Test can indeed be used as a constant $(\log n, \text{poly}(\log n))$-restricted extended verifier with the required properties. ∎

Proposition 5.29. *For every parameter $e \geqslant 2$, the polynomial coding scheme C_e^{poly} is $(\log n, \text{poly}(\log n))$-correctable.*

Proof. Assuming that a given function $\tilde{g} \colon \mathbb{F}_p^m \to \mathbb{F}_p$ is δ-close to some low degree polynomial g, we know that for all but a fraction δ of points $x \in \mathbb{F}_p^m$, the values $g(x)$ and $\tilde{g}(x)$ agree, but we do not know at *which* values x the function $\tilde{g}$ misrepresents the polynomial g. To show the correctability of the polynomial coding scheme, we need to give a verifier which checks probabilistically whether $g(x)$ and $\tilde{g}(x)$ agree *at some given point x*.

For this purpose, we use the same table T as used with the Low Degree Test, containing for each $x, h \in \mathbb{F}_p^m$ the coefficients of a univariate degree d polynomial that is supposed to represent the restriction of $\tilde{g}$ to the line $l_{x,h}$. The idea of the test is now to verify whether at the given point x, for some randomly chosen h, the corresponding table entry is identical to the restriction of the polynomial g to that line. Since both are univariate degree d polynomials, this can be checked by considering only a few points on the line (as we will see, one point is in fact enough). Now we do not know the polynomial g of course, but we can use the hypothesis that g is δ-close to $\tilde{g}$. This means that for that (uniformly chosen) point $y = x + th$ where we need to check $g(y)$, we can simply use $\tilde{g}(y)$ instead and will thus make an error with only a small probability.

Once we know that for the chosen h the corresponding table entry agrees with g, we know all values of g on the line $l_{x,h}$, especially at the point x. Thus we can simply check whether this value agrees with $\tilde{g}(x)$. The procedure just outlined can be stated as follows:

<u>Self-Correct</u>
Read input n and compute $p := p_{e,n}$, $m := m_n$, and $d := d_n$.
Read input i and determine the corresponding $x \in \mathbb{F}_p^m$.
Assume π_0 and π_1 encode $\tilde{g}$ and T as described above.

Choose $h \in \mathbb{F}_p^m$ and $t \in \mathbb{F}_p \setminus \{0\}$ uniformly at random.

if $T(x,h)_1 \neq \tilde{g}(x)$ **then reject**

if $\sum_{i=0}^{d} T(x,h)_{i+1} \cdot t^i \neq \tilde{g}(x + th)$ **then reject**

accept

This algorithm satisfies the requirements: If $\tilde{g}$ is a low degree polynomial, then we can simply choose the same table $T_{\tilde{g}}$ as above, and the procedure will never reject. It remains to show that if $\tilde{g}$ is δ-close to the low degree polynomial g, but $\tilde{g}(x) \neq g(x)$, then the algorithm rejects with probability $\geqslant 3/4$.

We consider now those values of h where $\tilde{P}_{x,h}(0) = \tilde{g}(x)$ holds. Since we have $g(x) \neq \tilde{g}(x)$, and thus $\tilde{P}_{x,h}(0) \neq g(x)$ follows, the two degree d polynomials $\tilde{P}_{x,h}$ and the restriction of g to $l_{x,h}$ can agree at most at d points $t \in \mathbb{F}_p \setminus \{0\}$. Since $p \geqslant 8d$ and hence $d/(p-1) \leqslant 1/7$ by construction of the polynomial coding scheme, we conclude that in this case $\mathrm{Prob}_{t \neq 0}[\tilde{P}_{x,h}(t) = g(x + th)] \leqslant 1/7$ holds, hence

$$\mathrm{Prob}_{h,t \neq 0}[\tilde{P}_{x,h}(0) \neq \tilde{g}(x) \vee \tilde{P}_{x,h}(t) \neq g(x + th)] \geqslant \frac{6}{7}.$$

Now if we could replace g by $\tilde{g}$, the above probability would just be equal to the reject probability of our algorithm. But since for $t \neq 0$ the expression $x + th$ is uniformly distributed over $\mathbb{F}_p^m$ iff h is, and by assumption $\tilde{g}$ is δ-close to g, we also have $\mathrm{Prob}_{h,t \neq 0}[\tilde{g}(x + th) \neq g(x + th)] \leqslant \delta$. Since obviously $\tilde{P}_{x,h}(t) \neq g(x + th)$ implies $\tilde{P}_{x,h}(t) \neq \tilde{g}(x + th)$ or $\tilde{g}(x + th) \neq g(x + th)$, we conclude that

$$\mathrm{Prob}_{h,t \neq 0}[\tilde{P}_{x,h}(0) \neq \tilde{g}(x) \vee \tilde{P}_{x,h}(t) \neq g(x + th)]$$
$$\leqslant \mathrm{Prob}_{h,t \neq 0}[\tilde{P}_{x,h}(0) \neq \tilde{g}(x) \vee \tilde{P}_{x,h}(t) \neq \tilde{g}(x + th)] + \delta.$$

But this means that the algorithm rejects with probability of at least $6/7 - \delta$, which is $\geqslant 3/4$ for some small enough δ.

We conclude the proof by noting that the algorithm SELF-CORRECT obviously satisfies the given resource constraints: the algorithm uses $\mathcal{O}(m \log p) = \mathcal{O}(\log n)$ random bits to choose h and t, and queries one entry of T and two values of $\tilde{g}$. ∎

As in the case of the linear code, we immediately obtain the following corollary.

Corollary 5.30. *For every parameter $e \geqslant 2$, the polynomial coding scheme $\mathcal{C}_e^{\mathrm{poly}}$ is $(\log n, \mathrm{poly}(\log n))$-robust.*

Remark 5.31. For the proof of Theorem 5.54 we need yet another property of the polynomial coding scheme: For all proof strings (π_0, π_1) that are not rejected by the low degree test given in the proof of Proposition 5.28, we want to efficiently find (and decode) the codeword nearest to π_0 (which is uniquely defined since π_0 is δ-close to some codeword, for some $\delta < 1/4$, otherwise it would have been rejected by the low degree test).

But we know from the proof of Proposition 5.28 that if the total rejection rate of the LOW DEGREE TEST procedure is less that $3/4$ (i.e. the proof string (π_0, π_1) is not rejected), then the rejection probability of a single test cannot be higher than $\delta/8$. (We suppose that we have the case $p \geqslant 1/\delta$ here; the case $p < 1/\delta$ can again be solved using brute force.) Now this means that the functions $\tilde{g}$ and T encoded by π_0 and π_1 satisfy the conditions of Theorem 5.69, but in this case — as we will show in Corollary 5.70 — we can compute the low degree polynomial g that is δ-close to $\tilde{g}$ in time polynomial in $p_{e,n}^{mn} = \mathcal{O}(\mathrm{poly}(n))$ (this holds for every parameter $e \geqslant 2$ of the polynomial coding scheme).

5.4 Efficient Solution Verifiers

5.4.1 Solution Verifiers

Solution verifiers are a special type of extended verifier which interpret their input x as Boolean formula and expect their input π_0 to be an encoded assignment for this formula. Before we give a more precise definition of solution verifiers we introduce some notation related to Boolean formulae.

Boolean formulae are built in the usual way from propositional variables and the logical connectives $\wedge$, $\vee$ and $\bar{\ }$, which are interpreted as conjunction, disjunction, and negation, respectively. A *literal* is a propositional variable or the negation of a propositional variable and a *clause* is a disjunction of literals. A Boolean formula is *in conjunctive normal form* iff it is a conjunction of clauses.

A *3-CNF formula* is a Boolean formula in conjunctive normal form with exactly three literals per clause. In order to fix an appropriate representation of 3-CNF formulae by binary strings observe that usually we are only interested in the satisfiability of a 3-CNF formula and thus the formula essentially remains the same under renaming of variables. Consequently we can assume that the variables of a 3-CNF formula φ in n variables are just $x_1, \ldots, x_n$. In this situation it is straightforward to represent a single literal of φ by $k = \lceil \log n \rceil + 1$ bits and to represent a single clause by $3k$ bits. In order to represent the whole formula φ we will specify the value of $3k$ by a length $3k$ prefix $0^{3k-1}1$ and we will use another $3km$ bits for representing the m clauses of φ, thus leaving us with a binary representation of length $3k(m + 1)$. We will usually identify a 3-CNF formula with its binary representation and thus for example the language 3SAT of all satisfiable 3-CNF formulae becomes a set of binary strings.

Remark 5.32. For technical reasons we will assume from now on that a 3-CNF formula with a binary representation of length n has exactly n variables in the sense that in an assignment for the formula we have to specify values for n variables (which do not necessarily all occur in φ). This convention is safe due to the fact that for a 3-CNF formula with a binary representation of length n

the number of variables occurring in the formula must be less than n because representing a single literal requires at least two bits.

Definition 5.33. *An extended verifier V is a solution verifier with coding scheme C iff for every input (x, π_0),*

V accepts in case x is a 3-CNF formula such that π_0 is a code w.r.t. C for a satisfying assignment of x and

V rejects in case π_0 is not $1/4$-close to a code for a satisfying assignment of x.

Remark 5.34. Every solution verifier can be viewed as a verifier for 3SAT. Let V be a solution verifier. Then each formula x either is satisfiable and in this case there is some π_0 on which V accepts, or x is not satisfiable and then V rejects for every proof π_0 because in this case no string is $1/4$-close to a code for a satisfying assignment. According to Remark 5.4 the solution verifier V can thus be considered as a verifier for 3SAT.

5.4.2 Arithmetization of 3-CNF Formulae

In Remark 5.35 we describe a method for arithmetizing 3-CNF formulae which will subsequently be used during the construction of solution verifiers.

Remark 5.35. We can assume that every clause of a 3-CNF formula belongs to one of the following types

$$\psi_1 := x_i \vee x_j \vee x_k, \qquad\qquad \psi_2 := x_i \vee x_j \vee \overline{x_k},$$
$$\psi_3 := x_i \vee \overline{x_j} \vee \overline{x_k}, \qquad\qquad \psi_4 := \overline{x_i} \vee \overline{x_j} \vee \overline{x_k}.$$

These possible types of clauses can be arithmetized by the polynomials

$$p_1 := (1 - x_i)(1 - x_j)(1 - x_k), \qquad p_2 := (1 - x_i)(1 - x_j)x_k,$$
$$p_3 := (1 - x_i)x_j x_k, \qquad\qquad p_4 := x_i x_j x_k,$$

respectively. More precisely we let V be the set of variables $\{x_0, \ldots, x_{n-1}\}$ and for ease of notation we identify V with the set $\{0, \ldots, n-1\}$. By means of this identification the usual strict ordering on the natural numbers induces a strict ordering $<$ on V. Recall from Section 5.3.4 that a function g from V to a field $\mathbb{F}_p$ corresponds to a binary string $s := \sigma(g, <)$ of length n. Here $s = b_0 \ldots b_{n-1}$ can be viewed as an assignment for the variables in V where the value for the variable x_j is equal to b_j. We leave it to the reader to show that now indeed for every function g from V to $\mathbb{F}_p$, the expression $p_r(g(i), g(j), g(k))$ evaluates to zero iff the assignment $\sigma(g, <)$ satisfies the clause $\psi_r(x_i, x_j, x_k)$.

Next we consider a 3-CNF formula φ in variables $x_0, \ldots, x_{n-1}$ and again we let $V = \{0, \ldots, n-1\}$. In order to obtain an arithmetization of φ we define indicator functions $\chi_1, \ldots, \chi_4$ from V^3 to $\{0, 1\}$ which correspond to the four types of clauses and where for example $\chi_1(i, j, k)$ is 1 iff φ contains a clause $x_i \vee x_j \vee x_k$. Then a function g from V to a field $\mathbb{F}_p$ corresponds to satisfying assignment $s = \sigma(g, <)$ for φ iff we have for all $r = 1, \ldots, 4$ and for all (i, j, k) in V^3

$$\chi_r(i, j, k) \cdot p_r(g(i), g(j), g(k)) = 0 \tag{5.13}$$

Thus we can for example test whether $\sigma(g, <)$ is a satisfying assignment for φ by testing whether for $r = 1, \ldots, 4$, the function shown in equation (5.13) is identically zero on V^3.

Note that the arithmetization by the polynomials $p_1, \ldots, p_4$ as given in Remark 5.35 differs from the more intuitive standard way of arithmetizing Boolean formulae where for example the formulae $\bar{x}$ and $x \wedge y$ are arithmetized as $1 - x$ and xy, respectively.

5.4.3 An $(n^3, 1)$-Restricted Solution Verifier

In this section, an $(n^3, 1)$-restricted solution verifier with linear function coding scheme will be constructed. As described in the previous section, we use the vector

$$\hat{\varphi}(x) = \hat{\varphi}(x_0, \ldots, x_{n-1}) = (\chi_r(i, j, k) \cdot p_r(x_i, x_j, x_k))_{(i,j,k,r) \in V^3 \times [1:4]}$$

of length $4n^3$ as an *arithmetization* of the 3-CNF formula φ on n variables in $V = \{x_0, \ldots, x_{n-1}\}$. Here and in the sequel we denote by $[m_1 : m_2]$ the set $\{n \in \mathbb{N} : m_1 \leqslant n \wedge n \leqslant m_2\}$ and we abbreviate the vector $(x_0, \ldots, x_{n-1})$ by x.

As usual, we identify a function $g : V \rightarrow \mathbb{F}_2$ with a binary string of length n, and hence with an assignment for a 3-CNF formula φ. Note that a satisfying assignment of the 3-CNF formula φ corresponds in a natural way to an assignment of $(x_0, \ldots, x_{n-1})$ for which the vector $\hat{\varphi}(x_0, \ldots, x_{n-1})$ is zero.

Observation 5.36. *A 3-CNF formula φ is satisfiable if and only if its arithmetization $\hat{\varphi}(x)$ is zero for some $x \in \mathbb{F}_2^n$.*

Testing whether such a vector $\hat{\varphi}(x)$ is zero can be done by testing whether the scalar product of $\hat{\varphi}(x)$ and a random vector $v \in \mathbb{F}_2^{4n^3}$ is zero. While $\hat{\varphi}(x)^T \cdot v = 0$ whenever $\hat{\varphi}(x) = 0$, it is nonzero with probability $\frac{1}{2}$ whenever $\hat{\varphi}(x) \neq 0$ as shown in the following lemma:

Lemma 5.37. *Let $0 \neq z \in \mathbb{F}_2^n$, then $\mathrm{Prob}_v[z^T v \neq 0] = \frac{1}{2}$.*

Proof. First note that due to Definition 5.15 $l_z(v) = z^T v$ and $l_0 \equiv 0$. Hence the claim follows from Lemma 5.16, which states that l_z and l_0 are exactly $\frac{1}{2}$-close. ∎

We now take a closer look at the scalar product $\hat{\varphi}(x)^T \cdot v = l_{\hat{\varphi}(x)}(v)$. A fundamental observation is that $l_{\hat{\varphi}(x)}(v)$ can be rewritten as follows:

$$l_{\hat{\varphi}(x)}(v) = \sum_{\substack{(i,j,k) \in V^3 \\ r \in [1:4]}} \chi_r(i,j,k) \cdot p_r(x_i, x_j, x_k) \cdot v_{(i,j,k,r)}$$

$$= c(v) + \sum_{i \in S_1(v)} x_i + \sum_{(i,j) \in S_2(v)} x_i x_j + \sum_{(i,j,k) \in S_3(v)} x_i x_j x_k, \quad (5.14)$$

where the constant $c(v)$ and the sets $S_1(v)$, $S_2(v)$, $S_3(v)$ depend only on v and the given 3-CNF formula φ, but they are independent from an assignment for x. Thus, if we have random access to the following sums

$$\sum_{i \in S} g(i) \quad \text{for all} \quad S \subseteq \{0, \ldots, n-1\}, \quad (5.15)$$

$$\sum_{(i,j) \in S} g(i) \cdot g(j) \quad \text{for all} \quad S \subseteq \{0, \ldots, n-1\}^2, \quad (5.16)$$

$$\sum_{(i,j,k) \in S} g(i) \cdot g(j) \cdot g(k) \quad \text{for all} \quad S \subseteq \{0, \ldots, n-1\}^3, \quad (5.17)$$

of a given assignment g for $x = (x_0, \ldots, x_{n-1})$, the values of these three sums will suffice to check the satisfiability of the given formula φ. Observe that each listing of all sums given in equation (5.15), (5.16), and (5.17) can be expressed as a codeword of the linear function code, where the sets S are interpreted as its arguments.

To be more precise, we need the following definitions. Let $x, x' \in \mathbb{F}_2^n$ be two vectors, then we denote by $x \circ x'$ the following vector $y \in \mathbb{F}_2^{n^2}$ given by $y_{i,j} = x_i \cdot x_j$. Here we identify the matrix element $y_{i,j}$ with the vector element $y_{(i-1)n+j}$. Similarly, let $x \in \mathbb{F}_2^n$ and $y \in \mathbb{F}_2^{n^2}$ be two vectors, then we denote by $x \circ y$ the following vector $z \in \mathbb{F}_2^{n^3}$ given by $z_{i,j,k} = x_i \cdot y_{j,k}$. Again, we identify the tensor element $z_{i,j,k}$ with the vector element $z_{(i-1)n^2+(j-1)n+k}$. For a fixed vector $a \in \mathbb{F}_2^n$, let $b = b_{i,j} = a \circ a \in \mathbb{F}_2^{n^2}$ and $c = c_{i,j,k} = a \circ b \in \mathbb{F}_2^{n^3}$. Using these vectors a, b, and c, we can define the following three codewords $A := l_a$, $B := l_b$, and $C := l_c$ of the linear function code with codeword length of 2^n, 2^{n^2}, and 2^{n^3}, respectively. If $a_i = g(i)$ for all $i \in \{0, \ldots, n-1\}$ then the codewords A, B, and C correspond exactly to the listing of all sums given in equation (5.15), (5.16), and (5.17), respectively.

In what follows we construct a solution verifier based on these three linear codes. Recall that for a solution verifier the input x is an encoding of a 3-CNF formula φ and that π_0 of the proof $\pi = (\pi_0, \pi_1)$ is an encoding using the linear function coding scheme of a satisfying assignment of φ. Note that if g is a satisfying assignment of the given 3-CNF formula encoded by x, then the listing of the values of all sums given in equation (5.15) is exactly the encoding using the linear function coding scheme required by the definition of a solution verifier.

In what follows, the solution verifier will use the following two pattern functions $\alpha_0(x) = (n, 1)$ and $\alpha_1(x) = (n^2, 1)(n^3, n^2 + 1)$. As we have seen in Corollary 5.20 the linear function coding scheme C_m^{lin} is $(m, 1)$-robust. Thus, by Proposition 5.14 we can assume that the codewords in the proof specified by the pattern function α_0 and α_1 are indeed codewords of the linear function coding scheme, i.e., our solution verifier assumes that for each input $(x, (\pi_0, \pi_1))$ the pattern function $\alpha_0(x)$ obeys π_0 and $\alpha_1(x)$ obeys π_1. Moreover, Corollary 5.20 implies that a $(n^3, 1)$-restricted solution subverifier suffices for this test. For ease of notation we denote by A, B, and C the codewords specified by the proof $\pi = (\pi_0, \pi_1)$ and the pattern functions α_0 and α_1, i.e., $\pi = (\pi_0, \pi_1) = (A, BC)$. Hence, the solution verifier has to check that:

1. A, B, C are consistently defined w.r.t. the same vector $a \in \mathbb{F}_2^n$,

2. vector a corresponds to a satisfying assignment.

In what follows we show how both conditions can be checked by a $(n^3, 1)$-restricted solution verifier with linear function coding scheme. Clearly, for codewords A, B, and C we have:

$$\forall x, x' \in \mathbb{F}_2^n : A(x) \cdot A(x') = B(x \circ x') \quad \Leftrightarrow \quad b = a \circ a$$

and

$$\forall x \in \mathbb{F}_2^n, y \in \mathbb{F}_2^{n^2} \quad A(x) \cdot B(y) = C(x \circ y) \quad \Leftrightarrow \quad c = a \circ b.$$

Together with the above equalities, it is possible to check whether the three linear function codes A, B, and C are consistently defined.

CONSISTENCY TEST
Interpret $\pi = (\pi_0, \pi_1)$ as (A, BC) as discussed above.
Choose $x, x' \in \mathbb{F}_2^n$ uniformly at random.
if $(A(x) \cdot A(x') \neq B(x \circ x'))$ **then reject**
Choose $x \in \mathbb{F}_2^n$ and $y \in \mathbb{F}_2^{n^2}$ uniformly at random.
if $(A(x) \cdot B(y) \neq C(x \circ y))$ **then reject**
accept

In CONSISTENCY TEST a constant number of positions in the proof will be queried and $\mathcal{O}(n^2)$ random bits are sufficient to determine the query positions. Again in the procedure CONSISTENCY TEST a correct proof π will never be rejected. On the other hand, an incorrect proof will be rejected with probability of at least $1/4$ as can be seen from the following lemma.

Lemma 5.38. *If there does not exist a vector $a \in \mathbb{F}_2^n$ such that $A(x) = a^T x$, $B(y) = (a \circ a)^T y$, and $C(z) = (a \circ a \circ a)^T z$, then* CONSISTENCY TEST *rejects with probability at least 1/4.*

Proof. If $b = a \circ a$ and $c = a \circ b$, there is nothing to prove. Hence, we first assume that $b \neq a \circ a$. Recall Lemma 5.37, i.e., for vectors $\alpha \neq \hat{\alpha} \in \mathbb{F}_2^n$ we have

$$\text{Prob}_x[\alpha^T x \neq \hat{\alpha}^T x] = \frac{1}{2}. \tag{5.18}$$

This implies that matrices $\beta \neq \hat{\beta} \in \mathbb{F}_2^{n^2}$ satisfy

$$\text{Prob}_x[\beta x \neq \hat{\beta} x] \geqslant \frac{1}{2}. \tag{5.19}$$

Combining these (in)equalities with the observation that

$$x^T(a \circ a)x' = \sum_i \sum_j x_i \cdot a_i \cdot a_j \cdot x'_j = A(x) \cdot A(x')$$

and

$$x^T b x' = \sum_i \sum_j x_i \cdot b_{i,j} \cdot x'_j = B(x \circ x'),$$

we obtain under the assumption that $b \neq a \circ a$ (note that $a \circ a$ and b are interpreted as matrices):

$\text{Prob}_{x,x'}[A(x) \cdot A(x') \neq B(x \circ x')]$

$= \quad \text{Prob}_{x,x'}[x^T(a \circ a)x' \neq x^T b x']$

$= \quad \text{Prob}_{x,x'}[(a \circ a)x' \neq bx' \wedge x^T((a \circ a)x') \neq x^T(bx')]$

$= \quad \text{Prob}_{x'}[(a \circ a)x' \neq bx'] \cdot \text{Prob}_{x,x'}[x^T((a \circ a)x') \neq x^T(bx')|(a \circ a)x' \neq bx']$

using equation (5.18) and (5.19)

$\geqslant \quad \dfrac{1}{4}.$

Now we assume that $c \neq a \circ b = a \circ a \circ a$. Similarly, we obtain for tensors $\gamma, \hat{\gamma} \in \mathbb{F}_2^{n^3}$

$$\text{Prob}_{y \in \mathbb{F}_2^{n^2}}[\gamma y \neq \hat{\gamma} y] \geqslant \frac{1}{2}. \tag{5.20}$$

With the following observation that

$$x^T(a \circ b)y' = \sum_i \sum_{j,k} x_i \cdot a_i \cdot b_{j,k} \cdot y_{j,k} = C(x \circ y)$$

we obtain under the assumption that $c \neq a \circ b$ (note that $a \circ b$ and c are interpreted as tensors):

$$\text{Prob}_{x,y}[A(x) \cdot B(y) \neq C(x \circ y)]$$
$$= \text{Prob}_{x,y}[x^T(a \circ b)y \neq x^T cy]$$
$$= \text{Prob}_{x,y}[(a \circ b)y \neq cy \wedge x^T((a \circ b)y) \neq x^T(cy)]$$
$$= \text{Prob}_y[(a \circ b)y \neq cy] \cdot \text{Prob}_{x,y}[x^T((a \circ b)y) \neq x^T(cy)|(a \circ b)y \neq cy]$$
$$\text{using equation (5.18) and (5.20)}$$
$$\geqslant \frac{1}{4}.$$

∎

Now we have checked whether the codewords in the given proof $\pi = (\pi_0, \pi_1) = (A, BC)$ are defined consistently w.r.t. some vector $a \in \mathbb{F}_2^n$. It remains to test whether this vector a corresponds to a satisfying assignment of the given 3SAT formula, i.e., whether $a^T v = 0$ for some vector v. Due to equation (5.14) the following procedure will suffice.

SATISFIABILITY TEST
Interpret x as an encoding of a 3-CNF formula φ.
Interpret $\pi = (\pi_0, \pi_1)$ as (A, BC) as discussed above.
Choose $v \in \mathbb{F}_2^n$ uniformly at random.
Compute $c := c(v) \in \mathbb{F}_2$ from φ and v.
Compute $S_1 := S_1(v) \in \mathbb{F}_2^n$ from φ and v.
Compute $S_2 := S_2(v) \in \mathbb{F}_2^{n^2}$ from φ and v.
Compute $S_3 := S_3(v) \in \mathbb{F}_2^{n^3}$ from φ and v.
if $(c + A(S_1) + B(S_2) + C(S_3) = 0)$
    then accept
    else reject

Note that both algorithms will never reject a correct proof, while both algorithms rejects a wrong proof with probability of at least 1/4. Repeating both test five times, the rejection probability can be amplified to at least 3/4 as required in the definition of a solution verifier. Altogether, we have proved the following fundamental theorem.

Theorem 5.39. *There exists an $(n^3, 1)$-restricted solution verifier with coding scheme $\mathcal{C}^{\text{lin}}$.*

By definition of a solution verifier, we also have established the first step of proving the PCP-Theorem as announced in Section 5.1.

Corollary 5.40. 3SAT $\in$ PCP$(n^3, 1)$.

Since 3SAT is $\mathcal{NP}$-complete, each problem in $\mathcal{NP}$ can be reduced to 3SAT while the size of the 3-CNF formula is polynomially related to the size of the original instance. Thus, we obtain the following corollary.

Corollary 5.41. $\mathcal{NP} \subseteq$ PCP$(\text{poly}(n), 1)$.

5.4.4 A $(\log n, \mathrm{poly}(\log n))$-Restricted Solution Verifier

Theorem 5.42. *There is a $(\log n, \mathrm{poly}(\log n))$-restricted solution verifier with coding scheme C^{poly}.*

Proof. Before giving the details of the solution verifier V satisfying the specified resource bounds, we recall the exact format of the input given to V. This input consists of the input string x encoding a 3-CNF formula φ according to the straightforward scheme given in Section 5.4.1 and the first part π_0 of the proof string that is supposed to encode a satisfying assignment of φ according to the polynomial coding scheme C^{poly}. Observe that since by Remark 5.32 an assignment of φ consists of the values of n propositional variables, where n is the length of the input string x, the proof string π_0 is supposed to contain a codeword of the code C_n^{poly}. Recall from Definition 5.25 that these codewords represent m_n-variate polynomials of degree at most h_n in each variable over the field $\mathbb{F}_{p_n}$, where p_n, h_n, and m_n are as defined in Section 5.3.4. Furthermore, if H_n denotes the set of the first h_n elements of $\mathbb{F}_{p_n}$, we can decode a code $\tilde{g}$ for an assignment g by noting that $g(i) = \tilde{g}(x)$ where x is the i-th element of $H_n^{m_n}$ according to the ordering $<_n$ on $\mathbb{F}_{p_n}^{m_n}$ defined by the polynomial coding scheme.

By Proposition 5.14, using a pattern function α_0 that maps every string of length n to the pattern $(n, 1)$ and an empty pattern function α_1, it suffices to construct a symbol constant $(\log n, \mathrm{poly}(\log n))$-restricted extended verifier V, such that V accepts all inputs (x, π_0) where π_0 is a code for a satisfying assignment of the 3-CNF formula φ encoded by x, and V rejects all inputs (x, π_0) where π_0 is a code for an assignment that does not satisfy φ. The assumptions of Proposition 5.14 are satisfied since the polynomial coding scheme is $(\log n, \mathrm{poly}(\log n))$-robust by Corollary 5.30, and furthermore the length of a codeword of C_n^{poly} is $k_n := p_n^{m_n} \lceil \log p_n \rceil$, and therefore $\log k_n$ is in $\mathcal{O}(m_n \log p_n + \log \log p_n) = \mathcal{O}(\log n)$.

Thus we will assume from now on that π_0 does indeed contain a code $\tilde{g}$ for some assignment g of φ; the task of the verifier V consists therefore solely of determining whether this assignment satisfies φ or not. Note that according to Remark 5.35 this is the case if and only if for all $r = 1, \ldots, 4$ and $0 \leqslant i, j, k < n$ we have

$$\chi_r(i, j, k) \cdot p_r(g(i), g(j), g(k)) = 0,$$

where the p_r are fixed polynomials that are linear in each variable and the χ_r can be computed from x in time polynomial in n. Since we have no immediate access to g but only to its low degree extension $\tilde{g}$, we also compute appropriate low degree extensions $\tilde{\chi}_r$ of the functions χ_r. Note that these can also be computed in time polynomial in n, and that each $\tilde{\chi}_r$ is a $3m_n$-variate polynomial over $\mathbb{F}_{p_n}$ of degree at most h_n in each variable. As a consequence, for $r = 1, \ldots, 4$ the functions

$$f_r(x, y, z) := \tilde{\chi}_r(x, y, z) \cdot p_r(\tilde{g}(x), \tilde{g}(y), \tilde{g}(z))$$

are $3m_n$-variate polynomials over $\mathbb{F}_{p_n}$ of degree at most $2h_n$ in each variable. Observe that each single value of f_r can be computed by our verifier using just three queries to the codeword $\bar{g}$. Furthermore, we know by construction that g is a satisfying assignment if and only if for all $r = 1, \ldots, 4$,

$$f_r(x, y, z) = 0 \quad \text{for all } x, y, z \in H_n^{mn}.$$

This means that the problem of verifying whether g is a satisfying assignment is reduced to the task of checking whether the low degree polynomials f_r are identically zero on a certain part of their domain. At this point we will use a test devised by Babai, Fortnow, Levin, and Szegedy to perform this check probabilistically. That test is an extension of the so-called LFKN test used to check whether the *sum* of the values of a polynomial over a certain part of its domain is zero, which is due to Lund, Fortnow, Karloff, and Nisan. Both of those tests will be described in detail in Section 5.6. Here we only state the result about the procedure EXTENDED LFKN-TEST, which takes a function f and an additional table T as inputs:

Theorem 5.56. *Let $m, d \in \mathbb{N}$ be constants and F be a finite field with $|F| \geqslant 4m(d+|H|)$, where $H \subseteq F$ denotes an arbitrary subset of F. Let $f : F^m \to F$ be a polynomial of degree at most d in every variable. Then the procedure* EXTENDED LFKN-TEST *has the following properties:*

- *If f satisfies the equation*

$$f(x) = 0 \quad \text{for all } x \in H^m, \tag{5.21}$$

  then there exists a table T_f such that EXTENDED LFKN-TEST(f, T_f) *always accepts.*

- *If f does not satisfy equation (5.21), then* EXTENDED LFKN-TEST(f, T) *rejects with probability at least $1/2$ for all tables T.*

- EXTENDED LFKN-TEST *queries f at one point, randomly chosen from F^m, reads m entries of size $\mathcal{O}(m(d + |H|) \log |F|)$ from T, and uses $\mathcal{O}(m \log |F|)$ random bits.*

Now we are able to construct a verifier V as desired: It simply uses four sections of its proof string π_1 as tables needed by the EXTENDED LFKN-TEST for checking whether the functions f_r vanish on the subcube H_n^{3mn}. The conditions of Theorem 5.56 are satisfied: Since f_r has degree at most $2h_n$ in each variable, we only need to ensure that $p_n \geqslant 36 m_n h_n$. But this is the case by definition of the parameters of the polynomial coding scheme. Thus we conclude that if π_0 encodes indeed a satisfying assignment for x, there is a proof string π_1 such that V accepts $(x, (\pi_0, \pi_1))$, and if π_0 encodes an assignment that does not satisfy x, for any proof string π_1 the verifier rejects with probability at least $1/2$. By repeating the procedure twice, the rejection probability can be amplified to the desired value $3/4$.

What remains to be checked are the resource constraints: The verifier makes two calls to EXTENDED LFKN-TEST, each of them using $\mathcal{O}(3m_n \log p_n) = \mathcal{O}(\log n)$ random bits, and querying a single point of f_r, which translates into three queries of $\tilde{g}$, and m_n entries of size $\mathcal{O}(9m_n h_n \log p_n) = \mathcal{O}(\text{poly}(\log n))$ in its table, which translates into queries of $\mathcal{O}(\text{poly}(\log n))$ many bits of π_1. Thus we conclude that V is indeed a symbol constant $(\log n, \text{poly}(\log n))$-restricted extended verifier. ∎

5.4.5 A Constant $(\log n, \text{poly}(\log n))$-Restricted Solution Verifier

Note that the solution verifier given in Theorem 5.42 is *not* constant, since the EXTENDED LFKN-TEST makes $m_n = \lceil \log n / \log \log n \rceil + 1$ queries to its additional table. As we will show in this section, that solution verifier can nevertheless be turned into a constant verifier. To this purpose we use a technique similar to one used in the proof of Proposition 5.29. Recall that the additional table used there for showing the correctability of the polynomial coding scheme contained polynomials $P_{x,h}$ that were supposed to describe the values of a low degree polynomial g along the line $\{x + th \mid t \in \mathbb{F}_p\}$. We then argued that — since $P_{x,h}$ and g were both polynomials of low degree — we could probabilistically check whether such a polynomial $P_{x,h}$ described g correctly at *all* points of that line, while actually querying g only at *one single* place during the test.

We now generalize this setting and consider not just lines, but general *curves* of the form $\{u_0 + tu_1 + t^2 u_2 + \cdots + t^c u_c \mid t \in \mathbb{F}_p\}$, where $u_0, \ldots, u_c \in \mathbb{F}_p^m$ are the *parameters* of the curve, and c is called its *degree*. Note that in the case $c = 1$ the notion of curve coincides with that of line. As in the case of lines, we consider polynomials $P_{u_0, \ldots, u_c}^g : \mathbb{F}_p \to \mathbb{F}_p$ describing the values of g along the curve C with parameters $u_0, \ldots, u_c$:

$$P_{u_0, \ldots, u_c}^g(t) := g(C(t)), \text{ where } C(t) := u_0 + tu_1 + \cdots + t^c u_c.$$

Note that if the total degree of g is at most d, the degree of $P_{u_0, \ldots, u_c}^g$ is at most cd. Thus, we will suppose that our table T contains for each $(c+1)$-tuple $u_0, \ldots, u_c \in \mathbb{F}_p^m$ a univariate degree cd polynomial $P_{u_0, \ldots, u_c}$, represented by its coefficients, that is supposed to be equal to $P_{u_0, \ldots, u_c}^g$.

We further observe that, since each component of $u_0 + tu_1 + \cdots + t^c u_c$ is a polynomial in t of degree at most c, we can for each choice of $c + 1$ points $x_1, \ldots, x_{c+1}$ find a uniquely defined curve $C\langle x_1, \ldots, x_{c+1}\rangle$ of degree c such that $C\langle x_1, \ldots, x_{c+1}\rangle(i - 1) = x_i$ for $1 \leqslant i \leqslant c + 1$. We let $P\langle x_1, \ldots, x_{c+1}\rangle$ denote the polynomial $P_{u_0, \ldots, u_c}$ where $u_0, \ldots, u_c$ are the parameters of the curve $C\langle x_1, \ldots, x_{c+1}\rangle$. Note that using e.g. Lagrange interpolation, those parameters can be computed in polynomial time if the points $x_1, \ldots, x_{c+1}$ are given.

The interesting point is now that if we are sure that $P := P\langle x_1, \ldots, x_{c+1}\rangle$ correctly describes g along the curve $C := C\langle x_1, \ldots, x_{c+1}\rangle$, we can infer all values $g(x_1), \ldots, g(x_{c+1})$ by simply evaluating $P(0), \ldots, P(c)$. Thus, all that we need are the coefficients of P, which we find by querying a *single* segment of our additional table. But now we can use the usual argument: since $P(\cdot)$ and $g(C(\cdot))$ are both polynomials of degree at most cd, they are either equal or agree at most at cd/p places. Thus we will be able to ensure probabilistically that P does indeed describe g along C correctly by simply verifying $P(t) = g(C(t))$ for a single point $t \in \mathbb{F}_p$, chosen uniformly at random. The details of this argument will be given below.

Theorem 5.43. *There is a constant* $(\log n, \operatorname{poly}(\log n))$*-restricted solution verifier with coding scheme* C^{poly}.

Proof. We denote the solution verifier from Theorem 5.42 by V_0. Our goal is now to turn that verifier into a constant verifier. Since V_0 queries only constantly many segments of π_0, we have to take care only of the queries to π_1. We alter V_0 into a solution verifier which works within the same resource bounds but expects the part π_1 of its proof to be suitably encoded. By using the technique discussed above, the new verifier will query only constantly many symbols of this codeword plus constantly many segments of length $\operatorname{poly}(\log n)$ in an additional part of the proof.

Before describing the operation of the new verifier in detail, we note that although the length of the proof string π_1 is not determined a priori, for any given input x of length n, the total number of places of π_1 the verifier V_0 could access for *any* random string τ is at most $2^{\mathcal{O}(\log n)} \cdot \mathcal{O}(\operatorname{poly}(\log n)) = \mathcal{O}(\operatorname{poly}(n))$. Thus, we can modify V_0 so that, given an input x, it first constructs in polynomial time a list of all places i of π_1 that might be accessed (by simply enumerating all possible random strings), sorts them into a list $i_0, i_1, \ldots$ in ascending order, and proceeds then with its normal operation while replacing any lookup of π_1 at place i_j with a lookup of π_1 at place j. This verifier V_0' is obviously also a solution verifier, and satisfies the same resource constraints as V_0.

Therefore, we can assume without loss of generality that V_0 expects a proof π_1 of length $l(n)$ where $l(n)$ is in $\mathcal{O}(\operatorname{poly}(n))$ when given an input x of length n. Furthermore we denote by $q(n)$ the number of queries to π_1 made by V_0. (Note that by adding dummy queries we can always assure that V_0 makes the same number $q := q(n)$ of queries to π_1 for any input of length n and any random string.) Since $q(n)$ is in $\mathcal{O}(\operatorname{poly}(\log n))$, there are constants e and N such that for all $n \geqslant N$ we have $q(n) \leqslant \lceil \log n \rceil^e$. We assume from now on that the new verifier will be called only with inputs of size $n \geqslant N$, since for $n < N$ we can simply consider the old verifier V_0 as constant verifier. The new verifier that we are going to construct will then expect its proof to contain a code g for π_1 according to the code $C^{\operatorname{poly}}_{e+2,l(n)}$, and in addition a table T containing the polynomials $P_{u_0,\ldots,u_{q-1}}$ supposed to describe the values of g along all curves with parameters $u_0, \ldots, u_{q-1} \in \mathbb{F}_p^m$, where $p := p_{e+2,l(n)}$ and $m := m_{l(n)}$ are the defining parameters of the code $C^{\operatorname{poly}}_{e+2,l(n)}$ and $q := q(n)$. Note that by definition of the polynomial coding scheme with parameter $e+2$, the field size p will be at least $36\lceil\log l(n)\rceil^{e+2}$. Since the degree $d := d_{l(n)}$ of g satisfies $d \leq \lceil\log l(n)\rceil^2$ according to that definition, and furthermore $q(n) \leq \lceil\log n\rceil^e \leq \lceil\log l(n)\rceil^e$ holds by definition of e, this implies $p \geq 36qd$.

By Proposition 5.14, using an empty pattern function α_0 and a pattern function α_1 that maps every string of length n to the pattern $(l(n), 1)$, it suffices to construct a symbol constant $(\log n, \text{poly}(\log n))$-restricted extended verifier V such that for all x and g where g is a code for a proof string π_1 such that V_0 accepts $(x, (\pi_0, \pi_1))$, there is a table T such that V accepts $(x, (\pi_0, g, T))$, and that for all x and g where g is a code for a proof string π_1 such that V_0 rejects $(x, (\pi_0, \pi_1))$, V rejects $(x, (\pi_0, g, T))$ for any table T. The assumptions of Proposition 5.14 are satisfied since the polynomial coding scheme is $(\log n, \text{poly}(\log n))$-robust by Corollary 5.30, and furthermore the length $k_{l(n)}$ of a code for a string of length $l(n)$ satisfies $\log k_{l(n)} = \mathcal{O}(\log l(n)) = \mathcal{O}(\log n)$.

The verifier V proceeds now as follows: It simulates the operation of V_0 on the given input x and random string τ to compute the places $a_1, \ldots, a_q$ of π_1 that V_0 would query, and the corresponding values $x_1, \ldots, x_q \in \mathbb{F}_p^m$ where each x_i is the a_i-th element of $\mathbb{F}_p^m$ according to the ordering $<_n$ defined by the polynomial coding scheme. (Recall that the decoding function was defined so that to get the a_i-th bit of the string encoded by the codeword g, we have to check whether the value $g(x_i)$ equals zero or not, where g is interpreted as function $\mathbb{F}_p^m \to \mathbb{F}_p$.) The verifier further computes the parameters of the curve $C := C\langle x_1, \ldots, x_q\rangle$, and looks up the polynomial $P := P\langle x_1, \ldots, x_q\rangle$ from its additional table T. It then chooses $t \in \mathbb{F}_p$ uniformly at random and checks whether $P(t) = g(C(t))$, rejecting if this test fails. It finally sets $v_i = 0$ if $P(i) = 0$ and $v_i = 1$ if $P(i) \neq 0$ for $1 \leq i \leq q$ and continues simulating what V_0 would do, if its queries of π_1 at places $a_1, \ldots, a_q$ had resulted in $v_1, \ldots, v_q$, respectively.

This verifier V has indeed the required properties: Suppose first that g is indeed code for a proof string π_1 such that V_0 accepts $(x, (\pi_0, \pi_1))$. If we then construct a table T_g by correctly tabulating all curve polynomials of g, the verifier V will correctly retrieve all values v_i, and thus accept as well. Now suppose on the other hand g is a code for a proof string π_1 such that V_0 rejects $(x, (\pi_0, \pi_1))$. This means that V_0 rejects $(x, (\pi_0, \pi_1))$ on at least 3/4 of all random strings, and thus V rejects $(x, (\pi_0, g, T))$ on at least 3/4 of all random strings where the polynomial P describes the values of g along the line C correctly, and thus the values v_i are retrieved correctly. Furthermore, since $P(\cdot)$ and $g(C(\cdot))$ are univariate polynomials of degree at most $(q-1)d$, in all cases where $P(\cdot) \not\equiv g(C(\cdot))$ they can agree at most at $(q-1)d$ points $t \in \mathbb{F}_p$, and it follows that

$$\text{Prob}_t[P(t) = g(C(t))] \leq \frac{(q-1)d}{p} \leq \frac{1}{4},$$

where the second inequality holds since $p \geq 36qd$ as remarked above. But this means that on at least 3/4 of all random strings where P does *not* describe g along C correctly, the verifier V rejects as well. Since we have shown that V rejects in at least 3/4 of all cases where $P(\cdot) \equiv g(C(\cdot))$, and it also rejects in at least $3/4$ of all cases where $P(\cdot) \not\equiv g(C(\cdot))$, we can conclude that the total rejection probability of V is at least $3/4$. Finally, we note that in addition to the random bits used by V_0, we only need $\mathcal{O}(\log p) = \mathcal{O}(\log \log n)$ random bits to choose t, and that V queries its proof string — apart from the queries of π_0 done by V_0 — only at a constant number of symbols of g and a constant number of segments from T, each of size $\mathcal{O}(qd \log p) = \mathcal{O}(\text{poly}(\log n))$.   ∎

5.5 Composing Verifiers and the PCP-Theorem

5.5.1 Splittable Coding Schemes

The proof of the PCP-Theorem works by composing the extended verifiers we have seen so far. In connection with the composition of extended verifiers we encounter a technical problem: assuming that a proof is partitioned into non-overlapping blocks of length l, for some fixed k and for every selection of k such blocks, we want to test efficiently whether some designated string of length $n = kl$ is equal to the concatenation of the k blocks selected. As usual we obtain an efficient probabilistic test as desired by working with codes instead of the strings themselves: we fix some appropriate coding scheme with codes $C_0, C_1 \ldots$ and we assume that firstly, instead of the designated string we are given a code for it in C_n and secondly, in case a block holds a binary string x then this block is given as a code z in C_n such that z decodes to a string which has x as a prefix.

Definition 5.44. *For a string w and natural numbers $i \leqslant j \leqslant |w|$ we denote by $w_{[i:j]}$ the substring of w which corresponds to the places i through j.*

With a code C and a decoding function dec understood, we say the sequence $z_1, \ldots, z_k$ of codewords in C is a split representation *of a codeword z_0 in C iff for some l, the string $\text{dec}(z_0)$ is equal to $\text{dec}(z_1)_{[1:l]}\text{dec}(z_2)_{[1:l]} \cdots \text{dec}(z_k)_{[1:l]}$.*

The coding scheme $C = (C_1, \text{dec}_1), (C_2, \text{dec}_2), \ldots$ is (r,q)-splittable *into k parts iff there is an (r,q)-restricted extended verifier V where for every natural number l and for $n = kl$ there is a subset L_n of C_n such that*

1. *for each string x of length l there is a codeword w in L_n where x is equal to $\text{dec}_n(w)_{[1:l]}$,*

2. *for all $z_1, \ldots, z_k$ in L_n there is some z_0 in C_n such that firstly, V accepts the input $(0^n, z_0 \ldots z_k)$ and secondly, $z_1, \ldots, z_k$ is a split representation of the codeword z_0,*

3. *V rejects all inputs $(0^n, z_0 \ldots z_k)$ where the strings $z_0, \ldots, z_k$ are all in C_n, but $z_1, \ldots, z_k$ is not a split representation of z_0,*

4. for each input $(0^n, \pi_0)$ *and random string* τ *the extended verifier* V *reads at most constantly many symbols from* π_0 *where* π_0 *is considered as a string over the alphabet* Σ_n *of the code* C_n.

A coding scheme C *is* (r, q)-*splittable iff for every natural number* $k \geqslant 1$, *the coding scheme is* (r, q)-*splittable into* k *parts.*

With the notation introduced in Definition 5.44 at hand we now reconsider the problem to check whether the concatenation of k binary strings $x_1, \ldots, x_k$ is equal to some designated string x_0. Given an (r, q)-splittable coding scheme C with codes $C_1, C_2, \ldots$ and an extended verifier V and subsets L_n of C_n as in Definition 5.44, we are able to construct a corresponding probabilistic test.

We assume that the blocks have length l and we let $n = kl$. The test will receive as input appropriate encodings $z_0, \ldots, z_k$ of the strings $x_0, \ldots, x_k$. In case a block holds a binary string x, we assume that the block is given as a code z in L_n according to Condition 1 in Definition 5.44, that is, as a code z which decodes to a string of the form xy. Further we assume that in case the string x_0 is in fact equal to the concatenation of x_1 through x_k, then x_0 is given as a code z_0 for x_0 in C_n according to Condition 2, that is, as a code z_0 which in particular possesses the split representation $z_1, \ldots, z_k$. We now obtain a probabilistic test as desired by running the extended verifier V on input $(0^n, z_0 \ldots z_k)$. In case the designated string is in fact equal to the concatenation of x_1 through x_k then the input is accepted by our choice of the z_j. On the other hand, according to Condition 3 the extended verifier V will not accept the input in case the z_i are all in C_n but z_1 through z_k are not a split representation of z_0.

By the preceding discussion an (r, q)-restricted extended verifier can use a subverifier in order to check probabilistically whether, intuitively speaking, k codes from C_n which figure at arbitrary places in the proof contain the same information as some code from C_n which occurs somewhere else as a segment of the proof. Remark 5.45 shows that for an (r, q)-robust coding scheme C this subverifier can in fact be chosen to work as intended for all possible inputs and not just for inputs which obey the pattern function which is implicit in our notation $(0^n, z_0 \ldots z_k)$.

Remark 5.45. The concept of a coding scheme which is (r, q)-splittable is designed to be used with coding schemes which are in addition (r, q)-robust. Given such a coding scheme and an extended verifier V as in Definition 5.44, then according to Proposition 5.14 for some positive rational δ there is an (r, q)-restricted extended verifier V_0 which basically works as V, but which in addition can be shown to reject every input $(0^n, y_0 \ldots y_k)$ for which either some of the strings y_i is not δ-close to C_n or the strings y_i are all δ-close to codewords z_i in C_n but the strings $z_1, \ldots, z_k$ are not a split representation of z_0.

Proposition 5.46. *The linear function coding scheme is* $(n, 1)$-*splittable.*

Proof. We construct an extended verifier V as required in the definition of $(n, 1)$-splittable. For every natural number k and for $n = kl$ we let L_n^{lin} be equal to C_n^{lin}. Condition 1 then is immediate because given a string x of length l, we can choose for w some code for $x0^{(k-1)l}$ in C_n.

Given an input $(0^n, z_0z_1 \ldots z_k)$ we can assume that the z_i are indeed in C_n^{lin} because only in this situation V is required to work correctly. Thus each z_i is a code for some binary string $x_i = x_{i,1} \ldots x_{i,n}$ and if we let $y = y_1 \ldots y_n$ be the the j-th element of $\mathbb{F}_2^n$ in lexicographical order, then the j-th bit of z_i is equal to

$$l_{x_i}(y) = \sum_{t=1}^{n} x_{i,t} y_t.$$

Then for

$$x := x_{1,1} \ldots x_{1,l} x_{2,1} \ldots x_{2,l} \quad \ldots \quad x_{k,1} \ldots x_{k,l}$$

the linear function l_x maps $y = y_1 \ldots y_n$ to

$$l_x(y) = \sum_{t=1}^{n} x_t y_t = \sum_{i=1}^{k} \sum_{t=1}^{l} x_{i,t} y_{(i-1)l+t} = \sum_{i=1}^{k} l_{x_i}(y_{[(i-1)l+1 \, : \, il]} 0^{(k-1)l}). \quad (5.22)$$

By our construction the functions l_x and l_{x_0} are equal iff the codewords $z_1, \ldots, z_k$ are a split representation of z_0. In order to check the latter property V chooses an argument y in $\mathbb{F}_2^n$ uniformly at random and checks whether both functions agree at place y. Observe that according to equation (5.22) in order to compute $l_x(y)$ it suffices to read one value each of l_{x_1} through l_{x_k}. If the two functions are not equal then by Lemma 5.16 they differ on half of their domain, and hence this check fails with probability $1/2$ in case the $z_1, \ldots, z_k$ are not a split representation of z_0, but succeeds always, otherwise. Then by repeating this check twice, while rejecting if one of the iterations fails, we can amplify the probability for rejection from $1/2$ to $3/4$.

By the preceding discussion V satisfies Conditions 3 and 4. Concerning Condition 2, it suffices to observe that for all strings $z_1, \ldots, z_k$ in C_n the function l_x corresponds to a codeword z_0 in C_n where V accepts the input $(0^n, z_0z_1 \ldots z_k)$. ∎

Recall in connection with Proposition 5.47 that the polynomial coding scheme was defined w.r.t. an admissible sequence $<_0, <_1, \ldots$ of orderings where admissible means that $<_n$ is a strict ordering on H_n^{mn} such that the relevant operations can be done in polynomial time.

Proposition 5.47. *For every natural number k, we can choose an admissible sequence $<_0, <_1, \ldots$ such that thereby the polynomial coding scheme becomes $(\log n, \text{poly}(\log n))$-splittable into k parts.*

Proof. Recall that each of the decoding functions $\text{dec}_n^{\text{poly}}$ depends on an ordering $<_n$ on H_n^{mn} which we have not yet specified in detail. Now we will specify an admissible sequence of orderings such that the polynomial coding scheme becomes $(\log n, \text{poly}(\log n))$-splittable. Here it suffices to specify the ordering $<_n$ in detail for the cases where n is a multiple of k, while for all other n, as before we might use arbitrary orderings such that the relevant operations can be done in polynomial time.

We fix a natural number l and for $n = lk$ we let $p = p_n$, $m = m_n$, $h = h_n$, $H = H_n$, $C_n = C_n^{\text{poly}}$, and $\text{dec}_n = \text{dec}_n^{\text{poly}}$. For all j in $\mathbb{F}_p$ we denote by S_j the slice of $\mathbb{F}_p^m$ defined by the equation $x_m = j$. In order to define the strict ordering $<_n$ on H^m we specify a partition $T_0, \ldots, T_{2p-1}$ of H^m: for each $j \leqslant p - 1$, the set T_j contains the first l elements of $S_j \cap H^m$ w.r.t. the lexicographical ordering $<_{\text{lex}}$ on $\mathbb{F}_p^m$, and the set T_{p+j} is the relative complement $(S_j \cap H^m) \setminus T_j$ of T_j in $S_j \cap H^m$. Then we specify $<_n$ by requiring that two elements in some set T_i are ordered according to the ordering $<_{\text{lex}}$, and that for $i < j$ all elements of T_i are below all elements of T_j.

By our construction the first kl elements of H^m w.r.t. $<_n$ are the first l elements of $S_0 \cap H^m$, followed by the first l elements of $S_1 \cap H^m$, and so on. Recalling that the decoding function dec_n maps a codeword w in C_n to the binary string $x = \sigma(w, <_n, n)$ we obtain that by our choice of $<_n$ we have $x = x_1 \ldots x_k$ where, intuitively speaking, x_j is decoded from the first l places of $S_j \cap H^m$.

For a function f on $\mathbb{F}_p^m$ we denote by $f|_{x_m=j}$ the function on $\mathbb{F}_p^{m-1}$ which is obtained by firstly restricting f to the slice S_j and then identifying in the natural way the elements of S_j and of $\mathbb{F}_p^{m-1}$. By definition the code C_n^{poly} contains all m-variate polynomials over $\mathbb{F}_p$ of degree at most mh. Now we let L_n^{poly} be the set of all polynomials q in C_n^{poly} such that in addition the polynomial $q|_{x_m=0}$ has degree at most $mh - k + 1$. Recall that for every function g from H^m to $\mathbb{F}_p$ there is a polynomial

$$\sum_{\{h_1,\ldots,h_m\} \in H^m} \prod_{i=1}^{m} \prod_{y \in H-\{h_i\}} \frac{x_i - y}{h_i - y} \cdot g(h_1, \ldots, h_m) \tag{5.23}$$

which by construction agrees with g on H^m and which has degree at most $m(h-1)$ and hence is in C_n^{poly}. By inspection we obtain that the restriction of the polynomial in (5.23) to the slice defined by $x_m = 0$ has degree at most

$$(m-1)(h-1) = mh - (m + h - 1).$$

As a consequence for all values of n which are so large that $m + h$ is greater or equal to k, the polynomial is in L_n^{poly} and Condition 1 in Definition 5.44 is satisfied. Here again we handle the finitely many remaining values of n by a brute force approach.

It remains to construct an extended verifier V as required in the definition of (r, q)-splittable coding scheme. On input $(0^n, z_0 z_1 \ldots z_k)$, the verifier V interprets z_0 through z_k as functions from $\mathbb{F}_p^m$ to $\mathbb{F}_p$, which we denote by $g_0, \ldots, g_k$, respectively. We can assume that the z_i are indeed in C_n^{poly}, because only in this situation V is required to work correctly. Thus the g_j are polynomials of degree at most mh, and then so are for $1 \leqslant j \leqslant k$, the functions $g_0|_{x_m=j-1}$ and $g_j|_{x_m=0}$. The extended verifier V now tries to verify that for $j = 1, \ldots, k$, the functions $g_0|_{x_m=j-1}$ and $g_j|_{x_m=0}$ are the same. In order to do so for each such pair of functions, V chooses an argument x in their common domain $\mathbb{F}_p^{m-1}$ uniformly at random and checks whether both functions agree at x. By the theorem of Schwartz stated as Theorem 5.22 two distinct $(m-1)$-variate polynomials of degree at most mh over $\mathbb{F}_p$ agree on at most a mh/p fraction of their domain. According to equation (5.11) mh/p is less than $1/36$ and hence in case the functions in one of the pairs are not the same, V will accept with probability at most $1/36$.

The extended verifier V obviously satisfies Condition 4 in Definition 5.44. Concerning Condition 3, observe that in case the codewords $z_1, \ldots, z_k$ are not a split representation of some codeword z_0, then by definition $\text{dec}_n(z_0)$ differs from the string $\text{dec}_n(z_1)_{[1:l]}, \ldots, \text{dec}_n(z_k)_{[1:l]}$. But by choice of the ordering $<_n$ this just means that for some j the functions $g_0|_{x_m=j-1}$ and $g_j|_{x_m=0}$ disagree, and consequently V rejects the input $(0^n, z_0 z_1 \ldots z_k)$.

Concerning Condition 2, we assume that codewords $z_1, \ldots, z_k$ in L_n^{poly} are given which correspond to polynomials $q_1, \ldots, q_k$, respectively. Then by definition of L_n^{poly} the functions $q_j|_{x_m=0}$ are polynomials of degree at most $mh - k + 1$ and consequently the function q_0 defined by

$$q_0(x_1, \ldots, x_m) = \sum_{j=0}^{k-1} \prod_{i \in \{0, \ldots, k-1\} - \{j\}} \frac{x_m - i}{j - i} \cdot [q_{j+1}|_{x_m=0}(x_1, \ldots, x_{m-1})]$$

is a m-variate polynomial over $\mathbb{F}_p$ of degree at most mh, that is, q_0 corresponds to a codeword z_0 in C_n^{poly}. This finishes our proof, because by construction, the codewords $z_1 \ldots, z_k$ are a split representation of z_0 and hence V accepts the input $(0^n, z_0 \ldots z_k)$. More precisely, in order to show that the $z_1, \ldots, z_k$ are a split representation of z_0 observe that for the lexicographical ordering $<_{\text{lex}}$ we have for all $a_1, \ldots, a_{m-1}, b_1, \ldots, b_{m-1}, i, j$ in $\mathbb{F}_p$

$$(a_1, \ldots, a_{m-1}, i) <_{\text{lex}} (b_1, \ldots, b_{m-1}, i) \quad \text{iff} \quad (a_1, \ldots, a_{m-1}, j) <_{\text{lex}} (b_1, \ldots, b_{m-1}, j).$$

As a consequence and according to the definition of the ordering $<_n$ in terms of the lexicographical ordering the restrictions of $<_n$ to the slices $S_1, \ldots, S_k$ order these slices in essentially the same way. Thus for $1 \leqslant i \leqslant k$, we will use essentially identical orderings while decoding the length l strings $\text{dec}(z_0)_{[(i-1)l+1:il]}$ and $\text{dec}(z_i)_{[1:l]}$ from the restriction of q to slice S_{i-1} and from the restriction of q_i to slice S_0, respectively. ∎

5.5.2 Decision Formulae and Composition of Extended Verifiers

In connection with Definition 5.48 recall our convention according to which an assignment for a 3-CNF formula of length n has length n, too.

Definition 5.48. *Let φ be a 3-CNF formula of length n.*

A partial assignment *for φ is a binary string of length less or equal to n. Given a partial assignment a for φ of length k, a* complementary assignment *for φ and a is a binary string of length $n - k$.*

The formula φ is satisfiable with partial assignment a *iff there is some complementary assignment b for φ and a such that ab is a satisfying assignment for φ.*

In the sequel we will use the following variant of Cook's theorem, for a proof see Section 5.8.

Theorem 5.71 (Cook-Levin). *Let L be a language which is recognized by a (nondeterministic) Turing machine in time $\mathcal{O}(n^c)$. Then for each input size n, we can construct in time $\mathcal{O}(n^{2c})$ a 3-CNF formula ψ of size $\mathcal{O}(n^{2c})$ such that for all binary strings x of length n, the string x is in L iff the formula ψ is satisfiable with partial assignment x.*

We will show in Lemma 5.51 that a pair of a verifier and a solution verifier which satisfy certain technical conditions can be composed to yield a more efficient verifier. Here the key idea is to transform, for a given verifier V_1, the evaluation of the bits read from the proof into a 3-CNF formula φ, then using the solution verifier V_2 for checking the satisfiability of the formula φ. If we assume that for some input x and some random string τ the answers to V_1 queries are given as a binary string a, then the evaluation of a amounts to a computation with input (x, τ, a). Here we can assume due to V_1 being polynomially time bounded that the running time of this computation, as well as the length of τ and a are polynomially bounded in n. Thus by Cook's theorem there is a formula φ of size polynomial in n, such that φ is satisfiable with partial assignment $x\tau a$ iff the evaluation of a results in acceptance. However even if the verifier V_1 is (r, q)-restricted for some sublinear function q, in general the formula obtained according to Cook's theorem will have size polynomial in n. In this situation in general we obtain formulae which are larger than the input formulae of V_1 and consequently the composed verifier might be less efficient than the verifier V_2 we have started with. So we are led to the consideration of verifiers where the behavior for given x and τ can be described by formulae of small size.

Definition 5.49. *Let V be an (r, q)-restricted verifier, and let x and τ be binary strings of length n and $r(n)$, respectively. Let $i_1 < \ldots < i_l$ be the positions of the proof which are queried by V while working on x and τ, and for a proof π of length at least i_l, let $a(x, \tau, \pi) = \pi_{i_1} \ldots \pi_{i_l}$ contain the corresponding bits of π.*

A decision formula *for V, x, and τ is a 3-CNF formula φ such that for every proof π of suitable length, V accepts (x, π) on random string τ iff the formula φ is satisfiable with partial assignment $a(x, \tau, \pi)$.*

An (r,q)-restricted verifier V has feasible decision *iff given strings x and τ of length n and $r(n)$, respectively, we can compute in time polynomial in n a decision formula for V, x and τ of size at most polynomial in $q(n)$.*

Observe in connection with Definition 5.49 that the decision formula depends on the verifier V, the argument x, and the random string τ, but not on the proof π. Observe further that during the evaluation of a decision formula φ we will use the bits read from the proof plus some complementary assignment which specifies values for some auxiliary variables. However we need not refer to x or τ, that is, intuitively speaking, the values of x and τ are hardcoded into the decision formula.

Proposition 5.50. *The constant $(n^3, 1)$-restricted solution verifier from Theorem 5.39 has feasible decision.*

The constant $(\log n, \mathrm{poly}(\log n))$-restricted solution verifier from Theorem 5.43 has feasible decision.

Proof. Concerning the first assertion about the $(n^3, 1)$-restricted extended verifier observe that an $(r(n), 1)$-restricted verifier always has feasible decision. In order to obtain a decision formula as desired we simply simulate the verifier for all possible assignments of answers to the constantly many queries to the proof in order to compute the Boolean function which maps such an assignment a to 0 in case V rejects, and to 1, otherwise. Given the truth table of this Boolean function we construct a formula φ_0 in conjunctive normal form such that the Boolean function under consideration maps a to 1 iff a is a satisfying assignment for φ_0. Finally we employ standard techniques in order to transform φ_0 into a 3-CNF formula φ such that an assignment a satisfies φ_0 iff φ is satisfiable with partial assignment a.

Concerning the second assertion let V be the $(\log n, \mathrm{poly}(\log n))$-restricted verifier from Theorem 5.43. For concreteness we assume that the number of bits V reads from its proof is bounded by a function q where $q(n)$ is in $\mathrm{poly}(\log n)$. Inspection of the construction of V reveals that for an input x of length n, a corresponding random string τ, and a proof π, the evaluation of the answer string $a(x, \tau, \pi)$ of length at most $q(n)$ can be viewed as a computation with running time in $\mathrm{poly}(q(n))$. Here this computation requires as input, besides $a(x, \tau, \pi)$, parameters such as $\log n$ or p_n of size in $\mathrm{poly}(\log n)$, but not the whole string x. If we denote the concatenation of these additional parameters as y, the input to the computation can be written as $a(x, \tau, \pi)y$ and has length in $\mathrm{poly}(\log n)$. By Cook's theorem we obtain a 3-CNF formula φ_0 of size in $\mathrm{poly}(\log n)$ such that φ_0 is satisfiable with partial assignment $a(x, \tau, \pi)y$ iff V accepts (x, π) on random string τ. In order to obtain a decision formula as required it suffices to hardcode y into φ_0 without destroying 3-CNF. In order to do so we let x_j be the variable which corresponds to the first bit of y and we replace each variable for which we want to hardcode an assignment 0, respectively 1, by $\overline{x_j}$, respectively by x_j. Finally we add a clause $(x_j \vee x_j \vee x_j)$ in order to ensure that in a satisfying assignment the variable x_j is assigned the value 1.   ∎

Lemma 5.51 (Composition Lemma). *Let V_1 be an (r_1, q_1)-restricted constant verifier with feasible decision for some language L where for all f in $\mathrm{poly}(q_1)$ and for all but finitely many n holds $f(n) \leqslant n$.*

For every natural number $k \geqslant 1$ let there be an (r_2, q_2)-restricted solution verifier V_2 with a coding scheme which is (r_2, q_2)-robust and is (r_2, q_2)-splittable into k parts.

Then there is an

$$(r_1(n) + r_2(q_1'(n)), q_2(q_1'(n)))\text{-restricted verifier } V_3 \qquad (5.24)$$

for L where q_1' is a function in $\mathrm{poly}(q_1)$.

In case V_2 is constant, has feasible decision, or satisfies both of these properties, respectively, then we can arrange that the same holds for V_3.

Observe that the bounds in equation (5.24) reflect that the composite verifier V_3 first uses r_1 random bits in order to simulate V_1 on an input of size n and to compute a decision formula φ of size $q_1'(n)$, and then consumes $\mathcal{O}(r_2(q_1'(n)))$ random bits and reads $\mathcal{O}(q_2(q_1'(n)))$ bits of the proof while simulating V_2 on φ.

Proof.   In the sequel we construct a verifier V_3 as required in the lemma and we leave it to the reader to check that V_3 satisfies the resource bounds required.

On input x of length n, the verifier V_3 basically proceeds as follows.

Step 1. V_3 simulates V_1 on input x and a random string τ of suitable length in order to compute the positions which V_1 reads from its proof.

Step 2. V_3 computes a decision formula φ^τ for V_1, x, and τ and assumes that the proof of V_3 contains an appropriate encoding π_0^τ which essentially contains an assignment for φ^τ. (Recall that an assignment for the decision formula φ^τ consist of a partial and a complementary assignment where the partial assignment corresponds to the positions computed during Step 1.)

Step 3. In order to ensure that the partial assignments given by the encodings π_0^τ are consistent for the various values of τ, the verifier V_3 assumes that some part of its proof contains a proof π^{V_1} for V_1. Here for technical reasons the proof π^{V_1} is not given directly, but is first partitioned into blocks of appropriate size. Then we choose a solution verifier V_2 with coding scheme $\mathcal{C}$ as in the assumption of the composition lemma and encode each block according to $\mathcal{C}$. By means of this encoding and by robustness and the splitting property of $\mathcal{C}$, the verifier V_3 can check within its resource bounds whether the partial assignment given by π_0^τ agrees with the bits read from π^{V_1}.

Step 4. Related to the tests in Step 3, in case π_0^τ encodes a string y then y is not an assignment for φ^τ, however the values for the variables of φ^τ are given by a proper subset of the bits of y. Accordingly V_3 transforms φ^τ into a basically equivalent formula ψ^τ such that (ψ^τ, π_0^τ) is an appropriate input for the solution verifier V_2.

Step 5. V_3 simulates V_2 on input (ψ^τ, π_0^τ).

Here V_3 rejects immediately in case one of the tests in Step 3 fails, and in case Step 5 is reached, then V_3 accepts iff the simulation of V_2 results in acceptance.

With the construction of V_3, Steps 1, 2, and 5 are straightforward, so it suffices to consider Steps 3 and 4. First we consider the encoding of a given proof π^{V_1} for V_1 related to Step 3. We partition π^{V_1} into blocks of equal length $b(n)$. Here, for the moment, b is an arbitrary function in $\mathrm{poly}(q_1(n))$ which is computable in polynomial time and which is larger than both the size of the decision formula φ^τ and the number of queries asked by V_1. While simulating V_1 we will query positions in at most $\mathcal{O}(1)$ such blocks because, firstly, V_1 is constant and hence queries only $\mathcal{O}(1)$ many segments of its proof and, secondly, by construction each such segment is contained in two successive blocks. Thus there is a constant k which is independent of x such that V_1 queries positions in exactly $k-1$ blocks. Next we choose some (r_2, q_2)-restricted verifier V_2 with a coding scheme $\mathcal{C}$ which is (r_2, q_2)-robust and (r_2, q_2)-splittable into k parts. We let $l = b(n)$ and let L_{kl} be a subset of the kl-th code C_{kl} of $\mathcal{C}$ as in the definition of splittable coding scheme. Then we replace each block of π^{V_1} by a code z in L_{kl} for some string y of length kl such that the block under consideration is equal to the length l prefix of y.

Further we assume that for every τ, the proof of V_3 contains at a designated position some code $z_k = z_k^\tau$ in L_{kl} for some block of length $b(n)$. Here we view z_k as code for a string y which contains as a prefix a complementary assignment for φ^τ. Consequently there is an assignment for the formula φ^τ which is given in the form of codewords $z_1, \ldots, z_{k-1}, z_k$ in L_{kl}. But if we want to use the solution verifier V_2 in order to check whether this assignment satisfies φ^τ, it is necessary that the assignment is given as a single code. So we assume that for every τ the proof of V_3 contains a code π_0^τ in C_{kl}. By choice of L and by assumption on the coding scheme $\mathcal{C}$, the extended verifier V_3 can now use an

$$(r_2(kl), q_2(kl))\text{-restricted extended verifier } \tilde{V} \tag{5.25}$$

in order to test probabilistically for some suitable $\delta < 1/4$ whether the alleged codes $\pi_0^\tau, z_1, \ldots, z_k$ are δ-close to C_{kl}, and if so, whether the nearest codewords of the z_j are a split representation of the nearest codeword of π_0^τ. Observe that by definition of splittable coding scheme for all choices of the k codes in L there is always at least one codeword z_0 in C such that the corresponding test accepts with probability 1 in case we let π_0^τ be equal to z_0. Observe further that by assumption on q_1 for almost all n we have $kl \leqslant n$ and that by convention the resource bounds r_2 and q_2 are non-decreasing. Thus for almost all n, the values $r_2(kl)$ and $q_2(kl)$ are less or equal to $r_2(n)$ and $q_2(n)$, respectively. As usual we handle the finitely many remaining values of n by a brute force approach.

Now Step 5 basically amounts to simulating the solution verifier V_2 on the formula φ^τ and the assignment as given by π_0^τ. However π_0^τ is not a code for an assignment of π_0^τ, but a code for a string of length kl which among its bits *contains* such an assignment. Thus we have to alter the formula φ^τ into a formula ψ^τ such that the new formula together with π_0^τ provides a reasonable input for V_2. In connection with this modification, recall the representation for 3-CNF formulae we have agreed on in Section 5.4.1. Now, the modification of φ^τ has to take care of the two following problems. First, in case V_1 queries positions $j_1 < \ldots < j_{t'}$ of its proof, then specifying a single variable of φ^τ requires $\lceil \log t \rceil$ bits. Here $t > t'$ is the number of variables in the formula φ^τ and thus also the length of its binary representation. With the new formula ψ^τ we will have to specify the corresponding t positions within an assignment of length kl and hence it requires $\lceil \log kl \rceil$ bits to specify a single variable of ψ^τ. Second, according to the convention in Remark 5.32, in case we want to present to V_2 an assignment of length kl, then also the corresponding input formula must have length kl. In order to handle these two problems we proceed as follows. Firstly, we replace in φ^τ the variable identifiers by length $\lceil \log kl \rceil$ variable identifiers w.r.t. the assignment encoded by π_0^τ. Secondly, we patch the formula thus obtained by dummy clauses in order to obtain a formula of length kl. In order to make this construction go through we ensure that

$$3(\lceil \log kl \rceil + 1) \text{ divides } kl, \tag{5.26}$$

$$\frac{t}{3(\lceil \log t \rceil + 1)} < \frac{kl}{3(\lceil \log kl \rceil + 1)}. \tag{5.27}$$

Then by equation (5.27) the replacement of the formula identifiers in the length n formula φ^τ results in an intermediate formula of length less than kl. Further we can patch this intermediate formula with dummy clauses in order to obtain the length kl formula ψ^τ according to equation (5.26) because by our representation of 3-CNF formulae the intermediate, as well as the patched formula consists of a prefix and several clauses which each have length $3(\lceil \log kl \rceil + 1)$. Observe that it is always possible to satisfy equations (5.26) and (5.27) by choosing an appropriate value for $l = b(n)$ in $\mathcal{O}(\text{poly}(q_1(n)))$, and that this choice is compatible with the previous conditions on b.

Concerning the simulation of V_2 on the pair (ψ^τ, π_0^τ) observe that ψ^τ has length kl in $\text{poly}(q_1(n))$ and thus the simulation queries $\mathcal{O}(q_2(\text{poly}(q_1(n))))$ places of the proof and uses a random string τ_2 of length in $\mathcal{O}(r_2(\text{poly}(q_1(n))))$. During the simulation of V_2 we use for each random string τ a different part π_1^τ of the proof of V_3 as proof π_1 of V_2. Thus in particular in case φ^τ is satisfiable we can choose π_1^τ such that the simulation results in acceptance with probability 1, and in case φ^τ is not satisfiable then for all choices of π_1^τ the simulation results in acceptance with probability at most $1/4$. Here the probabilities are w.r.t. τ_2.

In order to verify that V_3 recognizes L it suffices to show that for every fixed $\beta > 1$ and for all inputs x we can achieve that Claims 1 and 2 are satisfied.

Claim 1. If V_1 accepts $(x, \tilde{\pi})$ for some proof $\tilde{\pi}$, then there is a proof π such that V_3 accepts (x, π).

Claim 2. Given a proof π for V_3 we can construct a proof $\tilde{\pi}$ for V_1 such that for every rational $\varepsilon > 0$, if V_3 rejects (x, π) with probability at most ε then V_1 rejects $(x, \tilde{\pi})$ with probability at most $\beta\varepsilon$.

We show that Claims 1 and 2 together imply that V_3 recognizes L. For every x in L, the verifier V_1 accepts x for some proof, and then so does V_3 according to Claim 1. Conversely, let x be not in L and assume for a contradiction that V_3 does not reject x, that is, assume that there is a proof π where V_3 rejects (x, π) with probability at most $3/4$. By Remark 5.3 and because V_1 recognizes L we can assume that for every proof $\tilde{\pi}$, the verifier V_1 rejects $(x, \tilde{\pi})$ with probability at least $4/5$. We obtain a contradiction by choosing $\beta > 1$ so small that $(3/4)\beta$ is less than $4/5$ because according to Claim 2 then there is a proof $\tilde{\pi}$ where V_1 rejects $(x, \tilde{\pi})$ with probability at most $(3/4)\beta < 4/5$.

Claim 1 is more or less immediate from the construction of V_3. Concerning Claim 2, we fix some $\beta > 1$ and we let π be a proof for V_3. We assume that π contains an encoding of a proof π^{V_1} for V_1 which is given as the concatenation $z_1 \ldots z_t$ of alleged codes z_j in L_{kl} as above. In order to obtain a proof $\tilde{\pi}$ as required essentially we want to let $\tilde{\pi} = d(y_1) \ldots d(y_t)$ where $d(y_j)$ is some arbitrary string of length l in case y_j is not $1/4$-close to C_{kl}, and otherwise, $d(y_j)$ is the length l prefix of the binary string encoded by the nearest codeword for z_j. However, related to the proof of Theorem 5.54, we use a slightly more liberal definition of the function d. Recall that in (5.25) we have already fixed a verifier $\tilde{V}$ which for some suitable $\delta < 1/4$ rejects an input $(0^{kl}, y)$ in case y is not δ-close to C_{kl}. We now assume that we are given a function d where $d(y)$ is the length l prefix of the binary string encoded by the nearest codeword for y in case $(0^{kl}, y)$ is not rejected by $\tilde{V}$, and $d(y)$ is some arbitrary string of length l, otherwise. Then, similarly as above, we let $\tilde{\pi} = d(y_1) \ldots d(y_t)$. Observe that there is always some function d as required, because in case $\tilde{V}$ does not reject $(0^{kl}, y)$, then y is δ-close to C_{kl}. (In the proof of Theorem 5.54, we will show that for the polynomial coding scheme the function d can in fact be chosen to be computable in time polynomial in the length of the input y.)

In order to show Claim 2, by contraposition it is sufficient to show for every $\varepsilon > 0$, that in case V_1 rejects $\tilde{\pi}$ with probability greater than $\beta\varepsilon$, then V_3 rejects π with probability greater than ε. Now consider a random string τ such that V_1 rejects $(x, \tilde{\pi})$ on τ. We can assume that in case, firstly, π_0^τ or one of the strings z_j which correspond to the y_j which are accessed by V_1 on x and τ is rejected by $\tilde{V}$ or, secondly, the nearest codewords of these z_j are not a split representation of π_0^τ, then one of the tests performed by V_3 during Step 3 fails with probability greater than $1/\beta$. But otherwise, firstly, the subverifier $\tilde{V}$ will not reject the z_j, that is, by definition of d, the first l bits of $d(z_j)$ in fact relate to z_j and, secondly, the assignment given by π_0^τ, intuitively speaking, agrees with the strings $d(z_j)$. Consequently and by assumption on τ, in this case we can assume that the simulation of V_2 in Step 3 results in rejection with probability greater than $1/\beta$. Thus for every random string τ of V_1 such that V_1 rejects $(x, \tilde{\pi})$ on τ, the corresponding simulation by V_3 on (x, π) will result in rejection with conditional probability of at least $1/\beta$. By summing up these conditional probabilities over all such τ, we obtain that V_3 rejects (x, π) with probability greater than $(\beta\varepsilon)1/\beta = \varepsilon$.   ∎

5.5.3 The PCP-Theorem

We will obtain the PCP-Theorem as an easy consequence of Theorem 5.52.

Theorem 5.52. *There is a* $(\log n, 1)$*-restricted verifier for* 3SAT.

Proof. Consider the $(n^3, 1)$-restricted solution verifier V_{lin} from Theorem 5.39 and the $(\log n, \mathrm{poly}(\log n))$-restricted solution verifier V_{poly} from Theorem 5.43. Both solution verifiers are constant and by Remark 5.4 they can be viewed as verifiers which accept the language 3SAT. As a consequence both of them can play the role of the Verifier V_1 in the composition lemma with L equal to 3SAT. On the other hand they can also play the role of V_2 due to the properties of their coding schemes and because both have feasible decision according to Proposition 5.50. In this connection, recall that the linear function coding scheme is $(n, 1)$-splittable and that for every natural number k we can arrange that the polynomial coding scheme becomes $(\log n, \mathrm{poly}(\log n))$-splittable into k parts.

Applying Lemma 5.51 with V_1 and V_2 both equal to V_{poly} yields a constant

$$(\log n + \log \log n, \mathrm{poly}(\log \log n))\text{-restricted verifier } V'$$

which accepts the language 3SAT. Applying Lemma 5.51 again with V_1 equal to V' and V_2 equal to V_{lin} then yields a

$$(\log n + \mathrm{poly}(\log \log n), 1)\text{-restricted verifier } V$$

which accepts the language 3SAT. The verifier V is in fact $(\log n, 1)$-restricted because $\mathrm{poly}(\log \log n)$ is contained in $\mathcal{O}(\log n)$. Observe that in the above reasoning we have used several times that constant factors and exponents become absorbed due to equations such as

$$\mathcal{O}(\log(\mathrm{poly}(\log n))) \;=\; \mathcal{O}(\log \log n).$$

                                                                       ∎

Theorem 5.53 (PCP-Theorem). *The class* $\mathcal{NP}$ *is equal to* $\mathrm{PCP}(\log n, 1)$.

Proof. Recall that 3SAT is $\leqslant_m^P$-complete for $\mathcal{NP}$. By Theorem 5.52, the language 3SAT is in $\mathrm{PCP}(\log n, 1)$. Now the latter class is closed downwards under $\leqslant_m^P$-reductions and hence contains not only 3SAT, but also its lower $\leqslant_m^P$-cone $\mathcal{NP}$. In order to show the reverse containment let V be a $(\log n, 1)$-restricted verifier and observe that by the argument given in the proof of Theorem 5.43 we can assume without loss of generality that V expects its proof to be of polynomial length. As a consequence there is an $\mathcal{NP}$-machine which simulates V. The $\mathcal{NP}$-machine first guesses a proof of polynomial length, then simulates V for all random strings, and accepts iff V accepts for all random strings. ∎

Theorem 5.54 contains a version of Theorem 5.52 which is used in Chapter 4. A slightly stronger form of Theorem 5.54 has been stated in [KMSV94].

Theorem 5.54. *There is a $(\log n, 1)$-restricted verifier V for 3SAT and a rational $\varepsilon \geqslant 0$ such that for all x and π where V accepts (x, π) with probability at least $1 - \varepsilon$, we can in time polynomial in the length of x compute a proof π' where V accepts (x, π') with probability 1.*

Proof. We show that the $(\log n, 1)$-restricted verifier V for 3SAT constructed in the proof of Theorem 5.52 can be assumed to satisfy the additional conditions required in Theorem 5.54. Recall that the verifier V is obtained by successive applications of Lemma 5.51, the composition lemma. More precisely we construct verifiers $\tilde{V}_1, \tilde{V}_2$, and $\tilde{V}_3$ where $\tilde{V}_1$ is the $(\log n, \mathrm{poly}(\log n))$-restricted verifier from Theorem 5.43, $\tilde{V}_2$ is the $(\log n, \mathrm{poly}(\log \log n))$-restricted verifier V' from Theorem 5.52, and $\tilde{V}_3$ is equal to V. Observe that $\tilde{V}_{i+1}$ is obtained by applying the composition lemma to $\tilde{V}_i$ and to either again the $(\log n, \mathrm{poly}(\log n))$-restricted verifier from Theorem 5.43 or the $(n^3, 1)$-restricted verifier from Theorem 5.39.

For a given rational $\beta > 1$, we consider an application of the composition lemma to verifiers V_1 and V_2 in order to construct a composite verifier V_3. According to Claim 2 from the proof of the composition lemma we can arrange that given an input x and a proof π for V_3 which leads to rejection with probability at most ε, we can construct a proof $\tilde{\pi}$ such that V_1 rejects x on proof $\tilde{\pi}$ with probability at most $\beta \varepsilon$. Thus in particular we can arrange that for every x and for every proof π for $\tilde{V}_{i+1}$ which is rejected with probability at most ε, we can construct a proof $\tilde{\pi}$ such that $\tilde{V}_i$ rejects π with probability at most $\beta \varepsilon$. Then given x and a proof $\pi = \tilde{\pi}^3$ for $V = \tilde{V}_3$ where V rejects x on proof π with probability at most ε, by successive applications of the above construction we obtain proofs $\tilde{\pi}^2$ and $\tilde{\pi}^1 = \tilde{\pi}$ for $\tilde{V}_2$, and $\tilde{V}_1$, respectively, such that for $i = 1, 2$ the verifier $\tilde{V}_i$ rejects x on proof $\tilde{\pi}^i$ with probability at most $\varepsilon_i = \beta^{3-i}\varepsilon$. By choosing an arbitrary ε with $0 < \varepsilon < 1/4$ and some sufficiently small $\beta > 1$, we can achieve that ε_1 is less than $1/4$.

Now $\tilde{V}_1$ is in fact a solution verifier with polynomial coding scheme and thus in case $\tilde{V}_1$ rejects $(x, \tilde{\pi})$ with probability $\varepsilon_1 < 1/4$ then $\tilde{\pi}$ must have the form $(\tilde{\pi}_0, \tilde{\pi}_1)$ where, firstly, $\tilde{\pi}_0$ is $1/4$-close to a code in C_n^{poly} for a satisfying assignment

a of the 3-CNF formula x and, secondly, $\tilde{\pi}_0$ together with some tables in $\tilde{\pi}_1$ is not rejected by the standard low degree test. By the latter property and by Remark 5.31 we can compute this satisfying assignment a from $\tilde{\pi}$ in time polynomial in the length of x.

We leave it to the reader to show that given a satisfying assignment a for the 3-CNF formula x we can successively compute proofs for $\tilde{V}_1$ through $\tilde{V}_3$, respectively, which lead to acceptance with probability 1. Here in order to compute such a proof for $\tilde{V}_{i+1}$ basically we take the already computed proof for $\tilde{V}_i$, partition it into blocks as in the proof of the composition lemma, encode these blocks, and finally add some additional tables which are for example used by the verifier $\tilde{V}_{i+1}$ while testing whether the alleged codes for the blocks are in fact close to the code under consideration.

It remains to show that we can compute $\tilde{\pi}$ from π in time polynomial in x. We obtain the proof $\tilde{\pi}$ by successive applications of Claim 2 in the proof of the composition lemma to the verifiers $\tilde{V}_3$ and $\tilde{V}_2$. Here for $i = 3, 2$, during an application of Claim 2 we consider the encoded blocks $y_1, \ldots, y_t$ in the proof $\tilde{\pi}^i$ of $\tilde{V}_i$, and we apply to each block some function d. According to the proof of Claim 2, here it suffices to use a function d such that in case some encoded block y together with some designated parts of $\tilde{\pi}^i$ is not rejected by some fixed test for being close to a code in the coding scheme of the solution verifier $\tilde{V}_{i-1}$, then $d(y)$ is equal to a prefix of appropriate length of the string decoded from the nearest codeword of y.

In case $i = 3$, the inputs y for the function d are alleged codewords in the code C^{lin}_{kl} of the linear function coding scheme. Here kl is in $\text{poly}(\log \log n)$ which is a subset of $\mathcal{O}(\log n)$. Thus the codewords in C_{kl} have length in $2^{\mathcal{O}(\log n)}$ and hence in $\text{poly}(n)$. Consequently we obtain a function d as required by applying a majority operation as in equation (5.2). For $i = 2$, the inputs for the function d are alleged codewords in the code C^{poly}_{kl} of the polynomial coding scheme where kl is in $\text{poly}(\log n)$. Now, if we choose as test for being close to a code in the polynomial coding scheme the usual low degree test, then by the preceding remark the function d only has to work correctly in case its input y is not rejected by the low degree test. As a consequence we obtain a function d as required according to Remark 5.31. ∎

5.6 The LFKN Test

The goal of this section is to describe a method of verifying whether a given polynomial f is identically zero on a certain part of its domain. This test is an essential ingredient of the $(\log n, \text{poly}(\log n))$-restricted solution verifier that was constructed in Section 5.4.4.

Due to the restrictions that this verifier has to fulfill, we have to obey certain constraints while performing the test. In fact, we will be able to construct a test that evaluates the given polynomial f at just a constant number of points of its domain. Since this information is of course not sufficient to come to a correct decision with sufficiently high probability, we will make use of the fact that the verifier has access to a proof string containing additional information. We will thus require access to an table T in addition to the course-of-values of f. The specific format of that table will be described in detail below.

We can thus summarize the function of the test that we have to construct: The algorithm gets a pair (f, T) as input, where it is understood that f is a m-variate polynomial of total degree d over a finite field F, and tries to check whether $f(x) = 0$ for all $x \in H^m$ for some subset $H \subseteq F$. The output of the algorithm should fulfill the following requirement: If f actually is identically zero on H^m, then for some table T_f the algorithm should always accept, while if f is not identically zero in H^m, then for any table T the algorithm should reject with high probability.

Note that we will not solve that problem directly, but will reduce it to a similar problem, namely that of verifying whether $\sum_{x \in H^m} f(x) = 0$ under the same constraints. This test for verifying large sums, the so-called LFKN test, was invented by Lund, Fortnow, Karloff, and Nisan in [LFKN92]. The extension of that test to verify everywhere vanishing functions is due to [BFLS91].

A note on the use of finite fields: While the verifier from Section 5.4.4 uses only the fields $\mathbb{F}_p$ of prime cardinality, the reduction mentioned above will make it necessary to construct field extensions. Since we thus have to deal with general finite fields $\mathbb{F}_q$ for prime powers $q = p^k$ anyway, we will assume throughout this section that we are performing arithmetic in any finite field. Note that while the arithmetical operations in $\mathbb{F}_p$ can be simulated by simply using the natural numbers $\{0, 1, \ldots, p-1\}$ and performing the usual modular arithmetic mod p, the situation is not quite that easy in the general case of $\mathbb{F}_{p^k}$. But there are well-known methods of implementing the arithmetical operations of finite fields, so we will simply assume that some such method is used. Important for our purposes is that elements of $\mathbb{F}_{p^k}$ can be represented using $\log p^k$ bits, and that the operations can be performed in polynomial time. (We could e.g. use the fact that $\mathbb{F}_{p^k} \cong \mathbb{F}_p[x]/(m)$, where m is some irreducible polynomial of degree k over $\mathbb{F}_p$. Thus we can identify the elements of $\mathbb{F}_{p^k}$ with polynomials of degree less than k over $\mathbb{F}_p$, which can be represented by their coefficients. For details on arithmetic in finite fields see e.g. [LN87] and [MBG$^+$93].)

5.6.1 Verifying Large Sums

As noted above, we will first solve this problem: Suppose we have a polynomial $f: F^m \to K$ of degree at most d in every variable over a finite field F, where K is a field extension of F, and want to check whether

$$\sum_{x \in H^m} f(x) = 0, \tag{5.28}$$

where $H \subseteq F$ is an arbitrary subset of F, while evaluating f only at a constant number of points (in fact, at only *one* point).

Actually, we have access to some additional information from a table T. How could this information be used? The crucial idea of the LFKN test consists in considering partial sums of the sum in (5.28): We define for every tuple $a = (a_1, \ldots, a_{i-1})$ with $1 \leqslant i \leqslant m$ and $a_1, \ldots, a_{i-1} \in F$ the function $g_{a_1, \ldots, a_{i-1}} : F \to K$ by

$$g_{a_1, \ldots, a_{i-1}}(x) := \sum_{b_{i+1}, \ldots, b_m \in H} f(a_1, \ldots, a_{i-1}, x, b_{i+1}, \ldots, b_m).$$

Since f has degree at most d in each variable, each g_a can obviously be expressed as univariate polynomial of degree at most d. Observe further that for all $a_1, \ldots, a_i \in F$, the functions $g_{a_1, \ldots, a_{i-1}}$ and $g_{a_1, \ldots, a_i}$ are related by the equation

$$\sum_{x \in H} g_{a_1, \ldots, a_i}(x) = g_{a_1, \ldots, a_{i-1}}(a_i),$$

and that equation (5.28) can be written as

$$\sum_{x \in H} g_\varepsilon(x) = 0,$$

where ε denotes the tuple of length 0.

Now assume that the additional table T our test is allowed to use would consist of polynomials $\tilde{g}_a : F \to K$ of degree at most d that are supposed to be equal to the g_a. (Note that T need only contain the $d + 1$ coefficients $\in K$ of each such polynomial; each entry of T thus occupies only $(d+1)\lceil \log |K| \rceil$ bits.) Then, we could easily check whether $\sum_{x \in H} \tilde{g}_\varepsilon(x) = 0$ by simply evaluating $\tilde{g}_\varepsilon(x)$ at every point $x \in H$ and calculating the sum of these values. But even if this test holds, we can conclude (5.28) only if we know in addition that $\tilde{g}_\varepsilon$ is really equal to g_ε. At this point we use the fact that these two functions are univariate polynomials of degree at most d and are therefore equal if they agree at more than d points. In fact, since d is small compared to $|F|$, we will be able to show that even if we just know that $\tilde{g}_\varepsilon(a_1) = g_\varepsilon(a_1)$ for a *single* randomly chosen point $a_1 \in F$, we can conclude with high probability that $\tilde{g}_\varepsilon \equiv g_\varepsilon$. (The probabilistic argument will be carried out in detail below.)

By the argument just given we have reduced the question whether (5.28) holds to the problem whether $\tilde{g}_\varepsilon(a_1) = g_\varepsilon(a_1)$. But since $g_\varepsilon(a_1) = \sum_{x \in H} g_{a_1}(x)$, we can check this by verifying that $\tilde{g}_\varepsilon(a_1) = \sum_{x \in H} \tilde{g}_{a_1}(x)$, once we know that $\tilde{g}_{a_1} \equiv g_{a_1}$. By the same argument as above, that question reduces to the problem whether $\tilde{g}_{a_1}(a_2) = g_{a_1}(a_2)$ for some randomly chosen point $a_2 \in F$. By recursively applying the same argument, we finally get to the question whether $\tilde{g}_{a_1, \ldots, a_{m-1}}(a_m) = g_{a_1, \ldots, a_{m-1}}(a_m)$, where the a_i have all been chosen uniformly at random from F. But since $g_{a_1, \ldots, a_{m-1}}(a_m) = f(a_1, \ldots, a_{m-1}, a_m)$ by definition, we can verify this trivially by simply querying the given function f at this point. Note that this single point $(a_1, \ldots, a_m)$ is the only point where f has been evaluated during the entire procedure.

We can now state the procedure outlined above as the following algorithm LFKN-TEST. As noticed above, LFKN-TEST has access to f and a table T containing for all $1 \leqslant i \leqslant m$ and $a_1, \ldots, a_{i-1} \in F$ a polynomial $\tilde{g}_{a_1, \ldots, a_{i-1}}$, represented by $(d+1)$ coefficients $\in K$. Thus, T has $\sum_{i=1}^{m} |F|^{i-1} = (|F|^m - 1)/(|F| - 1) = \mathcal{O}(|F|^{m-1})$ many entries, each of size $(d+1)\lceil \log |K| \rceil$.

LFKN-TEST
Choose $a_1, \ldots, a_m \in F$ uniformly at random.
Access T to get the functions $\tilde{g}_{a_1, \ldots, a_{i-1}}$ for all $1 \leqslant i \leqslant m$.
if $\sum_{x \in H} \tilde{g}_\varepsilon(x) \neq 0$ **then reject**
for $i := 1$ **to** $m - 1$ **do**
    if $\sum_{x \in H} \tilde{g}_{a_1, \ldots, a_i}(x) \neq \tilde{g}_{a_1, \ldots, a_{i-1}}(a_i)$ **then reject**
if $f(a_1, \ldots, a_m) \neq \tilde{g}_{a_1, \ldots, a_{m-1}}(a_m)$ **then reject**
accept

What remains to be shown is that LFKN-TEST indeed satisfies all required conditions:

Theorem 5.55. *Let $m, d \in \mathbb{N}$ be constants and F be a finite field with $|F| \geqslant 4md$ and K be a field extension of F. Let $f : F^m \to K$ be a polynomial of degree at most d in every variable, and let $H \subseteq F$ denote an arbitrary subset of F. Then the procedure LFKN-TEST has the following properties:*

- *If f satisfies equation (5.28), then there exists a table T_f such that the procedure LFKN-TEST(f, T_f) always accepts.*

- *If f does not satisfy equation (5.28), then LFKN-TEST(f, T) rejects with probability at least $3/4$ for all tables T.*

- *LFKN-TEST queries f at only one point, randomly chosen from F^m, reads m entries of size $(d+1)\lceil \log |K| \rceil$ from T, and uses $\mathcal{O}(m \log |F|)$ random bits.*

Proof. Suppose first that equation (5.28) holds. If we build the table T_f by setting $\tilde{g}_a :\equiv g_a$ for all tuples a, it is obvious from the definition of the functions g_a that LFKN-TEST never rejects. Thus suppose on the contrary that (5.28) does not hold. We shall prove a lower bound on the rejection probability of the algorithm by induction on the number of steps performed before the procedure rejects. To this purpose we define a predicate $\varphi(a)$ expressing the condition that the procedure does not recognize a reason to reject by examining the tuple a:

$$\varphi(\varepsilon) \equiv \sum_{x \in H} \tilde{g}_\varepsilon(x) = 0$$

$$\varphi(a_1, \ldots, a_i) \equiv \varphi(a_1, \ldots, a_{i-1}) \wedge \sum_{x \in H} \tilde{g}_{a_1, \ldots, a_i}(x) = \tilde{g}_{a_1, \ldots, a_{i-1}}(a_i) \quad (i < m)$$

$$\varphi(a_1, \ldots, a_m) \equiv \varphi(a_1, \ldots, a_{m-1}) \wedge f(a_1, \ldots, a_m) = \tilde{g}_{a_1, \ldots, a_{m-1}}(a_m)$$

Note that the procedure rejects iff $\varphi(a_1, \ldots, a_m)$ is false. Since $f(a_1, \ldots, a_m) = g_{a_1, \ldots, a_{m-1}}(a_m)$ by definition, this means that the total rejection probability is given by $\text{Prob}_{a_1, \ldots, a_m}[\neg\varphi(a_1, \ldots, a_{m-1}) \vee g_{a_1, \ldots, a_{m-1}}(a_m) \neq \tilde{g}_{a_1, \ldots, a_{m-1}}(a_m)]$. We shall now be able to prove a lower bound of $1 - md/|F| \geqslant 3/4$ (since by assumption $|F| \geqslant 4md$) for this probability by showing that for every $1 \leqslant i \leqslant m$,

$$\text{Prob}_{a_1, \ldots, a_i}[\varphi(a_1, \ldots, a_{i-1}) \wedge g_{a_1, \ldots, a_{i-1}}(a_i) = \tilde{g}_{a_1, \ldots, a_{i-1}}(a_i)] \leqslant \frac{id}{|F|}. \quad (5.29)$$

The proof of (5.29) proceeds by induction on i. We first consider the case $i = 1$. Note that since $\varphi(\varepsilon)$ does not depend on a_1, we can simply consider the two cases $\varphi(\varepsilon)$ holds and $\varphi(\varepsilon)$ does not hold separately. If $\varphi(\varepsilon)$ does not hold, we obviously have $\text{Prob}_{a_1}[\varphi(\varepsilon) \wedge g_\varepsilon(a_1) = \tilde{g}_\varepsilon(a_1)] = 0$. If however $\varphi(\varepsilon)$ does hold, we conclude that $g_\varepsilon \not\equiv \tilde{g}_\varepsilon$ since by assumption (5.28) does not hold. But those two functions are both polynomials of degree at most d, hence we conclude that they can agree at most at d points, but that means $\text{Prob}_{a_1}[\varphi(\varepsilon) \wedge g_\varepsilon(a_1) = \tilde{g}_\varepsilon(a_1)] \leqslant d/|F|$.

Now assume that (5.29) holds for some $i < m$. Note that $g_{a_1, \ldots, a_i} \equiv \tilde{g}_{a_1, \ldots, a_i}$ and $\varphi(a_1, \ldots, a_i)$ imply $g_{a_1, \ldots, a_{i-1}}(a_i) = \sum_{x \in H} g_{a_1, \ldots, a_i}(x) = \sum_{x \in H} \tilde{g}_{a_1, \ldots, a_i}(x) = \tilde{g}_{a_1, \ldots, a_{i-1}}(a_i)$. Thus we conclude from the induction hypothesis that

$$\text{Prob}_{a_1, \ldots, a_i}[\varphi(a_1, \ldots, a_i) \wedge g_{a_1, \ldots, a_i} \equiv \tilde{g}_{a_1, \ldots, a_i}] \leqslant \frac{id}{|F|}.$$

On the other hand we conclude as above that for all $a_1, \ldots, a_i$ with $g_{a_1, \ldots, a_i} \not\equiv \tilde{g}_{a_1, \ldots, a_i}$ we have

$$\text{Prob}_{a_{i+1}}[g_{a_1, \ldots, a_i}(a_{i+1}) = \tilde{g}_{a_1, \ldots, a_i}(a_{i+1})] \leqslant \frac{d}{|F|}.$$

Combining both inequalities yields

$$\text{Prob}_{a_1, \ldots, a_{i+1}}[\varphi(a_1, \ldots, a_i) \wedge g_{a_1, \ldots, a_i}(a_{i+1}) = \tilde{g}_{a_1, \ldots, a_i}(a_{i+1})] \leqslant \frac{(i+1)d}{|F|}.$$

What remains to be shown is that the resource constraints are observed. But this is obvious: The only use of randomness is in the choice of the points $a_1, \ldots, a_m$ from F, which can be done using $\mathcal{O}(m \log |F|)$ random bits. The table T is queried exactly m times to get the polynomials $\tilde{g}_{a_1, \ldots, a_{i-1}}$ for $1 \leqslant i \leqslant m$, where the representation of each $\tilde{g}_a$ is of size $(d+1)\lceil \log |K| \rceil$, and f is evaluated only in the final test at exactly one point, chosen uniformly at random from F^m. ∎

5.6.2 Verifying Everywhere Vanishing Functions

We now go on to solve the problem that we are mainly interested in: Suppose we have a polynomial $f \colon F^m \to F$ of degree at most d over a finite field F and want to check whether f is identically zero at a certain part $H \subseteq F$ of its domain:

$$f(x) = 0 \quad \text{for all } x \in H^m. \tag{5.30}$$

Again, we want to perform this check under the same restrictions as in the previous section, that is, evaluating f at only a constant number of points. Note that if we were considering fields like $\mathbb{R}$, we could simply reduce the new problem to an application of Theorem 5.55, since over $\mathbb{R}$ equation (5.30) is equivalent to $\sum_{x \in H^m} f(x)^2 = 0$. But even over finite fields, where this simple trick does not work, we are able to reduce the test of (5.30) to an application of Theorem 5.55.

The basic idea behind this reduction is the following: We construct a univariate polynomial g whose coefficients are exactly the values $f(x)$ for all $x \in H^m$. Then equation (5.30) holds if and only if g is the zero polynomial. If we assume for a moment that the degree of g is not too high, this is the case iff g is identically zero. As in the previous section, we will try to verify whether this is the case by simply evaluating g at a single randomly selected point t of its domain. But now we switch our point of view and note that evaluating a polynomial is nothing else than calculating the sum of its evaluated monomials. Thus we will be able to use Theorem 5.55 to perform that check.

The construction of g yields of course a polynomial of degree at most $|H|^m$, since we want g to have $|H|^m$ many coefficients. However, $|H|^m$ might easily be greater than $|F|$, so if the domain of g were simply F, the degree argument would fail. Thus we let K be a field extension of F such that $4|H|^m \leqslant |K| \leqslant 4|H|^m|F|$ and enumerate H by an arbitrary bijection $\rho\colon H \to \{0, 1, \ldots, |H| - 1\}$. This can be extended to an enumeration σ of H^m by setting $\sigma(x) := \sum_{i=0}^{m-1} |H|^i \cdot \rho(x_i)$ for every $x = (x_0, \ldots, x_{m-1}) \in H^m$. Now we can define the function $g\colon K \to K$ by $g(t) := \sum_{x \in H^m} f(x) \cdot t^{\sigma(x)}$. Since now $|K| \geqslant 4|H|^m$, we have

$$\text{Prob}_{t \in K}[g(t) = 0] \leqslant \frac{1}{4} \quad \text{unless (5.30) holds.}$$

We now want to use Theorem 5.55 to test whether $g(t) = 0$ for a random point $t \in K$. To do so, we have to apply the theorem to the function $f_t\colon F^m \to K$ defined by $f_t(x) := f(x) \cdot t^{\sigma(x)}$. This is only possible if f_t can be expressed as a polynomial of low degree over F. But, since we have for all $x \in H^m$

$$f_t(x) = f(x) \cdot t^{\sigma(x)} = f(x) \cdot \prod_{i=0}^{m-1} t^{|H|^i \rho(x_i)} = f(x) \cdot \prod_{i=0}^{m-1} \sum_{h \in H} t^{|H|^i \rho(h)} \cdot L_h(x_i),$$

where $L_h(z)$ denotes the degree $|H|$ polynomial that is 1 if $z = h$ and 0 if $z \neq h$ for all $z \in H$, we see that $f_t(x)$ is indeed a polynomial of degree at most $d + |H|$ in every variable x_i.

Thus we can use the following algorithm to check whether (5.30) holds. Note that the algorithm has access to a table T that is supposed to be an array of $|K|$ many subtables, for each $t \in K$ one table for LFKN-TEST necessary to perform the check $\sum_{x \in H^m} f_t(x) = 0$. Since $|K| = \mathcal{O}(|F|^{m+1})$, the table T has therefore $|K| \cdot \mathcal{O}(|F|^{m-1}) = \mathcal{O}(|F|^{2m})$ entries, each of size $(d+|H|+1)\lceil\log|K|\rceil = \mathcal{O}(m(d+|H|)\log|F|)$.

EXTENDED LFKN-TEST
Construct the field extension K of F and choose $t \in K$ uniformly at random.
Let $T(t)$ denote the subtable of T corresponding to t.
Let f_t denote the function given by $f_t(x) := f(x) \cdot t^{\sigma(x)}$.
Perform the algorithm LFKN-TEST$(f_t, T(t))$.

The procedure EXTENDED LFKN-TEST indeed has the required properties:

Theorem 5.56. *Let $m, d \in \mathbb{N}$ be constants and F be a finite field with $|F| \geqslant 4m(d+|H|)$, where $H \subseteq F$ denotes an arbitrary subset of F. Let $f: F^m \to F$ be a polynomial of degree at most d in every variable. Then the procedure EXTENDED LFKN-TEST has the following properties:*

- *If f satisfies equation (5.30), then there exists a table T_f such that the procedure EXTENDED LFKN-TEST(f, T_f) always accepts.*

- *If f does not satisfy equation (5.30), then EXTENDED LFKN-TEST(f, T) rejects with probability at least $1/2$ for all tables T.*

- EXTENDED LFKN-TEST *queries f at one point, randomly chosen from F^m, reads m entries of size $\mathcal{O}(m(d+|H|)\log|F|)$ from T, and uses $\mathcal{O}(m\log|F|)$ random bits.*

Proof. We first note that since by assumption $|F| \geqslant 4m(d+|H|)$, all conditions of Theorem 5.55 are satisfied. We therefore conclude that if (5.30) holds, and therefore $\sum_{x \in H} f_t(x) = g(t) = 0$ for all $t \in K$, the algorithm always accepts if we set $T_f := \langle T_{f_t} \mid t \in K \rangle$.

If on the other hand (5.30) does not hold, then for any table T we conclude that EXTENDED LFKN-TEST accepts only if $g(t) = 0$ for the chosen t or if $g(t) \neq 0$ but the LFKN-TEST accepts anyway. The probability of the first condition is bounded by $1/4$ as noted above, and Theorem 5.55 guarantees that the probability of the second condition is bounded by $1/4$ as well, hence EXTENDED LFKN-TEST rejects with probability at least $1/2$.

What remains to be verified is that the resource constraints are satisfied. The table T is accessed only from within LFKN-TEST, thus at most m values are queried, each of size $\mathcal{O}(m(d+|H|)\log|F|)$ as noted above. Furthermore, LFKN-TEST evaluates f_t at only one point, thus f needs to be evaluated at only one point. We finally note that $\mathcal{O}(\log|K|) = \mathcal{O}(m\log|F|)$ random bits suffice for generating the random point $t \in K$, and that LFKN-TEST needs only an additional $\mathcal{O}(m\log|F|)$ random bits. ∎

5.7 The Low Degree Test

In the proof of Proposition 5.28, where we showed the $(\log n, \operatorname{poly}(\log n))$-checkability of the polynomial coding scheme, one essential ingredient was the Low Degree Test, a method of checking, given the course-of-values of some arbitrary function $f: F^m \to F$, whether it is δ-close to some polynomial of total degree at most d. This section constitutes the proof of Theorem 5.69, on which the low degree test is based.

The low degree test given here is due to Arora, Lund, Motwani, Sudan, and Szegedy [ALM+92], combining earlier results by Arora and Safra [AS92] and Rubinfeld and Sudan [RS92]. With the exception of the bivariate case, the presentation again follows [HPS94].

5.7.1 The Bivariate Case

Before treating the general case, we will first consider only polynomials in *two* variables. In addition, we will allow those polynomials to have degree at most d in *each* variable, while we will later consider polynomials of *total* degree at most d. The proof given here follows Polishchuk and Spielman [PS94] instead of using the so-called *Matrix Transposition Lemma* from [AS92] as in [HPS94], since their proof seems somewhat simpler, and their result improves the lower bound on the field size from $\mathcal{O}(d^3)$ to $\mathcal{O}(d)$.

The main idea for testing whether a given function is δ-close to a bivariate polynomial is based on the following property:

Lemma 5.57. *Let $d \in \mathbb{N}$ be constant and let F be a finite field such that $|F| > 2d$. Let furthermore $\langle r_s \mid s \in F \rangle$ and $\langle c_t \mid t \in F \rangle$ be families of polynomials over F of degree at most d such that*

$$r_s(t) = c_t(s) \quad \text{for all } s, t \in F.$$

Then there exists a bivariate polynomial $Q: F^2 \to F$ of degree at most d in each variable such that

$$Q(s, t) = r_s(t) = c_t(s) \quad \text{for all } s, t \in F.$$

Proof. Let $s_1, \ldots, s_{d+1}$ be distinct elements of F, and for each $1 \leqslant i \leqslant d+1$, let δ_i denote the degree d polynomial satisfying $\delta_i(s_i) = 1$ and $\delta_i(s_j) = 0$ for all $1 \leqslant j \leqslant d+1, j \neq i$. We can now define

$$Q(s, t) := \sum_{i=1}^{d+1} \delta_i(s) r_{s_i}(t).$$

It is clear that Q is a polynomial of degree at most d in each variable, and that $Q(s_i, t) = r_{s_i}(t) = c_t(s_i)$ for all $1 \leqslant i \leqslant d+1$, $t \in F$. But this means that for any $t \in F$, the degree d polynomials $Q(\cdot, t)$ and c_t agree at least at $d+1$ points, hence $Q(\cdot, t) \equiv c_t$. But this means $Q(s, t) = r_s(t) = c_t(s)$ for all $s, t \in F$. ∎

The lemma says that to check whether a list of values correspond to the values obtained from a bivariate degree d polynomial, it suffices to check that each row and column of the values can be obtained from a univariate degree d polynomial. Our goal is now to get a corresponding probabilistic result: If the rows and columns can be described at most points by univariate polynomials, then there is a bivariate polynomial describing the whole function at most points.

In fact, suppose we are given $\langle r_s \mid s \in F \rangle$ and $\langle c_t \mid t \in F \rangle$ as in the lemma that satisfy

$$\text{Prob}_{s,t}[r_s(t) \neq c_t(s)] \leqslant \delta^2.$$

The basic idea is to find a "error correcting" polynomial that is zero whenever $r_s(t)$ and $c_t(s)$ disagree.

Lemma 5.58. *Let F be a field and $S \subseteq F^2$ a set of size at most a^2 for some $a \in \mathbb{N}$. Then there exists a non-zero polynomial $E: F^2 \to F$ of degree at most a in each variable such that $E(x, y) = 0$ for all $(x, y) \in S$.*

Proof. The set of polynomials $E: F^2 \to F$ of degree at most a in each variable is a vector space of dimension $(a+1)^2$ over F. Consider the map that sends a polynomial to the vector of values that it takes for each element of S. That is, let $S = \{s_1, \ldots s_m\}$ and consider the map

$$\varphi : E(x, y) \mapsto (E(s_1), E(s_2), \ldots, E(s_m)).$$

This map is a homomorphism of a vector space of dimension $(a+1)^2$ into a vector space of dimension $m = |S| \leqslant a^2$ over F, which must have a non-trivial kernel. Thus, there exists a non-zero polynomial E such that $\varphi(E) = 0$, but that means $E(s) = 0$ for all $s \in S$. ∎

This means we can find a polynomial E such that $r_s(t)E(s, t) = c_t(s)E(s, t)$ for all $s, t \in F$, and we can apply Lemma 5.57 to get a polynomial P satisfying $P(s, t) = r_s(t)E(s, t) = c_t(s)E(s, t)$. Now we note that if P could be divided by E as formal polynomials, we could conclude the proof, since

$$\frac{P(s, t)}{E(s, t)} = r_s(t) = c_t(s) \quad \text{for all } s, t \in F \text{ with } E(s, t) \neq 0.$$

The rest of this section will be used to show that if $|F|$ is big enough, then E does in fact divide P. To do so, we will use Sylvester's Criterion that uses resultants to check whether two polynomials have a non-trivial common divisor.

Definition 5.59. *The* Sylvester matrix *of two polynomials* $P(x) = P_0 + P_1 x + \cdots + P_d x^d$ *and* $E(x) = E_0 + E_1 x + \cdots + E_e x^e$ *over a field* F *is defined to be the* $(d+e) \times (d+e)$ *matrix*

$$M(P,E) = \begin{pmatrix} P_d & \cdots & \cdots & P_0 & 0 & \cdots & 0 \\ 0 & P_d & \cdots & \cdots & P_0 & & \vdots \\ \vdots & & \ddots & & & \ddots & 0 \\ 0 & \cdots & 0 & P_d & \cdots & \cdots & P_0 \\ E_e & \cdots & \cdots & E_0 & 0 & \cdots & 0 \\ 0 & E_e & \cdots & \cdots & E_0 & & \vdots \\ \vdots & & \ddots & & & \ddots & 0 \\ 0 & \cdots & 0 & E_e & \cdots & \cdots & E_0 \end{pmatrix},$$

where the upper part of the matrix consists of e *rows of coefficients of* P *and the lower part consists of* d *rows of coefficients of* E.

The resultant $R(P,E)$ *of* P *and* E *is the determinant of the Sylvester matrix of* P *and* E:

$$R(P,E) = \det M(P,E).$$

Lemma 5.60 (Sylvester's Criterion). *Two polynomials* $P(x)$ *and* $E(x)$ *over a field* F *have a non-trivial common divisor if and only if the resultant* $R(P,E)$ *is zero.*

Proof. If we write $P(x) = P_0 + P_1 x + \cdots + P_d x^d$ and $E(x) = E_0 + E_1 x + \cdots + E_e x^e$, we note first that $P(x)$ and $E(x)$ have a non-trivial common divisor if and only if there exist non-trivial polynomials $A(x)$ of degree at most $e - 1$ and $B(x)$ of degree at most $d - 1$ such that $P(x)A(x) = E(x)B(x)$. One direction of this is obvious: If there exists a non-trivial common divisor $D(x)$ such that $P(x) = D(x)\hat{P}(x)$ and $E(x) = D(x)\hat{E}(x)$, we simply set $A(x) := \hat{E}(x)$ and $B(x) := \hat{P}(x)$. On the other hand, assume that such $A(x)$ and $B(x)$ exist. Since the degree of $P(x)$ is greater than the degree of $B(x)$, $P(x)$ and $E(x)$ must share a common divisor.

We can now reformulate this as a system of linear equations in the coefficients of A and B by comparing coefficients of the same power of x:

$$\begin{aligned} P_d A_{e-1} &= E_e B_{d-1} \\ P_{d-1} A_{e-1} + P_d A_{e-2} &= E_{e-1} B_{d-1} + E_e B_{d-2} \\ \vdots \quad \vdots \quad \vdots & \\ P_0 A_1 + P_1 A_0 &= E_0 B_1 + E_1 B_0 \\ P_0 A_0 &= E_0 B_0 \end{aligned}$$

If we treat the coefficients of A and B as the variables of a system of linear equations, then we find that this system has a non-trivial solution if and only if the Sylvester matrix $M(P,E)$ hat determinant zero, that is iff $R(P,E) = 0$. ∎

In addition, we will need the following fact about the derivatives of the determinant of a matrix of polynomials:

Lemma 5.61. *Let $M(x) = (p_{ij}(x))_{i,j}$ be a $n \times n$ matrix of polynomials in x over a field F, let $R(x) = \det M(x)$ denote the determinant of $M(x)$, and let $R^{(k)}(x)$ denote the k-th formal derivative of the polynomial $R(x)$. Let $x_0 \in F$ be such that $M(x_0)$ is a matrix of rank less than $n - k$ for some $k \geqslant 0$. Then $R^{(k)}(x_0) = 0$.*

Proof. The derivative $R'(x)$ of $R(x)$ can be expressed as

$$R'(x) = \begin{vmatrix} p'_{1,1}(x) & \cdots & p'_{1,n}(x) \\ p_{2,1}(x) & \cdots & p_{2,n}(x) \\ \vdots & \ddots & \vdots \\ p_{n,1}(x) & \cdots & p_{n,n}(x) \end{vmatrix} + \cdots + \begin{vmatrix} p_{1,1}(x) & \cdots & p_{1,n}(x) \\ p_{2,1}(x) & \cdots & p_{2,n}(x) \\ \vdots & \ddots & \vdots \\ p'_{n,1}(x) & \cdots & p'_{n,n}(x) \end{vmatrix}.$$

This means that $R'(x_0)$ is the sum of determinants of matrices of rank less than $n - k + 1$. By induction we conclude that $R^{(k)}(x_0)$ is the sum of determinants of $n \times n$ matrices of rank less than n. But then $R^{(k)}(x_0) = 0$. ∎

We have just one more problem to overcome: The results given above apply only to univariate polynomials over a field, while we are considering bivariate polynomials. We could of course simply treat a polynomial in $F[x,y]$ as univariate polynomial in y over $F[x]$, but this is no field. So, we treat the polynomials P and E as univariate polynomials in y over $F(x)$, the field of rational functions over F, and note that by Gauss' Lemma, E divides P over $F(x)$ iff E divides P over $F[x]$.

Lemma 5.62. *Let $E(x,y)$ be a polynomial of degree at most a in x and b in y and let $P(x,y)$ be a polynomial of degree at most $a+d$ in x and $b+d$ in y over a field F. If there exist distinct $x_1, \ldots, x_n \in F$ such that $E(x_i,y)$ divides $P(x_i,y)$ (in $F[y]$) for $1 \leqslant i \leqslant n$, where $n > 2(a+d)$, and furthermore there exist distinct $y_1, \ldots, y_m \in F$ such that $E(x,y_i)$ divides $P(x,y_i)$ (in $F[x]$) for $1 \leqslant i \leqslant m$, where $m > 2(b+d)$, then $E(x,y)$ divides $P(x,y)$ (in $F[x,y]$).*

Proof. We let $R(x,y)$ denote a greatest common divisor of $P(x,y)$ and $E(x,y)$ (which is uniquely determined up to multiplication with elements in $F\setminus\{0\}$), and assume by way of contradiction that E/R has degree $\geqslant 1$. Now if R is not constant, we let r (resp. s) denote its degree in x (resp. y), and set $\hat{P} := P/R$ and $\hat{E} := E/R$ and apply the lemma to these polynomials. We note that the conditions of the lemma are indeed satisfied for $\hat{P}$ and $\hat{E}$: For every $1 \leqslant i \leqslant n$ with

$R(x_i, y) \not\equiv 0$ we know that $\hat{E}(x_i, y)$ divides $\hat{P}(x_i, y)$, but $R(x_i, y)$ can be the zero polynomial for at most r values x_i. Thus we can find values $\hat{x}_1, \ldots, \hat{x}_{n-r}$ where $E(\hat{x}_i, y)$ divides $P(\hat{x}_i, y)$. Similarly we find values $\hat{y}_1, \ldots, \hat{y}_{m-s}$ with $E(x, \hat{y}_i)$ dividing $P(x, \hat{y}_i)$. Since $\hat{E}$ has degree at most $\hat{a} := a - r$ in x and $\hat{b} := b - s$ in y, and $\hat{P}$ has degree at most $\hat{a} + d$ in x and $\hat{b} + d$ in y, and obviously $n - r > 2(a + d) - r \geqslant 2(\hat{a} + d)$ and $m - s > 2(\hat{b} + d)$, we conclude that the conditions of the lemma are therefore fulfilled.

Thus we can assume without loss of generality that $P(x, y)$ and $E(x, y)$ have no non-trivial common divisor. We furthermore assume without loss of generality from now on that $b \geqslant a$. If we treat P and E as polynomials in y over $F(x)$ by writing

$$
\begin{aligned}
P(x, y) &= P_0(x) + P_1(x)y + \cdots + P_{b+d}(x)y^{b+d} \\
E(x, y) &= E_0(x) + E_1(x)y + \cdots + E_b(x)y^b,
\end{aligned}
$$

we can apply Gauss' Lemma and conclude that P and E have no non-trivial common divisor in $F(x)[y]$ either. But since we are now considering *univariate* polynomials over the field $F(x)$, we can apply Lemma 5.60 to conclude that the resultant $R(P, E)(x) \in F(x)$ is non-zero. Since $R(P, E)(x)$ is the determinant of the $(2b + d) \times (2b + d)$ Sylvester matrix

$$
M(P, E)(x) = \begin{pmatrix}
P_{b+d}(x) & \cdots & & P_0(x) & \cdots & 0 \\
\vdots & \ddots & & & \ddots & \vdots \\
0 & \cdots & P_{b+d}(x) & \cdots & \cdots & P_0(x) \\
E_b(x) & \cdots & E_0(x) & 0 & \cdots & 0 \\
\vdots & \ddots & & & \ddots & \vdots \\
0 & \cdots & 0 & E_b(x) & \cdots & E_0(x)
\end{pmatrix},
$$

the resultant can in fact even be viewed as a *polynomial* in x over F. In fact, since $M(P, E)(x)$ has b rows of coefficients of P, each of degree at most $a + d$, and $b + d$ rows of coefficients of E, each of degree at most a, the degree of $R(P, E)(x)$ is bounded by $b(a + d) + (b + d)a$, and hence $R(P, E)(x)$ has at most $b(a + d) + (b + d)a$ roots.

Now we will arrive at a contradiction by showing that in fact each x_i, $1 \leqslant i \leqslant n$, is a root of $R(P, E)(x)$ of multiplicity b, thus $R(P, E)(x)$ has at least $nb > 2(a + d)b \geqslant 2ab + (b + a)d = b(a + d) + (b + d)a$ roots by the assumptions of the lemma. Indeed, for each $1 \leqslant i \leqslant n$, $E(x_i, y)$ divides $P(x_i, y)$, which means $P(x_i, y) \equiv K_i(y)E(x_i, y)$ for some polynomial $K_i(y) = K_{i,0} + \cdots + K_{i,d}y^d$. By comparing coefficients of the same power of y, we conclude that

$$
P_z(x_i) = \sum_{r=0}^{d} K_{i,r} \cdot E_{z-r}(x_i) \quad \text{for all } 0 \leqslant z \leqslant b + d,
$$

if we set $E_s(x_i) = 0$ for $s < 0$ or $s > b$. But now we see that each of the first b rows of $M(P, E)(x_i)$ can be expressed as a linear combination of the last $b + d$

rows: indeed, if we denote the j-th row of $M(P, E)(x_i)$ by $M_j(P, E)(x_i)$ we find that

$$M_j(P, E)(x_i) = \sum_{r=0}^{d} K_{i,r} \cdot M_{(j-1+b)+(d-r)}(P, E)(x_i).$$

This implies that $M(P, E)(x_i)$ is matrix of rank at most $b + d$, and we conclude from Lemma 5.61 that $R^{(k)}(P, E)(x_i)$ is zero for all $k < b$. But that means that $R(P, E)(x)$ has a root of multiplicity b at x_i.   ∎

We can now finally state the main result on bivariate polynomials:

Theorem 5.63. *Let $0 < \varepsilon < 3/8$ and $d \in \mathbb{N}$ be constants and let F be a finite field such that $|F| \geqslant 8d$ and $|F| \geqslant 1/(3/8 - \varepsilon)$. Let furthermore $\langle r_s \mid s \in F \rangle$ and $\langle c_t \mid t \in F \rangle$ be families of polynomials over F of degree at most d and $A = (a_{st})$ be an $|F| \times |F|$ matrix such that*

$$\mathrm{Prob}_{s,t}[a_{st} = r_s(t) \wedge a_{st} = c_t(s)] \geqslant 1 - \varepsilon^2.$$

Then there exists a bivariate polynomial $Q \colon F^2 \to F$ of degree at most d in each variable such that the sets $S = \{s \in F \mid r_s \equiv Q(s, \cdot)\}$ and $T = \{t \in F \mid c_t \equiv Q(\cdot, t)\}$ satisfy

$$|S| \geqslant (1 - 2\varepsilon^2)|F| \qquad and \qquad |T| \geqslant (1 - 2\varepsilon^2)|F|.$$

Proof. We call a pair (s, t) an *error point* if $r_s(t) \neq a_{st}$ or $c_t(s) \neq a_{st}$. If we set $a := \lceil \varepsilon|F| \rceil$, we know by assumption that the total number of error points is bounded by $\varepsilon^2|F|^2 \leqslant a^2$. Hence by Lemma 5.58 there exists a bivariate polynomial $E \colon F^2 \to F$ of degree at most a in each variable such that $E(s, t) = 0$ for each error point (s, t). This implies that $r'_s(t) := r_s(t)E(s, t)$ and $c'_t(s) = c_t(s)E(s, t)$ are polynomials of degree $a + d$ satisfying $r'_s(t) = c'_t(s)$ for all $s, t \in F$. Therefore, by Lemma 5.57 there exists a polynomial $P \colon F^2 \to F$ of degree $a + d$ in each variable such that $P(s, t) = r_s(t)E(s, t) = c_t(s)E(s, t)$ for all $s, t \in F$.

Note that the assumptions imply $\varepsilon|F| + 1 \leqslant 3/8|F|$ and $d \leqslant 1/8|F|$, hence $a + d < \varepsilon|F| + 1 + d \leqslant 1/2|F|$. Furthermore, for every fixed $s_0 \in F$, $P(s_0, t)$ and $r_{s_0}(t)E(s_0, t)$ are univariate degree $a + d$ polynomials that agree on all $|F| > a + d$ points of their domain, and hence are the same polynomial. Thus, for each $s_0 \in F$, $E(s_0, t)$ divides $P(s_0, t)$, and we conclude similarly that for each $t_0 \in F$, $E(s, t_0)$ divides $P(s, t_0)$. Therefore, by Lemma 5.62 there exists a polynomial $Q \colon F^2 \to F$ of degree at most d in each variable such that $P(s, t) = Q(s, t)E(s, t) = r_s(t)E(s, t) = c_t(s)E(s, t)$ for all $s, t \in F$. This implies that in any row s where $E(s, \cdot)$ is not the zero polynomial (and is thus zero at most at a points), $Q(s, t) = r_s(t)$ for at least $|F| - a > d$ columns $t \in F$, and therefore $Q(s, \cdot) \equiv r_s$. But since E can be identically zero on at most a rows, it follows that $|S| \geqslant |F| - a$. We show similarly that $|T| \geqslant |F| - a$.

But we can show even more: If we consider an arbitrary column $t \in F$, we have $Q(s, t) = r_s(t)$ for all $s \in S$, and unless (s, t) is an error point, this implies

$Q(s,t) = c_t(s)$ as well. But if this is the case for at least $d+1$ rows $s \in S$, we have $Q(\cdot,t) \equiv c_t$, that is $t \in T$. Conversely, all columns $t \notin T$ must contain at least $|S| - d \geqslant |F| - (a+d) > 1/2|F|$ error points, since $a+d < 1/2|F|$. Therefore, the total number of error points must be greater than $1/2|F|(|F| - |T|)$, but since this number is bounded by $\varepsilon^2 |F|^2$, we conclude $|T| > (1 - 2\varepsilon^2)|F|$. Analogously we conclude $|S| > (1 - 2\varepsilon^2)|F|$. ∎

5.7.2 The General Case

We will now generalize the result to the case of functions of arbitrary arity. The presentation will from here on again follow [HPS94], which is based on [ALM+92]. The basic idea is here that similarly to considering the restrictions of a bivariate function to rows and columns, we will consider the restrictions of a multivariate function $f: F^m \to F$ to all "lines" of the form $\{x + th \mid t \in F\}$ for $x, h \in F^m$.

Our goal will be to show that if all these restrictions agree with high probability with univariate degree d polynomials, then the given function f is δ-close to some multivariate polynomial of total degree d. This will enable us to construct a low degree tester that gets as inputs the table of values of the function f that is to be tested, and additionally a list of degree d polynomials (specified by $d+1$ coefficients each) that are supposed to be equal to the restriction of f to every line $l_{x,h}$. The tester then simply proceeds to test this fact by computing the values of some randomly chosen polynomials from that list at some randomly chosen points and comparing them with the corresponding values of f.

Definition 5.64. *For $x, h \in F^m$, the set of points $l_{x,h} := \{x + th \mid t \in F\}$ is called the* line *through x with slope h.*

For a function $f: F^m \to F$ and points $x, h \in F^m$ the degree d line polynomial *for f on the line $l_{x,h}$ is the univariate degree d polynomial $P_{x,h}^{f,d}$ which maximizes the number of points $t \in F$ for which $P_{x,h}^{f,d}(t)$ equals $f(x + th)$.*

Note: If there are several degree d polynomials that agree on the same (maximal) number of points with the restriction of f to the line $l_{x,h}$, we arbitrarily select one of them as line polynomial. This selection must be done consistently, though: if for two pairs (x, h) and (x', h'), $l_{x,h}$ is the same set as $l_{x',h'}$, then for a point $y = x + th = x' + t'h'$, the line polynomials must agree at y, i.e. $P_{x,h}^{f,d}(t) = P_{x',h'}^{f,d}(t')$.

Note that although the line polynomial is not always defined in a deterministic way, under certain circumstances we know that a given polynomial *must* be the line polynomial:

Lemma 5.65. *For a function $f: F^m \to F$ and points $x, h \in F^m$, if any degree d polynomial $p: F \to F$ satisfies*

$$|\{t \in F \mid p(t) = f(x + th)\}| > 1/2(|F| + d),$$

then p is the line polynomial on $l_{x,h}$.

Proof. Let q denote the line polynomial on $l_{x,h}$. By definition we have

$$|\{t \in F \mid q(t) = f(x + th)\}| \geqslant |\{t \in F \mid p(t) = f(x + th)\}|.$$

Since $p(t) \neq q(t)$ implies $p(t) \neq f(x + th)$ or $q(t) \neq f(x + th)$, we can conclude

$$|\{t \in F \mid p(t) \neq q(t)\}| \leqslant |\{t \in F \mid p(t) \neq f(x+th)\}| + |\{t \in F \mid q(t) \neq f(x+th)\}|,$$

and therefore

$$|\{t \in F \mid p(t) \neq q(t)\}| \leqslant 2 \cdot |\{t \in F \mid p(t) \neq f(x + th)\}| < |F| - d,$$

hence

$$|\{t \in F \mid p(t) = q(t)\}| > d.$$

But since p and q are both degree d polynomials, we conclude $p \equiv q$. ∎

In that case, the line polynomial can even be efficiently computed. (Note that this problem is well-known in the theory of error-correcting codes: the decoding/correction of so-called Reed-Solomon codes is essentially the same as computing what is called a line polynomial in our setting. The algorithm given in the following lemma is in fact the so-called Berlekamp-Welch decoder used for this purpose. For more information on Reed-Solomon codes see e.g. [MS77]. The description of the Berlekamp-Welch decoder can be found in [BW86]; the proof given here follows [GS92] and [Sud92].)

Lemma 5.66. *There is an algorithm that given a function $f: F^m \to F$ and points $x, h \in F^m$ and a number $d < |F|$ finds a degree d polynomial $p: F \to F$ satisfying*

$$|\{t \in F \mid p(t) = f(x + th)\}| > 1/2(|F| + d),$$

provided any such polynomial exists. The running time of the algorithm is polynomial in $|F|$. (Note that according to Lemma 5.65, p must indeed be the line polynomial on $l_{x,h}$.)

Proof. Suppose there exists a polynomial p satisfying the condition stated in the lemma. This means that $p(t)$ and $f(x + th)$ disagree at most at $a := \lfloor 1/2(|F| - d - 1) \rfloor$ places, and thus there exists a degree a polynomial e with $e(t) = 0$ if and only if $p(t) \neq f(x + th)$. Thus we conclude $q(t) = e(t)f(x + th)$ for all $t \in F$, where $q(t) := p(t)e(t)$ is a polynomial of degree at most $a + d$.

We will show now that the reverse direction of this implication holds as well: Suppose that there exist polynomials e' of degree at most a and q' of degree at most $a + d$ with $q'(t) = e'(t)f(x + th)$ for all $t \in F$, where e' is not the zero polynomial. If furthermore e' divides q', we can obviously find a polynomial $p := q'/e'$ satisfying the required conditions. On the other hand, if e' does not divide q', we claim that such a polynomial p does not exist: Indeed, if there were one, we could find q and e as above, and conclude that $q'(t)e(t)f(x + th) = q(t)e'(t)f(x + th)$ for all $t \in F$, and since $f(x + th) = 0$ implies $q(t) = q'(t) = 0$ by choice of q and q', we have even $q'(t)e(t) = q(t)e'(t)$ for all $t \in F$. But since $q'(t)e(t)$ and $q(t)e'(t)$ are both polynomials of degree at most $2a + d$, and they agree at $|F|$ points, where $|F| > 2a + d$ by definition of a, we conclude that $q'e$ and qe' are the same polynomial. As $q = pe$ by definition, we see that qe' and hence $q'e$ is divisible by ee', and thus e' divides q', in contradiction to our assumption.

Thus the required algorithm can proceed as follows: It first tries to find polynomials q of degree at most $a+d$ and e of degree at most a with $q(t) = e(t)f(x+th)$ for all $t \in F$, where $e \not\equiv 0$. Note that this is equivalent to finding a nontrivial solution to the system of $|F|$ linear equations

$$\sum_{i=0}^{a+d} q_i t^i = f(x + th) \sum_{i=0}^{a} e_i t^i \qquad \text{for all } t \in F$$

in the $2a + d < |F|$ variables q_i and e_i. Obviously, if such a solution exists, it can be found (using e.g. Gauss elimination) in time polynomial in $|F|$. If no solution is found, or if the resulting polynomial e does not divide q, no polynomial p with the required properties exists, as was shown above. Otherwise, the procedure returns the polynomial $p := q/e$. ∎

As in the bivariate case, we begin by stating the deterministic result that builds the basis of the randomized version we will prove later: To check whether a function $f: F^m \to F$ can be described by a polynomial of degree d, it suffices to test whether all restrictions of f to some line $l_{x,h}$ agree with some univariate polynomial of degree at most d.

Lemma 5.67. *Let $m, d \in \mathbb{N}$ be constants and F be a finite field with $|F| > 2d$. A function $f: F^m \to F$ is a polynomial of total degree at most d if and only if for all $x, h \in F^m$ the restriction $f_{x,h}$ defined by $f_{x,h}(t) := f(x + th)$ is a univariate polynomial of degree at most d.*

Proof. One direction is obvious. We will prove the other direction by induction on m. The case $m = 1$ is trivial, so let $m > 1$. Let $a_0, \ldots, a_d$ be distinct points in F. According to the induction hypothesis, for all $i \leqslant d$ the function f_i defined by $f_i(x_2, \ldots, x_m) := f(a_i, x_2, \ldots, x_m)$ is described by a $(m - 1)$-ary polynomial of total degree d.

Let L_i be the univariate degree d polynomial which is 1 at a_i and 0 at all a_j for $j \neq i$. Consider the polynomial g defined by

$$g(x_1, \ldots, x_m) := \sum_{i=0}^{d} L_i(x_1) f_i(x_2, \ldots, x_m).$$

Note that for every $b_2, \ldots, b_m \in F$ the degree d polynomials $x \mapsto g(x, b_2, \ldots, b_m)$ and $x \mapsto f(x, b_2, \ldots b_m)$ coincide on the points $a_0, \ldots, a_d$ and hence on all $x \in F$. Thus we have $g(b_1, \ldots, b_m) = f(b_1, \ldots, b_m)$ for all $b_1, \ldots, b_m \in F$.

On the other hand we claim that g has total degree at most d. By construction the total degree of g is bounded by $2d$. Hence we can write g as

$$g(x_1, \ldots, x_m) = \sum_{i_1 + \cdots + i_m \leqslant 2d} \alpha_{i_1, \ldots, i_m} x_1^{i_1} \cdots x_m^{i_m},$$

and the restriction $g_{0,h} : t \mapsto g(th_1, \ldots, th_m)$ as

$$g_{0,h}(t) = \sum_{i=0}^{2d} t^i \sum_{i_1 + \cdots + i_m = i} \alpha_{i_1, \ldots, i_m} h_1^{i_1} \cdots h_m^{i_m}.$$

But by the assumptions of the lemma we know that $g_{0,h} = f_{0,h}$ is a polynomial of degree at most d in t for every choice of $h_1, \ldots, h_m \in F$, hence for all $i > d$ the degree i polynomial

$$p_i(h_1, \ldots, h_m) := \sum_{i_1 + \cdots + i_m = i} \alpha_{i_1, \ldots, i_m} h_1^{i_1} \cdots h_m^{i_m}$$

must be identically zero. Since $|F| > 2d \geqslant i$, this in turn implies $\alpha_{i_1, \ldots, i_m} = 0$ for all $i_1, \ldots, i_m$ with $i_1 + \cdots + i_m > d$, but that means that g has no term of total degree greater than d. ∎

An essential tool in the proof of the randomized version will be the following lemma, which is a corollary to Theorem 5.63: If families of matrices (A_i) and polynomials $(r_s(t)_i)$ and $(c_t(s)_i)$ satisfy the assumptions of Theorem 5.63 with high probability, then the degree d line polynomials for a fixed row and a fixed column coincide with high probability at their intersection.

Lemma 5.68. *Let $\delta > 0$ and $d, m \in \mathbb{N}$ be constants, let F be a finite field such that $|F| \geqslant 8d$ and $|F| \geqslant 24$ and let $s_0, t_0 \in F$ be fixed. Assume that for every pair $h_1, h_2 \in F^m$ there exist matrices $A = (a_{st})$ and families $\langle r_s \mid s \in F \rangle$ and $\langle c_t \mid t \in F \rangle$ of polynomials over F of degree at most d such that the sets*

$$\tilde{S} := \{s \in F \mid \mathrm{Prob}_{h_1, h_2}[r_s(t) = a_{st}] \geqslant 1 - \frac{\delta}{2} \text{ for all } t \in F\}$$

and

$$\tilde{T} := \{t \in F \mid \mathrm{Prob}_{h_1, h_2}[c_t(s) = a_{st}] \geqslant 1 - \frac{\delta}{2} \quad \text{for all } s \in F\}$$

satisfy

$$|\tilde{S}| \geqslant (1 - \delta)|F| \quad \text{and} \quad |\tilde{T}| \geqslant (1 - \delta)|F|.$$

Let P_{row} denote the degree d line polynomial for A on the s_0-th row and P_{col} the degree d line polynomial on the t_0-th column (where the matrix A is interpreted as function $F^2 \to F$). Then

$$\mathrm{Prob}_{h_1, h_2}[P_{row}(t_0) = P_{col}(s_0)] \geqslant 1 - 54\delta.$$

(Note that A, r_s, c_t, P_{row} and P_{col} all depend on h_1 and h_2, although that dependency is not reflected in the notation.)

Proof. Note that the assumptions of the lemma imply

$$\mathrm{Prob}_{s,t,h_1,h_2}[r_s(t) \neq a_{st}]$$
$$\leqslant \mathrm{Prob}_{s,t,h_1,h_2}[s \in \tilde{S} \to r_s(t) \neq a_{st}] + \mathrm{Prob}_{s,t,h_1,h_2}[s \notin \tilde{S}] \leqslant \frac{3}{2}\delta$$

and analogously

$$\mathrm{Prob}_{s,t,h_1,h_2}[c_t(s) \neq a_{st}] \leqslant \frac{3}{2}\delta,$$

hence

$$\mathrm{Prob}_{s,t,h_1,h_2}[r_s(t) \neq a_{st} \vee c_t(s) \neq a_{st}] \leqslant 3\delta.$$

Thus if we set $\varepsilon := 1/3$ we can conclude that

$$\mathrm{Prob}_{h_1,h_2}[\mathrm{Prob}_{s,t}[r_s(t) \neq a_{st} \vee c_t(s) \neq a_{st}] > \varepsilon^2] \leqslant \frac{3\delta}{\varepsilon^2}$$

and therefore

$$\mathrm{Prob}_{h_1,h_2}[\mathrm{Prob}_{s,t}[r_s(t) = a_{st} \wedge c_t(s) = a_{st}] \geqslant 1 - \varepsilon^2] \geqslant 1 - \frac{3\delta}{\varepsilon^2}.$$

From this we see that the matrix A satisfies the requirements of Theorem 5.63 with probability at least $1 - 3\delta/\varepsilon^2$, since $|F| \geqslant 8d$ and $|F| \geqslant 1/(3/8 - \varepsilon) = 24$ by assumption. If we let Q, S, and T be defined as in Theorem 5.63 (for those A that do not satisfy the requirements, let $Q \equiv 0$ and $S = T = \emptyset$), we conclude

$$\mathrm{Prob}_{h_1,h_2}[|S| \geqslant (1 - 2\varepsilon^2)|F| \wedge |T| \geqslant (1 - 2\varepsilon^2)|F|] \geqslant 1 - \frac{3\delta}{\varepsilon^2},$$

that is

$$\mathrm{Prob}_{h_1,h_2}[|F\backslash S| > 2\varepsilon^2|F| \vee |F\backslash T| > 2\varepsilon^2|F|] \leqslant \frac{3\delta}{\varepsilon^2},$$

which we can write as

$$\mathrm{Prob}_{h_1,h_2}[\mathrm{Prob}_s[r_s \not\equiv Q(s,\cdot)] > 2\varepsilon^2 \vee \mathrm{Prob}_t[c_t \not\equiv Q(\cdot,t)] > 2\varepsilon^2] \leqslant \frac{3\delta}{\varepsilon^2}. \quad (5.31)$$

On the other hand, if we apply the assumptions of the lemma in the special cases $t = t_0$ and $s = s_0$, respectively, we obtain

$$\text{Prob}_{s,h_1,h_2}[r_s(t_0) \neq a_{st_0}] \leqslant \frac{3}{2}\delta \quad \text{and} \quad \text{Prob}_{t,h_1,h_2}[c_t(s_0) \neq a_{s_0t}] \leqslant \frac{3}{2}\delta,$$

hence

$$\text{Prob}_{h_1,h_2}[\text{Prob}_s[r_s(t_0) \neq a_{st_0}] > \varepsilon^2 \vee \text{Prob}_t[c_t(s_0) \neq a_{s_0t}] > \varepsilon^2] \leqslant \frac{3\delta}{\varepsilon^2}. \quad (5.32)$$

Since $Q(s,t_0) \neq a_{st_0}$ implies $r_s \not\equiv Q(s,\cdot) \vee r_s(t_0) \neq a_{st_0}$, we conclude

$$\text{Prob}_s[Q(s,t_0) \neq a_{st_0}] \leqslant \text{Prob}_s[r_s \not\equiv Q(s,\cdot)] + \text{Prob}_s[r_s(t_0) \neq a_{st_0}],$$

hence $\text{Prob}_s[Q(s,t_0) \neq a_{st_0}] > 3\varepsilon^2$ implies that either $\text{Prob}_s[r_s \not\equiv Q(s,\cdot)] > 2\varepsilon^2$ or $\text{Prob}_s[r_s(t_0) \neq a_{st_0}] > \varepsilon^2$, and therefore

$$\text{Prob}_{h_1,h_2}[\text{Prob}_s[Q(s,t_0) \neq a_{st_0}] > 3\varepsilon^2]$$
$$\leqslant \text{Prob}_{h_1,h_2}[\text{Prob}_s[r_s \not\equiv Q(s,\cdot)] > 2\varepsilon^2 \vee \text{Prob}_s[r_s(t_0) \neq a_{st_0}] > \varepsilon^2].$$

As we can similarly deduce

$$\text{Prob}_{h_1,h_2}[\text{Prob}_t[Q(s_0,t) \neq a_{s_0t}] > 3\varepsilon^2]$$
$$\leqslant \text{Prob}_{h_1,h_2}[\text{Prob}_t[c_t \not\equiv Q(\cdot,t)] > 2\varepsilon^2 \vee \text{Prob}_t[c_t(s_0) \neq a_{s_0t}] > \varepsilon^2],$$

combining this observation with (5.31) and (5.32) yields

$$\text{Prob}_{h_1,h_2}[\text{Prob}_s[Q(s,t_0) \neq a_{st_0}] > 3\varepsilon^2 \vee \text{Prob}_t[Q(s_0,t) \neq a_{s_0t}] > 3\varepsilon^2] \leqslant 6\delta/\varepsilon^2. \quad (5.33)$$

By Lemma 5.65, $Q(\cdot,t_0) \not\equiv P_{row}$ implies

$$\text{Prob}_s[Q(s,t_0) = a_{st_0}] \leqslant \frac{1}{2}(1 + \frac{d}{|F|}) \leqslant \frac{1}{2}(1 + \frac{1}{8}) \leqslant \frac{9}{16},$$

since $|F| \geqslant 8d$. This in turn yields $\text{Prob}_s[Q(s,t_0) \neq a_{st_0}] \geqslant 7/16 > 1/3 = 3\varepsilon^2$. Analogously, we conclude that $Q(s_0,\cdot) \not\equiv P_{col}$ implies $\text{Prob}_t[Q(s_0,t) \neq a_{s_0t}] > 3\varepsilon^2$. Therefore we deduce from (5.33) that

$$\text{Prob}_{h_1,h_2}[Q(\cdot,t_0) \not\equiv P_{row} \vee Q(s_0,\cdot) \not\equiv P_{col}] \leqslant \frac{6\delta}{\varepsilon^2} = 54\delta,$$

which in turn implies

$$\text{Prob}_{h_1,h_2}[P_{row}(s_0) \neq P_{col}(t_0)] \leqslant 54\delta.$$

■

Now we can state the main theorem of this section.

Theorem 5.69. *Let $0 < \delta < 1/6066$ and $d, m \in \mathbb{N}$ be constants, and let F be a finite field with $|F| \geq 8d$ and $|F| \geq 1/\delta$. Let $\tilde{g}: F^m \to F$ be a function, and let $T: F^{2m} \to F^{d+1}$ be a function such that the degree d polynomials $\tilde{P}_{x,h}$ over F given by $\tilde{P}_{x,h}(t) = \sum_{i=0}^{d} T(x, h)_{i+1} \cdot t^i$ satisfy*

$$\text{Prob}_{x,h,t}[\tilde{P}_{x,h}(t) = \tilde{g}(x + th)] \geq 1 - \frac{\delta}{8}.$$

Then there exists a (unique) polynomial $g: F^m \to F$ of total degree d so that

$$\text{Prob}_x[g(x) = \tilde{g}(x)] \geq 1 - \delta.$$

Proof. For $x, h \in F^m$ we will denote the degree d line polynomial $P_{x,h}^{\tilde{g},d}$ by $P_{x,h}$. The function g is then given explicitly by

$$g(x) = \text{majority}_h\{P_{x,h}(0)\},$$

that is, $g(x)$ is a value so that $|\{h \mid P_{x,h}(0) = g(x)\}| \geq |\{h \mid P_{x,h}(0) = a\}|$ for every choice of $a \in F$. In case there should be several candidates equally suited for choice as $g(x)$, we choose one arbitrarily (we will see later that this cannot happen and g is indeed uniquely defined).

By assumption we have

$$\text{Prob}_{x,h,t}[\tilde{P}_{x,h}(t) \neq \tilde{g}(x + th)] \leq \frac{\delta}{8}, \tag{5.34}$$

hence

$$\text{Prob}_{x,h}[\text{Prob}_t[\tilde{P}_{x,h}(t) \neq \tilde{g}(x + th)] > \frac{1}{3}] \leq \frac{3\delta}{8}.$$

By Lemma 5.65 we know that $P_{x,h} \not\equiv \tilde{P}_{x,h}$ implies $\text{Prob}_t[\tilde{P}_{x,h}(t) = \tilde{g}(x + th)] \leq 1/2(1 + d/|F|) \leq 1/2(1 + 1/8) = 9/16$ and hence $\text{Prob}_t[\tilde{P}_{x,h}(t) \neq \tilde{g}(x + th)] \geq 7/16 > 1/3$. Thus we conclude

$$\text{Prob}_{x,h}[\tilde{P}_{x,h} \not\equiv P_{x,h}] \leq \frac{3\delta}{8}$$

and together with (5.34) also

$$\text{Prob}_{x,h,t}[P_{x,h}(t) \neq \tilde{g}(x + th)] \leq \frac{3\delta}{8} + \frac{\delta}{8} = \frac{\delta}{2}.$$

We can even strengthen the result: If we fix a $t \in F$, we can replace in the above equation t by $t + t'$ and x by $x - t'h$ to get

$$\text{Prob}_{x,h,t'}[P_{x-t'h,h}(t+t') \neq \tilde{g}(x - t'h + (t+t')h)] = \text{Prob}_{x,h,t'}[P_{x,h}(t) \neq \tilde{g}(x+th)],$$

since $P_{x-t'h,h}(t + t') = P_{x,h}(t)$ according to the consistency criterion for line polynomials. Thus we conclude

$$\text{Prob}_{x,h}[P_{x,h}(t) \neq \tilde{g}(x + th)] \leqslant \frac{\delta}{2} \quad \text{for all } t \in F. \tag{5.35}$$

Note that this equation is no longer a statement about averages taken over x, h, t, but one about averages taken over x, h that holds for every t. By an application of Lemma 5.68 we can strengthen the result even more, yielding a statement where averages are taken only over h: For every $h_1, h_2 \in F^m$, let $A = (a_{st})$ denote the matrix given by $a_{st} = \tilde{g}(x + sh_1 + th_2)$, and let $r_s(t) = P_{x+sh_1,h_2}(t)$ and $c_t(s) = P_{x+th_2,h_1}(s)$. If we replace x by $x + sh_1$ for $s \neq 0$ and h by h_2, (5.35) yields

$$\text{Prob}_{h_1,h_2}[r_s(t) \neq a_{st}] = \text{Prob}_{h_1,h_2}[P_{x+sh_1,h_2}(t) \neq \tilde{g}(x + sh_1 + th_2)] \leqslant \frac{\delta}{2}$$

for all $t \in F$, hence $|\tilde{S}| \geqslant |F \backslash \{0\}| = |F| - 1 \geqslant (1 - \delta)|F|$ since $|F| \geqslant 1/\delta$. As we similarly conclude $|\tilde{T}| \geqslant (1 - \delta)|F|$, all assumptions of Lemma 5.68 are satisfied, and the lemma yields

$$\text{Prob}_{h_1,h_2}[P_{x+s_0h_1,h_2}(t_0) \neq P_{x+t_0h_2,h_1}(s_0)] \leqslant 54\delta \quad \text{for all } x \in F^m, \ s_0, t_0 \in F, \tag{5.36}$$

since by definition $P_{x+s_0h_1,h_2}$ is the line polynomial for the s_0-th row and $P_{x+t_0h_2,h_1}$ the line polynomial for the t_0-th column of A. If we fix now $x \in F^m$ and $t \in F \backslash \{0\}$ and consider the special case $s_0 = 0$ and $t_0 = t$ of (5.36), we deduce that

$$\text{Prob}_{h_1,h_2}[P_{x+th_2,h_1}(0) \neq P_{x,h_2}(t)] \leqslant 54\delta.$$

On the other hand, replacing x by $x + t(h_2 - h_1)$ (this is possible since $t \neq 0$) and h by h_1 and using $P_{x+t(h_2-h_1),h_1}(t) = P_{x+th_2,h_1}(0)$ (by consistency), (5.35) yields

$$\text{Prob}_{h_1,h_2}[P_{x+th_2,h_1}(0) \neq \tilde{g}(x + th_2)] \leqslant \frac{\delta}{2}.$$

Combining those two inequalities we conclude

$$\text{Prob}_h[P_{x,h}(t) \neq \tilde{g}(x + th)] \leqslant 54.5\delta \quad \text{for all } x \in F^m \text{ and } t \in F \backslash \{0\}. \tag{5.37}$$

Furthermore we can use the special case $s_0 = t_0 = 0$ of (5.36) to prove the above claim that $g(x)$ is uniquely defined: Indeed, for every $x \in F^m$ we have

$$
\begin{aligned}
1 - 54\delta \ &\leqslant \ \text{Prob}_{h_1,h_2}[P_{x,h_1}(0) = P_{x,h_2}(0)] \\
&= \ \sum_{a \in F} \text{Prob}_{h_1}[P_{x,h_1}(0) = a] \cdot \text{Prob}_{h_2}[P_{x,h_2}(0) = a] \\
&\leqslant \ \text{Prob}_h[P_{x,h}(0) = g(x)] \cdot \sum_{a \in F} \text{Prob}_h[P_{x,h}(0) = a],
\end{aligned}
$$

where the last inequality holds since $\text{Prob}_h[P_{x,h}(0) = g(x)] \geqslant \text{Prob}_h[P_{x,h}(0) = a]$ follows from the definition of g for any $a \in F$. Since we furthermore know that $\sum_{a \in F} \text{Prob}_h[P_{x,h}(0) = a] = 1$, we conclude

$$\mathrm{Prob}_h[P_{x,h}(0) = g(x)] \geqslant 1 - 54\delta > \frac{1}{2} \quad \text{for all } x \in F^m. \tag{5.38}$$

We proceed now to prove the δ-closeness of g to $\tilde{g}$. First note that $g(x) \neq \tilde{g}(x)$ implies $\tilde{g}(x) \neq P_{x,h}(0)$ or $g(x) \neq P_{x,h}(0)$ for every $h \in F^m$. We can write this as

$$\mathrm{Prob}_h[\tilde{g}(x) \neq P_{x,h}(0)] + \mathrm{Prob}_h[g(x) \neq P_{x,h}(0)] \geqslant 1,$$

which implies

$$\mathrm{Prob}_h[\tilde{g}(x) \neq P_{x,h}(0)] \geqslant \mathrm{Prob}_h[g(x) = P_{x,h}(0)] > \frac{1}{2}$$

by (5.38). On the other hand, the special case $t = 0$ of (5.35) yields

$$\mathrm{Prob}_x[\mathrm{Prob}_h[\tilde{g}(x) \neq P_{x,h}(0)] > \frac{1}{2}] \leqslant \delta,$$

and thus

$$\mathrm{Prob}_x[g(x) \neq \tilde{g}(x)] \leqslant \delta. \tag{5.39}$$

Note that this allows us to extend (5.37) to the case $t = 0$: as (5.39) and (5.38) imply

$$\mathrm{Prob}_h[P_{x,h}(0) \neq \tilde{g}(x)] \leqslant 54\delta + \delta = 55\delta \quad \text{for all } x \in F^m,$$

we combine this with (5.37) to get

$$\mathrm{Prob}_h[P_{x,h}(t) \neq \tilde{g}(x + th)] \leqslant 55\delta \quad \text{for all } x \in F^m \text{ and } t \in F. \tag{5.40}$$

What remains to be shown is that g is actually a degree d polynomial. This requires another application of Lemma 5.68: Let $x, h \in F^m$ be fixed. For every $h_1, h_2 \in F^m$ let $A' = (a'_{st})$ denote the matrix given by

$$a'_{st} = \begin{cases} g(x + sh) & \text{if } t = 0 \\ \tilde{g}(x + sh + t(h_1 + sh_2)) & \text{otherwise} \end{cases}$$

and let $r'_s(t) = P_{x+sh,h_1+sh_2}(t)$ and $c'_t(s) = P_{x+th_1,h+th_2}(s)$. Note that replacing x with $x + th_1$ and h with $h + th_2$ for any $t \in F \backslash \{0\}$, (5.35) yields

$$\mathrm{Prob}_{h_1,h_2}[c'_t(s) \neq a'_{st}] = \mathrm{Prob}_{h_1,h_2}[P_{x+th_1,h+th_2}(s) \neq \tilde{g}(x+th_1+s(h+th_2))] \leqslant \frac{\delta}{2}$$

for all $s, t \in F$, $t \neq 0$. On the other hand, setting x to $x + sh$ for some $s \in F$ and replacing h with $h_1 + sh_2$ in (5.37) we get

$$\mathrm{Prob}_{h_1,h_2}[r'_s(t) \neq a'_{st}] =$$
$$\mathrm{Prob}_{h_1,h_2}[P_{x+sh,h_1+sh_2}(t) \neq \tilde{g}(x + sh + t(h_1 + sh_2))] \leqslant 54.5\delta$$

for all $t \in F \backslash \{0\}$. But for $t = 0$ we can use (5.38) to get

$$\text{Prob}_{h_1,h_2}[r'_s(0) \neq a'_{s,0}] = \text{Prob}_{h_1,h_2}[P_{x+sh,h_1+sh_2}(0) \neq g(x+sh)] \leqslant 54\delta, \quad (5.41)$$

concluding

$$\text{Prob}_{h_1,h_2}[r'_s(t) \neq a'_{st}] \leqslant 54.5\delta \quad \text{for all } s, t \in F.$$

Hence, if we replace δ by 109δ, the assumptions of Lemma 5.68 are again satisfied. Applying the lemma in the case $s_0 = t_0 = 0$ we therefore conclude

$$\text{Prob}_{h_1,h_2}[P_{row}(0) \neq P_{col}(0)] \leqslant 5886\delta. \qquad (5.42)$$

Furthermore, we note that by Lemma 5.65 we know that $P_{row} \not\equiv r'_0$ implies $|\{t \in F \mid r'_0(t) = a'_{0,t}\}| \leqslant 1/2(|F| + d)$, and hence $|\{t \in F \mid P_{x,h_1}(t) = \tilde{g}(x+th_1)\}| \leqslant 1/2(|F| + d) + 1 \leqslant 1/2(1 + 1/8)|F| + \delta|F| < 1153/2048|F|$ since $\delta < 1/2048$, and hence $\text{Prob}_t[P_{x,h_1}(t) \neq \tilde{g}(x+th_1)] > 895/2048$. Since (5.40) yields

$$\text{Prob}_{h_1}[\text{Prob}_t[P_{x,h_1}(t) \neq \tilde{g}(x+th_1)] > \frac{895}{2048}] \leqslant \frac{2048}{895} 55\delta,$$

we conclude

$$\text{Prob}_{h_1,h_2}[P_{row} \not\equiv r'_0] \leqslant 126\delta.$$

This combined with (5.41) for $s = 0$ and (5.42) yields

$$\text{Prob}_{h_1,h_2}[P_{col}(0) \neq g(x)] \leqslant (54 + 126 + 5886)\delta = 6066\delta < 1$$

by choice of δ. On the other hand, we see that since the first columns of the matrices A' do not depend on h_1 or h_2, P_{col} is in fact identical to $P_{x,h}^{g,d}$, the degree d line polynomial for g. Since $g(x)$ does not depend on h_1 or h_2 either, we conclude

$$P_{x,h}^{g,d}(0) = g(x) \quad \text{for all } x, h \in F^m.$$

Using the consistency condition $P_{x+th,h}^{g,d}(0) = P_{x,h}^{g,d}(t)$ we even have

$$P_{x,h}^{g,d}(t) = g(x+th) \quad \text{for all } x, h \in F^m, \, t \in F,$$

but then we can conclude from Lemma 5.67 that g is a degree d polynomial. ∎

For the proof of Theorem 5.54 we need the even stronger property that the polynomial g, whose existence is guaranteed by Theorem 5.69, is even efficiently computable. This follows as an immediate corollary of the proof just given and Lemma 5.66, though.

Corollary 5.70. *In the situation of Theorem 5.69, there exists an algorithm that takes the course-of-values of $\tilde{g}$ as input and computes the course-of-values of g as output, provided the conditions of the theorem are satisfied. The running time of the algorithm is polynomial in the size of its input (i.e. polynomial in $|F^m|$).*

Proof. Since the function g was explicitly given by

$$g(x) = \text{majority}_h\{P_{x,h}(0)\}$$

in the proof of Theorem 5.69, where $P_{x,h}$ was defined to be the degree d line polynomial for $\tilde{g}$ on the line $l_{x,h}$, we want to make use of this definition in our algorithm. Since there are only $|F^m|$ many values of h, we can evaluate the majority quantifier in the definition by brute force. The only remaining problem is to compute the line polynomials $P_{x,h}$.

Recall from Lemma 5.66 that we can compute all line polynomials satisfying

$$\text{Prob}_t[P_{x,h}(t) = \tilde{g}(x + th)] > 1/2(1 + \frac{d}{|F|})$$

in polynomial time. Since $1/2(1 + \frac{d}{|F|}) \leqslant 1/2(1 + 1/8) = 9/16$, and since

$$\text{Prob}_h[\text{Prob}_t[P_{x,h}(t) \neq \tilde{g}(x + th)] > 7/16] \leqslant 16/7 \cdot 55\delta < 1/40$$

follows from equation (5.40) and the fact that $\delta < 1/6066$, we conclude that for every $x \in F^m$, we can compute $P_{x,h}$ for at least $39/40$ of all $h \in F^m$.

On the other hand, we know from equation (5.38) that

$$\text{Prob}_h[P_{x,h}(0) = g(x)] \geqslant 1 - 54\delta > 99/100.$$

This means that even if get wrong results for $1/40$ of all $h \in F^m$ by using the algorithm of Lemma 5.66 for computing the $P_{x,h}(0)$, we will still receive the correct result $g(x)$ in the (overwhelming) majority of cases. ∎

5.8 A Proof of Cook's Theorem

Theorem 5.71 (Cook-Levin). *Let L be a language which is recognized by a (nondeterministic) Turing machine in time $\mathcal{O}(n^c)$. Then for each input size n, we can construct in time $\mathcal{O}(n^{2c})$ a 3-CNF formula ψ of size $\mathcal{O}(n^{2c})$ such that for all binary strings x of length n, the string x is in L iff the formula ψ is satisfiable with partial assignment x.*

Proof. To fix the notation, suppose that the machine $\mathcal{M}$ consists of the set of states $Q = \{q_0, q_+, q_-, q_3, q_4, \ldots, q_k\}$ — containing the starting state q_0, the accepting state q_+, and the rejecting state q_- — the tape alphabet $\Sigma = \{\diamond, 0, 1, \sigma_3, \sigma_4, \ldots, \sigma_l\}$, and the transition relation $\delta \subseteq Q \times \Sigma \times Q \times \Sigma \times \{-1, 0, +1\}$, where $(q, \sigma, q', \sigma', m) \in \delta$ is supposed to mean that if $\mathcal{M}$ is in state q while scanning the tape symbol σ at the current head position, it is allowed to change into state q', write σ' into the current tape cell, and move the head according to m (left, right, or no movement).

An instantaneous description of the configuration of the machine $\mathcal{M}$ must contain the state of $\mathcal{M}$, the position of the head, and the contents of the tape. We assume without loss of generality that the machine $\mathcal{M}$ is such that for every input of size n there is a $t(n) \in \mathcal{O}(n^c)$ such that every computation of $\mathcal{M}$ is of length exactly $t(n)$, and that $\mathcal{M}$ never tries to "run off the tape". In this case, the machine can only visit the tape cells at positions $0, \ldots, t(n)$, and thus the size of a configuration is limited. Furthermore, a whole computation of $\mathcal{M}$ can be represented by a sequence of $t(n)$ configurations. The basic idea is now to encode this sequence of configurations into those propositional variables that may occur in ψ in addition to the variables $x_1, \ldots, x_n$ that will always be assigned according to the input string x. For better readability, we give special names to those additional variables, according to what they are supposed to encode (note that the number of variables used is in $\mathcal{O}(n^{2c})$):

- Q_s^t: at time t, $\mathcal{M}$ is in state q_s $(0 \leqslant t \leqslant t(n), 0 \leqslant s \leqslant k)$
- H_r^t: at time t, the head scans a tape cell at position $\leqslant r$ $(0 \leqslant t, r \leqslant t(n))$
- $T_{r,s}^t$: at time t, the tape cell at position r contains σ_s $(0 \leqslant t, r \leqslant t(n), 0 \leqslant s \leqslant l)$

The formula ψ that we want to construct can be described as the conjunction of the following statements (note that we will sometimes use the shortcuts $P_0^t := H_0^t$ and $P_r^t := H_r^t \wedge \overline{H_{r-1}^t}$ $(0 < r \leqslant t(n))$ expressing that at time t, the head scans exactly the cell at position r):

- At any time, the machine is in exactly one state:

$$\bigwedge_{0 \leqslant t \leqslant t(n)} \left[\bigvee_{0 \leqslant s \leqslant k} Q_s^t \wedge \bigwedge_{0 \leqslant s_1 < s_2 \leqslant k} (\overline{Q_{s_1}^t} \vee \overline{Q_{s_2}^t}) \right]$$

- At any time, the head position flags are consistent:

$$\bigwedge_{0 \leqslant t \leqslant t(n)} H_{t(n)}^t \wedge \bigwedge_{0 \leqslant t \leqslant t(n)} \bigwedge_{0 \leqslant r < t(n)} \left[\overline{H_r^t} \vee H_{r+1}^t \right]$$

- At any time, every tape cell contains exactly one symbol:

$$\bigwedge_{0 \leqslant t \leqslant t(n)} \bigwedge_{0 \leqslant r \leqslant t(n)} \left[\bigvee_{0 \leqslant s \leqslant l} T_{r,s}^t \wedge \bigwedge_{0 \leqslant s_1 < s_2 \leqslant l} (\overline{T_{r,s_1}^t} \vee \overline{T_{r,s_2}^t}) \right]$$

- At time $t = 0$, the machine is in state q_0, scans the cell at position 0, and the tape is inscribed as follows: for $0 \leqslant r < n$, the cell at position r contains the symbol 1 if x_{r+1} is true and the symbol 0 otherwise, and all other tape cells contain the symbol $\diamond$:

$$Q_0^0 \wedge P_0^0 \wedge \bigwedge_{0 \leqslant r < n} \left[(x_{r+1} \vee T_{r,1}^0) \wedge (\overline{x_{r+1}} \vee T_{r,2}^0) \right] \wedge \bigwedge_{n \leqslant r \leqslant t(n)} T_{r,0}^0$$

– At time $t = t(n)$, the machine is in state q_+:

$$Q_1^{t(n)}$$

– For any $t < t(n)$, the tape contains the same symbol in cell r at time $t + 1$ as it contains at time t, unless the machine scans cell r at time t:

$$\bigwedge_{0 \leqslant t < t(n)} \bigwedge_{0 \leqslant r \leqslant t(n)} \bigwedge_{0 \leqslant s \leqslant l} \left[P_r^t \vee \overline{T_{r,s}^t} \vee T_{r,s}^{t+1} \right]$$

– For any $t < t(n)$, if the head scans cell r, there is a transition $(q, \sigma, q', \sigma', m) \in \delta$ such that the machine is in state q at time t and in state q' at time $t + 1$, the cell r contains σ at time t and σ' at time $t + 1$, and the head scans cell $r + m$ at time $t + 1$:

$$\bigwedge_{0 \leqslant t < t(n)} \bigwedge_{0 \leqslant r \leqslant t(n)} \left[\overline{P_r^t} \vee \bigvee_{(q_a, \sigma_b, q_c, \sigma_d, m) \in \delta} (Q_a^t \wedge T_{r,b}^t \wedge Q_c^{t+1} \wedge T_{r,b}^{t+1} \wedge P_{r+m}^{t+1}) \right]$$

If we call the conjunction of the formulae given above ψ', it should be obvious that ψ' is satisfiable with partial assignment x if and only if $\mathcal{M}$ accepts the input string x. But note that ψ' is no 3-CNF formula. This poses no problem though, since in each of the formulae above, the part enclosed in $[\cdots]$ is an expression of constant size (independent of n). Therefore, each of those parts can be replaced by an equivalent 3-CNF formula of constant size, using some additional variables (where the number of new variables is also a constant). Since the total number of parts to be translated is $\mathcal{O}(n^{2c})$, we conclude that we can construct a 3-CNF formula ψ with the following properties:

– For all x, ψ is satisfiable with partial assignment x iff ψ' is satisfiable with partial assignment x.

– ψ uses $\mathcal{O}(n^{2c})$ distinct variables (apart from $x_1, \ldots, x_n$).

– ψ consists of $\mathcal{O}(n^{2c})$ clauses of length 3.

To conclude the proof we note that the construction of ψ can obviously be done in time linear in its length.   ∎

Exercises

Exercise 5.1. Let $g : \mathbb{F}_2^n \to \mathbb{F}_2$ be a linear function and let $\tilde{g} : \mathbb{F}_2^n \to \mathbb{F}_2$ be a function which is δ-close to g.

a) Show that
$$\text{Prob}_y[g(x) \neq \tilde{g}(x+y) - \tilde{g}(y)] \leqslant 2\delta.$$

Note that this generalizes the first part in the proof of Corollary 5.19.

b) Is $\text{Prob}_y[g(x) = \tilde{g}(x+y) - \tilde{g}(y)]$ independent of x?

Exercise 5.2. In the proof of Theorem 5.39 we used three codewords A, B, and C of codeword length 2^n, 2^{n^2}, and 2^{n^3}, respectively, from the linear function coding scheme. Recall that these codewords encode an assignment for the variables in a given 3-CNF formula. In fact these codewords listed all values of the sums given in equation (5.15), (5.16), and (5.17).

a) Show that codeword A can always be efficiently extracted from codeword B or codeword C, and that codeword B can always be efficiently constructed from codeword C.

b) Is it possible to construct an $(n^3, 1)$-restricted solution verifier, which queries only bits from codeword C? If so, could the CONSISTENCY TEST be removed?

6. Parallel Repetition of MIP(2,1) Systems

Clemens Gröpl, Martin Skutella

6.1 Prologue

Two men standing trial for a joint crime try to convince a judge of their innocence. Since the judge does not want to spend too much time on verifying their joint alibi, he selects a pair of questions at random. Then each of the suspects gets only one of the questions and gives an answer. Based on the coincidence of the two answers, the judge decides on the guilt or innocence of the men.

The judge is convinced of the fairness of this procedure, because once the questioning starts the suspects can neither talk to each other as they are kept in separate rooms nor anticipate the randomized questions he may ask them. If the two guys are innocent an optimal strategy is to convince the judge by telling the truth. However, if they have actually jointly committed the crime, their answers will agree with probability at most ε, regardless of the strategy they use.

To reduce the error ε the judge decides to repeat the random questioning k times and to declare the two men innocent if all k pairs of answers coincide. This obviously reduces the error to ε^k. So he randomly chooses k pairs of questions and writes them on two lists, one for each of the accused.

The judge does not want to ask the questions one after another but, to make things easier, hands out the two lists to the accused and asks them to write down their answers. Of course he does not allow them to communicate with each other while answering the questions. Waiting for the answers, he once more thinks about the error probability. Can the two men benefit from knowing all the questions in advance?

6.2 Introduction

In mathematical terms, the situation described in the prologue fits into the context of *two-prover one-round proof systems*. In this chapter, we give an introduction into basic definitions and characteristics of two-prover one-round proof systems and the complexity class MIP(2, 1). (The letters MIP stand for *multi-prover interactive proof system*). Furthermore, we illustrate the central ideas of the proof of the Parallel Repetition Theorem. Our approach is mainly based on the papers of Raz [Raz95, Raz97] and Feige [Fei95].

Multi-prover interactive proofs were introduced by Ben-Or, Goldwasser, Kilian, and Wigderson in [BGKW88], motivated by the necessity to find new foundations to the design of secure cryptography. The story told in our prologue is an adaption of an example given in [BGKW88].

The chapter is organized as follows: We formally introduce MIP(2, 1) proof systems in Section 6.3. Then, in Section 6.4, we explain the problem of error reduction by parallel repetition of MIP(2, 1) proof systems and state Raz' Parallel Repetition Theorem. Its proof is based on the investigation of coordinate games, which are introduced in Section 6.5. Sections 6.6 and 6.7 give an overview of the proof of the Parallel Repetition Theorem.

6.3 Two-Prover One-Round Proof Systems

In a MIP(2, 1) proof system G, two computationally unbounded *provers* P_1, P_2 try to convince a probabilistic polynomial time *verifier* that a certain input I belongs to a pre-specified language L. The proof system has one round: Based on I and a random string τ, the verifier randomly generates a pair of questions (x, y) from a pool $X \times Y$ which depends on the input I and sends x to prover P_1 and y to prover P_2. The first prover responds by sending $u = u(x) \in U$ to the verifier, the second prover responds with $v = v(x) \in V$. Here U and V are pre-specified sets of possible answers to the questions in X resp. Y. Since the verifier is polynomially bounded, only a polynomial number of bits can be exchanged. In particular, the size of all possible questions and answers must be polynomially bounded in the size of the input I.

The two provers know the input I and can agree upon a strategy (i. e., mappings $u : X \to U$ and $v : Y \to V$) in advance, but they are not allowed to communicate about the particular random choice of questions the verifier actually asks. Based on x, y, u, and v, the verifier decides whether to accept or reject the input I. This is done by evaluating an *acceptance predicate* $Q : X \times Y \times U \times V \to \{0, 1\}$ where output 1 means acceptance and 0 rejection. We think of Q also as a subset of $Z := X \times Y \times U \times V$, i.e., $Q = \{(x, y, u, v) \in Z \mid Q(x, y, u, v) = 1\}$. The probability distribution of the question pairs (x, y) chosen from $X \times Y$ by the verifier is denoted by μ. The strategy of the verifier (i. e., his choice of μ and Q based on I) is called the *proof system*, while the strategy of the provers (i. e., their choice of u and v, based on the knowledge of I, μ and Q) is called the *protocol*.

For fixed input I, a proof system G can be interpreted as a *game* for two players (provers). Thus, if we talk of $G = G(I)$ as a game we implicitly consider a fixed input I. A game G is described by sets X, Y, U, V, a predicate Q, a probability measure μ (which makes use of a random string τ), and mappings u and v. We will write $\mu(x, y)$ for $\mu\big((x, y)\big)$. Through their choices of u and v the two players aim to maximize the μ-probability that Q is satisfied, which is given by

$$\sum_{X \times Y} \mu(x, y) \, Q\big(x, y, u(x), v(y)\big) \,.$$

This probability is called the *value of the protocol* (u, v). The *value* $w(G)$ *of the game* G is defined as the maximal value of a protocol, i. e.,

$$w(G) := \max_{\substack{u : X \to U \\ v : Y \to V}} \sum_{X \times Y} \mu(x, y) \, Q\big(x, y, u(x), v(y)\big) \,.$$

A strategy (u, v) of the two players for which the maximum is attained is called an *optimal protocol*. A protocol with value 1 is also called a *winning strategy*.

We can now formally define $\mathrm{MIP}(2, 1)$ proof systems.

Definition 6.1. *A verifier and two provers build a* $\mathrm{MIP}(2, 1)$ *proof system for a language L with* error probability ε *if:*

1. Perfect completeness: *For all $I \in L$, there exists a protocol with value 1 (i. e., the provers have a winning strategy);*

2. Soundness: *For all $I \notin L$, the value of the proof system is $w(G) \leqslant \varepsilon$ (i. e., the provers can only succeed with probability at most ε).*

(The largest value c such that for all $I \in L$, the value of the corresponding game is at least c, is called *completeness*.) The following example is one of the most important applications of $\mathrm{MIP}(2, 1)$ proof systems in the context of non-approximability results. It will be applied in Chapters 7, 9, and 10.

Example 6.2. Consider an RobE3Sat formula $\varphi = C_1 \wedge \ldots \wedge C_m$. Let z_{a_j}, z_{b_j} and z_{c_j} denote the variables in clause C_j, $j = 1, \ldots, m$. Then we can define the following two-prover interactive proof system:

1. The verifier chooses a clause with index $x \in \{1, \ldots, m\}$ and a variable with index $y \in \{a_x, b_x, c_x\}$ at random. Question x is sent to prover P_1 and question y is sent to prover P_2.

2. Prover P_1 answers with an assignment for the variables z_{a_x}, z_{b_x}, and z_{c_x}, and prover P_2 answers with an assignment for z_y.

3. The verifier accepts if and only if C_x is satisfied by P_1's assignment and the two provers return the same value for z_y.

For each protocol, the strategy v of prover P_2 induces an assignment to the variables in the obvious way. Moreover, to get an optimal protocol prover P_1 must answer in accordance with this assignment unless the given clause C_x is not satisfied. If C_x is not satisfied, P_1 must change the value of at least one variable in the clause. Thus, if φ is satisfiable the value of the proof system is 1. Otherwise, with probability at least ε the verifier chooses a clause C_x not satisfied by P_2's assignment and P_1 has to change the value of a variable $z_{y'}$ where $y' \in \{a_x, b_x, c_x\}$. Since $z_y = z_{y'}$ with conditional probability at least $1/3$ the value of the protocol is at most $1 - \varepsilon/3$.

As a result of the PCP-Theorem 4.1 we know that we can translate an arbitrary instance of a problem in $\mathcal{NP}$ to an instance of RoBE3SAT. The previous example is a MIP(2,1) proof system for RoBE3SAT with error probability $1 - \varepsilon/3$.

We briefly mention some other results on MIPs. The classes MIP(k, r) are generalizations of MIP$(2, 1)$ with k provers and r rounds of questioning. Both the verifier and the provers are allowed to use their knowledge of earlier questions and answers. Obviously, MIP$(k, r) \subseteq$ MIP(k', r') holds, if $k \leqslant k'$ and $r \leqslant r'$. Feige and Lovász [FL92] proved that MIP$(2, 1) =$ MIP$(k, r) = \mathcal{NEXP}$ for $k \geqslant 2$ and $r \geqslant 1$. The result that MIP$(1, \text{poly}(n)) = \mathcal{PSPACE}$ is known as Shamir's Theorem, see [Pap94]. It follows from results of [BGKW88] that multi-prover interactive proof systems with two-sided error (i.e., completeness < 1) behave similar to MIPs with one-sided error.

6.4 Reducing the Error Probability

The main application of MIPs in the context of approximability results is the construction of probabilistically checkable proofs with small error probability for $\mathcal{NP}$-hard optimization problems, see Chapters 7, 9, and 10. Without giving details, we shortly describe the basic idea behind this construction. One forces the two provers to write down their answers to all possible questions in a special encoding, the so-called long code. The resulting string serves as a proof in the context of PCPs. Thereby the soundness of the PCP directly depends on the error probability of the MIP. Thus, reducing the error in MIP$(2, 1)$ proof systems is an important, however subtle issue.

A straightforward approach is to repeat an MIP$(2, 1)$ protocol k times and to accept only if all executions are accepting. Ideally one would hope that this method reduces the error to ε^k. This is indeed true if the executions are performed sequentially and in an oblivious way, i.e., each prover must answer each question online before seeing the question for the next execution, and is not allowed to use his knowledge about the questions he was asked before.

Instead, we will consider parallel repetition, where each prover sends out its answers only after receiving all its questions. Such a k-fold parallel repetition of

G can be seen as a new two-prover one-round proof system with k *coordinates*, denoted by $G^{\otimes k}$. The verifier treats each coordinate of $G^{\otimes k}$ as an independent copy of the original game G and accepts in $G^{\otimes k}$ only if it would have accepted all the k copies of G. We will elaborate on the decomposition into coordinates in Section 6.5.

The *k-fold parallel repetition* $G^{\otimes k}$ of the game G consists of the four sets X^k, Y^k, U^k, and V^k, the probability measure $\mu^{\otimes k}$ defined on $\Omega^k := X^k \times Y^k$, and the acceptance predicate $Q^{\otimes k}$. For $\bar{x} \in X^k$ and $\bar{y} \in Y^k$ we write $\bar{x} = (x_1, \ldots, x_k)$ resp. $\bar{y} = (y_1, \ldots, y_k)$. The probability measure $\mu^{\otimes k}$ is defined by

$$\mu^{\otimes k}(\bar{x}, \bar{y}) := \prod_{i=1}^{k} \mu(x_i, y_i)\,.$$

The acceptance predicate is

$$Q^{\otimes k}(\bar{x}, \bar{y}, \bar{u}, \bar{v}) := \prod_{i=1}^{k} Q(x_i, y_i, u_i, v_i)\,.$$

A protocol for $\bar{G}$ consists of two mappings $\bar{u} : X^k \to U^k$ and $\bar{v} : Y^k \to V^k$. The value of $G^{\otimes k}$ is denoted by $w(G^{\otimes k})$. For simplicity, we will also use the notations $\bar{G} := G^{\otimes k}$, $\bar{X} := X^k$, $\bar{Y} := Y^k$, $\bar{U} := U^k$, $\bar{V} := V^k$, $\bar{\mu} = \mu^{\otimes k}$, $\bar{\Omega} := \bar{X} \times \bar{Y}$, and $\bar{Q} = Q^{\otimes k}$. As before, let $\bar{Z} := \bar{X} \times \bar{Y} \times \bar{U} \times \bar{V}$.

Note that, in contrast to the case of sequential repetition, prover P_1 is allowed to take all the questions $x_1, \ldots, x_k$ into account (and not only x_i), when it responds to question x_i. The same holds for prover P_2. If the input I belongs to the language L, the provers already have a winning strategy in G, so knowing also the questions of other coordinates they can do no better. The problem with parallel repetition is therefore, to which amount this side information can help the provers to cheat, if the input I does *not* belong to the language L.

At first, it was believed that, as in the sequential case, repeating a proof system k times in parallel reduces the error to ε^k, see [FRS88]. Later, Fortnow constructed an example where $w(G^{\otimes 2}) > w(G)^2$, showing that the two provers can benefit from their knowledge of all the questions in advance, see [FRS90]. Feige [Fei91] presented the following game G which has the same value as its 2-fold parallel repetition $G^{\otimes 2}$, see also Exercise 6.2.

Example 6.3. The verifier selects two integers x and y independently at random from $\{0, 1\}$ and sends x to prover P_1 and y to P_2. The provers have to reply with numbers u resp. v from $\{0, 1\}$. Moreover each of them must either point at itself or at the other prover. The verifier accepts if

1. both point at the same prover and

2. the prover which both point at replies with the number it was asked and

3. $u + v \equiv 0 \mod 2$.

For some years, it was an open conjecture whether parallel repetition is sufficent for every games G to reduce the error probability below arbitrarily small constants. This was proved by Verbitsky in [Ver94, Ver96] for the case where μ is a *uniform* measure supported by a subset of Ω. He pointed out that parallel repetition is connected to a density version of Hales-Jewett's theorem in Ramsey theory, proved by Furstenberg and Katznelson [FK91]. Unfortunately, the proof technique used gives no constructive bound on the number of required repetitions.

Recently, Raz [Raz95] showed that parallel repetition reduces the error probability of any MIP$(2,1)$ system even at an exponential rate, thus proving the so-called *Parallel Repetition Conjecture* [FL92]. The constant in the exponent (in his analysis) depends only on $w(G)$ and the *answer size* $s = s(G) := |U| \cdot |V|$ (of the original problem). His result is now known as the *Parallel Repetition Theorem*:

Theorem 6.4 (Parallel Repetition Theorem [Raz95]). *There exists a function* $W : [0,1] \to [0,1]$ *with* $W(z) < 1$ *for all* $z < 1$, *such that for all games* G *with value* $w(G)$ *and answer size* $s = s(G) \geq 2$ *and all* $k \in \mathbb{N}$ *holds*

$$w\left(G^{\otimes k}\right) \leq W\left(w(G)\right)^{k/\log s} .$$

For some applications, only the fact that parallel repetition reduces the error probability below arbitrarily small *constants* is used. Let us refer to this fact as the *weak parallel repetition theorem*.

In the meantime, Feige [Fei95] showed that Raz' theorem is nearly best possible. He proved that there is no universal constant $\alpha > 0$ such that

$$w\left(G^{\otimes k}\right) \leq w(G)^{\alpha k}$$

for all MIP$(2,1)$ proof systems. Hence, the exponent in Theorem 6.4 must depend on G. Its inverse logarithmic dependency on $s(G)$ was shown to be almost best possible by [FV96].

In the remaining sections of this chapter, we give an overview of the proof of the Parallel Repetition Theorem, based on [Raz95, Raz97, Fei95].

6.5 Coordinate Games

In the last section, we introduced $\bar{G}$ as the parallel repetition of G. In fact, $\bar{G}$ is best viewed as not being simply a repetition of the game G, but as a "simultaneous execution" of its *coordinates*, which will be defined next.

The *coordinate game* $\bar{G}^i$, where $i \in \{1, \ldots, k\}$, consists of:

– the sets $\bar{X}, \bar{Y}, U, V$;

– the probability measure $\bar{\mu}$;

– and the acceptance predicate $\bar{Q}^i$ defined by $\bar{Q}^i(\bar{x}, \bar{y}, u, v) := Q(x_i, y_i, u, v)$.

The value of $\bar{G}^i$ will be denoted by $w_i(\bar{G})$.

Thus, in the parallel repetition game $\bar{G}$ the verifier accepts if and only if all the predicates $\bar{Q}^i$, $i = 1, \ldots, k$, of the coordinate games $\bar{G}^i$ are satisfied. We will also say that $\bar{G}^i$ is the restriction of $\bar{G}$ to coordinate i. A protocol $\bar{u}, \bar{v}$ for the game $\bar{G}$ induces a protocol u_i, v_i for the game $\bar{G}^i$, where $u_i(\bar{x})$ and $v_i(\bar{y})$ are the i-th coordinates of the vectors $\bar{u}(\bar{x})$ resp. $\bar{v}(\bar{y})$.

Note that it is not required that the verifier will ask every question pair in $X \times Y$ with positive probability. In the proof of Theorem 6.4, we will consider restrictions of the game $\bar{G}$ to a question set $A \subseteq \bar{\Omega}$, which has the form of a cartesian product $A = A_X \times A_Y$, where $A_X \subseteq \bar{X}$ and $A_Y \subseteq \bar{Y}$. For $\mu(A) > 0$, we define

$$\bar{\mu}_A(\bar{x}, \bar{y}) := \begin{cases} \dfrac{\bar{\mu}(\bar{x}, \bar{y})}{\bar{\mu}(A)}, & (\bar{x}, \bar{y}) \in A; \\ 0, & (\bar{x}, \bar{y}) \notin A. \end{cases}$$

The game $\bar{G}_A$ is defined similarly to the game $\bar{G}$, but with the probability measure $\bar{\mu}_A$ instead of $\bar{\mu}$. The game $\bar{G}_A$ will be called the *restriction* of $\bar{G}$ to the question set A and is sometimes also denoted by $\bar{G}_{\bar{\mu}_A}$. If $\bar{\mu}(A) = 0$, we set $\bar{\mu}_A := 0$.

The following main technical theorem says that, provided $\bar{\mu}(A)$ is not "too" small, there always exists a coordinate i such that the value $w(\bar{G}_A^i)$ of the coordinate game $\bar{G}_A^i$ is not "too" large, compared with $w(G)$.

Theorem 6.5. *There exists a function $W_2 : [0,1] \to [0,1]$ with $W_2(z) < 1$ for all $z < 1$ and a constant c_0 such that the following holds. For all games G, dimensions k, and $A = A_X \times A_Y$, where $A_X \subseteq \bar{X}$ and $A_Y \subseteq \bar{Y}$, such that $-\log \bar{\mu}(A)/k \leqslant 1$ (i. e., $\bar{\mu}(A) \geqslant 2^{-k}$), there exists a coordinate i of $\bar{G}_A$ such that*

$$w(\bar{G}_A^i) \leqslant W_2\big(w(G)\big) + c_0 \left(\frac{-\log \bar{\mu}(A)}{k} \right)^{\frac{1}{16}}.$$

In Section 6.7 we will give some remarks on the proof of Theorem 6.5. But first let us see how Theorem 6.5 is applied in the proof of the Parallel Repetition Theorem 6.4

6.6 How to Prove the Parallel Repetition Theorem (I)

In order to prove Theorem 6.4, we will apply an inductive argument that yields a slightly stronger result. Let us define

$$C(k,r) := \max_{\substack{A=A_X \times A_Y \\ -\log \bar{\mu}(A) \leqslant r}} w(G_A^{\otimes k})$$

for $k \in \mathbb{N}$ and $r \in \mathbb{R}_+$. It will be convenient to define $C(0,r) := 1$. $C(k,r)$ is an upper bound for the value (i. e., error probability) of any restriction of k parallel repetitions of G to a set A, that has the form of a cartesian product $A = A_X \times A_Y$ and whose size $\bar{\mu}(A)$ is not very small, where "small" is quantified by r. Observe that $\bar{G} = \bar{G}_{\bar{\Omega}}$ and $\bar{\mu}(\bar{\Omega}) = 1$, so $C(k,r)$ is an upper bound for $w(\bar{G})$, and in particular, $C(k,0) = w(\bar{G})$.

For technical reasons, we will not apply Theorem 6.5, but the following re-statement of it.

Theorem 6.6. *There exists a function $W_2 : [0,1] \to [0,1]$ with $W_2(z) < 1$ for all $z < 1$ and a constant c_0 such that the following holds. For all games G, dimensions k, and $A = A_X \times A_Y$, where $A_X \subseteq \bar{X}$ and $A_Y \subseteq \bar{Y}$, such that for all Δ with $-\log \bar{\mu}(A)/k \leqslant \Delta \leqslant 1$ (i. e., $\bar{\mu}(A) \geqslant 2^{-\Delta k}$), there exists a coordinate i of $\bar{G}_A$ such that*

$$w(\bar{G}_A^i) \leqslant W_2\big(w(G)\big) + c_0 \Delta^{\frac{1}{16}}.$$

It will be necessary to choose Δ carefully such that certain assumptions made in the proof are satisfied.

Now we show how Raz derives Theorem 6.4 from Theorem 6.6. Let us denote the upper bound from Theorem 6.6 by

$$\hat{w} := W_2\big(w(G)\big) + c_0 \Delta^{\frac{1}{16}}.$$

and assume that $r \leqslant \Delta k$. We choose an $A = A_X \times A_Y \subseteq \bar{\Omega}$ with $-\log \bar{\mu}(A) \leqslant r$ that maximizes $w(\bar{G}_A)$, i. e., $C(k,r) = w(\bar{G}_A)$. (So $\bar{\mu}(A) > 0$.) By Theorem 6.6, we know that there must be a coordinate i (without loss of generality, we will assume that $i = 1$) whose value is not too big, namely $w(\bar{G}_A^i) \leqslant \hat{w}$. We will use this "good" coordinate in order to make an induction step in the following way. Recall that the value $w(\bar{G}_A)$ is defined as the maximal expected error probability with respect to the probability measure $\bar{\mu}_A$, taken over all protocols $\bar{u} : \bar{X} \to \bar{U}$, $\bar{v} : \bar{Y} \to \bar{V}$ the provers can agree upon. This means that an optimal protocol $\bar{u}$, $\bar{v}$ (for the provers) satisfies

$$w(\bar{G}_A) = \sum_{(\bar{x},\bar{y}) \in \bar{\Omega}} \bar{\mu}_A(\bar{x},\bar{y}) \, \bar{Q}\big(\bar{x},\bar{y},\bar{u}(\bar{x}),\bar{v}(\bar{y})\big).$$

Let us fix such an optimal protocol $\bar{u}$, $\bar{v}$. Next we partition A according to the behavior of $\bar{u}$ and $\bar{v}$ on the first coordinate. For every point $z = (x, y, u, v) \in Z$, we define a partition class $A(z) \subseteq A$ by

$$A(x, y, u, v) := \{(\bar{x}, \bar{y}) \in A \mid x_1 = x \wedge y_1 = y \wedge u_1(\bar{x}) = u \wedge v_1(\bar{y}) = v\}.$$

Note that the partition classes $A(z)$ again have the form of cartesian products as required for the application of Theorem 6.6. We denote the subset of questions of $A(z)$ such that the acceptance predicate $\bar{Q}$ is satisfied by

$$B(x, y, u, v) := \{(\bar{x}, \bar{y}) \in A(x, y, u, v) \mid \bar{Q}(\bar{x}, \bar{y}, \bar{u}(\bar{x}), \bar{v}(\bar{y})) = 1\}.$$

The sets $B(z)$ will be useful later. The size of the partition $\{A(z) \mid z \in Z\}$ is not too big, and therefore the average size of a partition class is not too small. Also, in many of these subsets $A(z)$ of A the protocol $\bar{u}$, $\bar{v}$ fails to satisfy the predicate $\bar{Q}^1$, because $i = 1$ is a good coordinate. If $\bar{Q}^1$ is not satisfied, then $\bar{Q}$ is also not satisfied, so we can forget about these subsets. This is a consequence of the fact that $\bar{Q}$ can be viewed as the product of $\bar{Q}^1$ and a $(k-1)$-dimensional predicate and that $\bar{Q}^1$ is constant on each $A(z)$. Let us deduce this fact formally. If $Q(z) = 0$, then $B(z) = \emptyset$, because $(\bar{x}, \bar{y}) \in A(z)$ implies

$$\bar{Q}(\bar{x}, \bar{y}, \bar{u}(\bar{x}), \bar{v}(\bar{y})) \leqslant Q(x_1, y_1, u_1(\bar{x}), v_1(\bar{y})) = Q(z) = 0.$$

If $Q(z) = 1$, then $(\bar{x}, \bar{y}) \in A(z)$ implies

$$\bar{Q}(\bar{x}, \bar{y}, \bar{u}(\bar{x}), \bar{v}(\bar{y})) = \underbrace{Q(z)}_{=1} [t] \prod_{j=2}^{k} Q(x_j, y_j, u_j(\bar{x}), v_j(\bar{y})).$$

These facts enable us to carry out an inductive argument.

We denote the conditional probability with which a random $(\bar{x}, \bar{y}) \in A$ that was chosen according to the distribution $\bar{\mu}_A$ lies in $A(z)$, depending on $z \in Z$, by

$$\alpha(z) := \bar{\mu}_A(A(z)) = \frac{\bar{\mu}(A(z))}{\bar{\mu}(A)}.$$

Of course, $\alpha(Z) = 1$, because $\{A(z) \mid z \in Z\}$ is a partition of A. Note that the event $(\bar{x}, \bar{y}) \in A(z)$ depends only on the first coordinate of $\bar{G}_A$, which is $(x_1, y_1, u_1(\bar{x}), v_1(\bar{y}))$. For some of the partition classes $A(z)$, the predicate $\bar{Q}^1(x_1, y_1, u_1(\bar{x}), v_1(\bar{y}))$ is not satisfied. Summing the α-probability of the other partition classes gives us the value of the first coordinate game. Thus the following claim holds.

Claim 6.1:

$$\alpha(Q) = w(\bar{G}_A^1) \leqslant \hat{w}.$$

This is the point where Theorem 6.6 is applied. But to get an assertion about the parallel repetition game $\bar{G}_A$, we also have to consider the probability that a random $(\bar{x}, \bar{y}) \in A(z)$ leads to acceptance in the game $\bar{G}_{A(z)}$. Let us denote this probability by

$$\beta(z) := \bar{\mu}_{A(z)}(\bar{Q}) = \frac{\bar{\mu}(B(z))}{\bar{\mu}(A(z))}.$$

If $\bar{\mu}(A(z)) = 0$, we set $\beta(z) := 0$. Now a moment's thought shows that the following claim is true.

<u>Claim 6.2:</u>

$$\sum_{z \in Q} \alpha(z)\,\beta(z) = w(\bar{G}_A) = C(k, r).$$

Next, we deduce an upper bound for $\beta(z)$. Observe that in fact, $A(z)$ and $B(z)$ are only $(k-1)$-dimensional sets, so it makes good sense to define

$$A'(z) := \left\{ ((x_2, \ldots, x_k), (y_2, \ldots, y_k)) \,\middle|\, (\bar{x}, \bar{y}) \in A(z) \right\}.$$

and

$$B'(z) := \left\{ ((x_2, \ldots, x_k), (y_2, \ldots, y_k)) \,\middle|\, (\bar{x}, \bar{y}) \in B(z) \right\}.$$

In a similar way, we define a protocol for the game $\bar{G}_{A'(z)}^{\otimes(k-1)}$ by

$$u'(x_2, \ldots, x_k) := \left(u_j(x, x_2, \ldots, x_k) \,\middle|\, j = 2, \ldots, k \right)$$

and

$$v'(y_2, \ldots, y_k) := \left(v_j(y, y_2, \ldots, y_k) \,\middle|\, j = 2, \ldots, k \right).$$

Since the predicate Q^{k-1} of the game $G_{A'(z)}^{\otimes(k-1)}$ is satisfied precisely at the elements of $B'(z)$ and the protocol $\bar{u}, \bar{v}$ was chosen optimal for $\bar{G}_A$, it follows that

$$w\left(G_{A'(z)}^{\otimes(k-1)} \right) \geqslant \frac{\mu^{\otimes(k-1)}(B'(z))}{\mu^{\otimes(k-1)}(A'(z))} = \frac{\bar{\mu}(B(z))}{\bar{\mu}(A(z))} = \beta(z).$$

A trivial upper bound for $w\left(G_{A'(z)}^{\otimes(k-1)} \right)$ that holds by definition is

$$w\left(G_{A'(z)}^{\otimes(k-1)} \right) \leqslant C\left(k-1, \; -\log \mu^{\otimes(k-1)}(A'(z)) \right).$$

Also, we have

$$\mu^{\otimes(k-1)}(A'(z)) = \frac{\bar{\mu}(A(z))}{\mu(x,y)} = \frac{\alpha(z)\,\bar{\mu}(A)}{\mu(x,y)} \geqslant 2^{-r}\,\frac{\alpha(z)}{\mu(x,y)}.$$

If we put these inequalities together, using the fact that the function $C(\cdot, \cdot)$ is monotone increasing in the second argument, we see that the following claim is true.

Claim 6.3:

For all $z = (x, y, u, v) \in Q$ with $\alpha(z) > 0$,

$$\beta(z) \leqslant C\left(k - 1, \; r - \log \frac{\alpha(z)}{\mu(x, y)}\right).$$

The claims 6.1, 6.2, and 6.3 lead to a recursive inequality for C. Let $T := \{z \in Q \mid \alpha(z) > 0\}$ be the set of all realizations of the coordinate game $\tilde{G}_A^1$ that occur with positive probability. Then it holds

$$C(k, r) = \sum_{z \in T} \alpha(z) \, \beta(z) \leqslant \sum_{z \in T} \alpha(z) \, C\left(k - 1, \; r - \log \frac{\alpha(z)}{\mu(x, y)}\right).$$

It remains to carry out a recursive estimation, using the bound for $C(k - 1, \cdot)$ to prove the bound for $C(k, \cdot)$. Define

$$c := \hat{w}^{\frac{1}{2(\log s + \Delta)}}$$

for abbreviation. (So c is a function of $w(G)$, $s(G)$, and Δ.) The heart of the proof of the following claim is a clever application of Jensen's inequality.

Claim 6.4:

$$C(k, r) \leqslant c^{\Delta k - r}.$$

provided $0 \leqslant \hat{w} < 1$ and $\frac{1}{\sqrt{2}} < c < 1$.

Since $0 < W_2\big(w(G)\big) < 1$ and $\hat{w}$ is monotone in Δ, we can find an appropriate Δ that satisfies the assumptions we made during the proof by starting from $\Delta = 0$ and increasing Δ until the conditions hold.

Finally, we show how Theorem 6.4 follows from claim 6.4. Note that $\Delta < 2 \log s$ (because $\Delta < 1$ and $s > 2$). Therefore, under the assumptions made above,

$$w(\bar{G}) = C(k, 0) \leqslant c^{\Delta k} = \hat{w}^{\frac{1}{2(\log s + \Delta)} \Delta k} \leqslant \hat{w}^{\frac{1}{4 \log s} \Delta k} = \left(\hat{w}^{\Delta/4}\right)^{k/\log s}.$$

So Theorem 6.4 holds with

$$W\big(w(G)\big) := \inf\left(\hat{w}^{\Delta/4}\right),$$

where the infimum is taken over all Δ satisfying the assumptions.

6.7 How to Prove the Parallel Repetition Theorem (II)

In this section we give an overview of basic ideas and techniques that are used in the proof of Theorem 6.6. We do not try to explain how to get the exact quantitative result of the theorem but rather aim to motivate the qualitative statement. The entire proof covers more than 30 pages and can be found in an extended version [Raz97] of [Raz95]. Our description is also based on the short overview of the proof given by Feige in [Fei95]. In particular we use Feige's notion of "good" coordinates and their characterization by certain properties, see Properties 1–3 below.

For the following considerations we keep X, Y, U, V, Q, μ, k and the corresponding game G fixed. Moreover, we will consider games consisting of X, Y, U, V, Q resp. $\bar{X}, \bar{Y}, \bar{U}, \bar{V}, \bar{Q}$ together with probability distributions other than μ resp. $\bar{\mu}$. For arbitrary probability measures $\alpha : \Omega \to \mathbb{R}$ and $\bar{\alpha} : \bar{\Omega} \to \mathbb{R}$ we denote the corresponding games by G_α resp. $\bar{G}_{\bar{\alpha}}$. Thus, we can denote the restricted game $\bar{G}_A$ alternatively by $\bar{G}_{\bar{\mu}_A}$ where $A = A_X \times A_Y$ is a fixed subset of $\bar{\Omega}$. To simplify notation we denote the probability measure $\bar{\mu}_A$ by $\bar{\pi}$.

We think of w as a function from the set of all probability measures on Ω resp. $\bar{\Omega}$ to $\mathbb{R}$ and denote the values of the games G_α and $\bar{G}_{\bar{\alpha}}$ by $w(\alpha)$ resp. $w(\bar{\alpha})$. In the following lemma we state two basic properties of the function w. The proof is left to the reader, see exercise 6.5.

Lemma 6.7. *The function w is continuous and concave.*

For $\bar{\alpha} : \bar{\Omega} \to \mathbb{R}$ define $\bar{\alpha}^i$ to be the projection of $\bar{\alpha}$ on the i-th coordinate, i. e., for $(x, y) \in \Omega$ let

$$\bar{\alpha}^i(x, y) := \sum_{\substack{(\bar{x}, \bar{y}) \in \bar{\Omega}: \\ (x_i, y_i) = (x, y)}} \bar{\alpha}(\bar{x}, \bar{y}) .$$

In particular $\bar{\alpha}^i$ is a probability measure on Ω. We will consider the following games:

- The game $\bar{G}_{\bar{\alpha}}$ as introduced above consisting of $\bar{X}, \bar{Y}, \bar{U}, \bar{V}, \bar{Q}$, with the measure $\bar{\alpha}$ and value $w(\bar{\alpha})$.

- The coordinate game $\bar{G}_{\bar{\alpha}}^i$ of $\bar{G}_{\bar{\alpha}}$ as defined in Section 6.5 consisting of the four sets $\bar{X}, \bar{Y}, U, V$, with the acceptance predicate $\bar{Q}^i$ and with the measure $\bar{\alpha}$. We denote the value of this game by $w_i(\bar{\alpha})$.

- The one-dimensional game $G_{\bar{\alpha}^i}$ induced by the measure $\bar{\alpha}^i$ on X, Y, U, V, Q with value $w(\bar{\alpha}^i)$.

Theorem 6.6 claims that for "large" $A = A_X \times A_Y$ (with respect to the measure $\bar{\mu}$) there exists a coordinate i such that the provers succeed in the corresponding coordinate game $\bar{G}_{\bar{\pi}}^i$ with probability "not much higher" than $w(G)$. We keep such a large subset A fixed for the rest of this section. To prove Theorem 6.6 Raz considers coordinates that satisfy certain properties which lead to the property stated in the theorem. As proposed by Feige we call coordinates with these properties "good". In the rest of this section we characterize good coordinates in an informal way.

A natural requirement on a good coordinate i seems to be:

Property 1: The projection $\bar{\pi}^i$ of $\bar{\pi}$ on coordinate i is very close to the original probability distribution μ.

To be more precise we say that $\bar{\pi}^i$ is close to μ if the *informational divergence* $D(\bar{\pi}^i \| \mu)$ of $\bar{\pi}^i$ with respect to μ is small. The informational divergence (or *relative entropy*) is a basic tool of information theory and is defined by

$$D(\bar{\pi}^i \| \mu) := \sum_{z \in \Omega} \bar{\pi}^i(z) \log \frac{\bar{\pi}^i(z)}{\mu(z)} .$$

The informational divergence is always non-negative and if it is small then the L_1 distance between the two measures is also small. A short discussion of basic properties can be found in [Raz95], for further information see [Gra90, CK81].

It is easy to show that for large A Property 1 holds for many coordinates i. As a consequence of Lemma 6.7 we know that the value $w(\bar{\pi}^i)$ of the game $G_{\bar{\pi}^i}$ is not much larger than $w(\mu)$ if the coordinate i satisfies Property 1. Thus it suffices to show that for one such coordinate i, the value $w_i(\bar{\pi})$ of the coordinate game $\bar{G}_{\bar{\pi}}^i$ is bounded by some "well behaved" function of $w(\bar{\pi}^i)$. On the other hand it is always true that

$$w(\bar{\pi}^i) \leqslant w_i(\bar{\pi}) , \tag{6.1}$$

because any protocol for the one-dimensional game $G_{\bar{\pi}^i}$ induces in a canonical way a protocol with the same value for the coordinate game $\bar{G}_{\bar{\pi}}^i$. Remember that in the one-dimensional game $G_{\bar{\pi}^i}$ on X, Y, U, V each prover is asked only one question. However, in the game $\bar{G}_{\bar{\pi}}^i$, each prover gets k questions $\bar{x}$ resp. $\bar{y}$ but only has to answer the i-th question x_i resp. y_i. Thus, ignoring all questions but the i-th, the provers of the game $\bar{G}_{\bar{\pi}}^i$ can apply the optimal strategy of the one-dimensional game $G_{\bar{\pi}^i}$. Moreover, since by definition of the projection $\bar{\pi}^i$ the probability distribution of questions (x, y) in the one-dimensional game exactly equals the distribution of the i-th question (x_i, y_i) in the game $\bar{G}_{\bar{\pi}}^i$, the value of this protocol is the same for both games.

The following lemma describes a special case where an optimal protocol for the one-dimensional game $G_{\bar{\pi}^i}$ is also optimal for the game $\bar{G}_{\bar{\pi}}^i$ (see [Raz95, Lemma 4.1]).

Lemma 6.8. *If there exist functions $\alpha_1 : X \times Y \to \mathbb{R}$, $\alpha_2 : \bar{X} \to \mathbb{R}$, $\alpha_3 : \bar{Y} \to \mathbb{R}$ such that for all $(\bar{x}, \bar{y}) \in \bar{X} \times \bar{Y}$*

$$\bar{\pi}(\bar{x}, \bar{y}) = \alpha_1(x_i, y_i)\, \alpha_2(\bar{x})\, \alpha_3(\bar{y})$$

then

$$w(\bar{\pi}^i) = w_i(\bar{\pi}) .$$

Sketch of Proof. Because of (6.1), it remains to show that $w(\bar{\pi}^i) \geqslant w_i(\bar{\pi})$. Inverting the argument given above we have to show that a protocol for $\bar{G}^i_{\bar{\pi}}$ can be used to define a corresponding protocol with the same value for $G_{\bar{\pi}^i}$. The basic idea is that in this special case the provers of the game $G_{\bar{\pi}^i}$ can simulate a k-dimensional input $(\bar{x}, \bar{y})$ with distribution $\bar{\pi}$ from the given input (x_i, y_i) with distribution $\bar{\pi}^i$. Thus, Exercise 6.4 completes the proof. ∎

Corollary 6.9. *If μ is a product measure, i. e.,*

$$\mu(x, y) = \beta(x)\, \gamma(y)$$

where $\beta : X \to \mathbb{R}$ and $\gamma : Y \to \mathbb{R}$ are arbitrary probability measures, then Property 1 suffices to define good coordinates and to prove Theorem 6.6 for this special case.

Proof. If μ is a product measure then the same holds for $\bar{\pi}$ by definition. Moreover, it is easy to see that the condition of Lemma 6.8 is satisfied in this case. This, together with the remarks after Property 1, completes the proof of Theorem 6.6. ∎

Unfortunately, in general we do not get an optimal protocol for $\bar{G}^i_{\bar{\pi}}$ by taking one for $G_{\bar{\pi}^i}$. The reason is that in the game $\bar{G}^i_{\bar{\pi}}$ a prover can loose important information by ignoring all but the i-th question. Recall that in the game $\bar{G}$ with measure $\bar{\mu}$ the question pairs selected at different coordinates are independent. This is no longer true in the restricted game $\bar{G}_{\bar{\pi}}$ with measure $\bar{\pi}$. Thus the question which a prover receives on a coordinate j different from i may already provide information on its i-th question. Feige defines the side information for coordinate i as the questions a prover receives on coordinates other than i.

The situation gets even worse if the question a prover receives in coordinate j is correlated with the question that the other prover receives on this coordinate. Then the side information for coordinate i may help him to guess the i-th question of the other prover. As a consequence the side information can obviously help the provers to succeed on coordinate i. Thus a good coordinate should not only meet Property 1, in addition we should require that the side information for good coordinates are in a way useless:

Property 2: The side information for coordinate i available to each prover conveys almost no information on the question that the other prover receives on this coordinate (beyond the direct information available to the prover through its own question on coordinate i and the description of the underlying probability measure used by the verifier).

Unfortunately, even this condition does not suffice to define good coordinates and to get the desired result. Consider the following example which is given in [Fei95]:

Example 6.10. Define a game G by $X = Y = U = V = \{0,1\}$. For $x \in X, y \in Y, u \in U$, and $v \in V$ the acceptance predicate Q is given by $x \wedge y = u \oplus v$ (where $\oplus$ denotes *exclusive or*). The probability distribution μ is defined by $\mu(x,y) = 1/4$ for all $(x,y) \in X \times Y$, i.e., the two questions are drawn independently from each other and all choices are equally likely.

This game is not trivial, i.e., its value is not 1, see Exercise 6.1. However, for the game $\bar{G} = G^{\otimes 2}$ consider the subset A of $X^2 \times Y^2$ that includes eight question pairs, written as $(x_1 x_2, y_1 y_2)$:

$$A = \{(00,00),(01,01),(00,10),(01,11),(10,00),(11,01),(10,11),(11,10)\}$$

It is an easy observation that the projection of A on its first coordinate gives the uniform distribution over question pairs (x,y). Moreover, the question y_1 is independent of the two questions $x_1 x_2$ and x_1 is independent of $y_1 y_2$, see [Fei95, Proposition 4]. Thus Properties 1 and 2 hold for coordinate 1. Nevertheless, there exists a perfect strategy for the provers on the first coordinate. Just observe that $x_1 \wedge y_1 = x_2 \oplus y_2$ for all question pairs in A. Thus we get a perfect strategy for the first coordinate if the first prover gives the answer x_2 while the second prover gives the answer y_2.

Example 6.10 seems to contradict the statement of Corollary 6.9. The reason is that the set A given above cannot be written as $A_X \times A_Y$ with $A_X \subseteq X^2$ and $A_Y \subseteq Y^2$. Nevertheless, the game G can be extended to a new game G' which, together with a subset $A' = A'_X \times A'_Y$, shows that Property 2 together with Property 1 does not suffice for a coordinate i to meet the property claimed in Theorem 6.6.

To overcome this drawback one has to substitute Property 2 by a somewhat stronger condition on good coordinates. The idea is to reduce the problem in a way to probability measures as considered in Lemma 6.8. One represents $\bar{\pi}$ as a convex combination of measures that satisfy the condition of Lemma 6.8. Therefore we need the following definition which Raz considers as "probably the most important notion" for his proof.

Definition 6.11. *For a fixed coordinate $i \in \{1, \ldots, k\}$ a set M of type $\mathcal{M}^i$ is given by*

1. *a partition of the set of coordinates $\{1, \ldots, k\} - \{i\}$ into $J \cup L$ and*

2. *values $a_j \in X$, for all $j \in J$, and $b_\ell \in Y$, for all $\ell \in L$.*

Then M is given by

$$M = \{(\bar{x}, \bar{y}) \in \bar{X} \times \bar{Y} \mid x_j = a_j \text{ for all } j \in J, y_\ell = b_\ell \text{ for all } \ell \in L\}.$$

$\mathcal{M}^i$ denotes the family of all sets M of type $\mathcal{M}^i$.

As a simple application of Lemma 6.8 one can prove that

$$w_i(\bar{\pi}_M) = w(\bar{\pi}_M^i) \tag{6.2}$$

for a set M of type $\mathcal{M}^i$. Moreover, the probability measure $\bar{\pi}$ induces in a natural way a measure $\rho_i : \mathcal{M}^i \to \mathbb{R}$ by

$$\rho_i(M) = \frac{\bar{\pi}(M)}{2^{k-1}} .$$

Since $\mathcal{M}^i$ is a cover of $\bar{X} \times \bar{Y}$ and each element $(\bar{x}, \bar{y})$ is covered exactly 2^{k-1} times, ρ_i is in fact a probability measure for $\mathcal{M}^i$ and we can write $\bar{\pi}$ as the convex combination

$$\bar{\pi} = \sum_{M \in \mathcal{M}^i} \rho_i(M) \bar{\pi}_M \tag{6.3}$$

At this point we can imagine the importance of the set $\mathcal{M}^i$. It finally enables us to write the probability measure $\bar{\pi}$ as the convex combination (6.3) of measures $\bar{\pi}_M$ with the nice property (6.2).

Moreover, we can now bound the value $w_i(\bar{\pi})$ of the coordinate game $\bar{G}_{\bar{\pi}}^i$. First note that a protocol for $\bar{G}_{\bar{\pi}}^i$ is also a protocol for each game $\bar{G}_{\bar{\pi}_M}^i$. Thus the concavity of the function w_i (see Exercise 6.5) yields

$$w_i(\bar{\pi}) \leqslant \sum_{M \in \mathcal{M}^i} \rho_i(M) w_i(\bar{\pi}_M) .$$

The right hand side of this inequality can be interpreted as the expectation $\mathbf{E}_{\rho_i}(w_i(\bar{\pi}_M))$. This together with equation (6.2) yields the upper bound

$$w_i(\bar{\pi}) \leqslant \mathbf{E}_{\rho_i}(w(\bar{\pi}_M^i)) .$$

We need the following notation. For a probability measure $\alpha : X \times Y \to \mathbb{R}$, define $\alpha(x, \cdot) : Y \to \mathbb{R}$ to be the induced measure on Y for fixed $x \in X$, i.e., for $y \in Y$

$$\alpha(x, \cdot)(y) = \frac{\alpha(x, y)}{\alpha(x)}$$

where $\alpha(x) = \sum_{y \in Y} \alpha(x, y)$. If $\alpha(x) = 0$ set $\alpha(x, \cdot)$ to be identically 0. Define $\alpha(\cdot, y)$ and $\alpha(y)$ in the same way.

The following lemma is a qualitative version of [Raz95, Lemma 4.3]. It defines a condition on coordinate i that suffices to upper bound $\mathbf{E}_{\rho_i}(w(\bar{\pi}_M^i))$.

Lemma 6.12. *Let coordinate i satisfy Property 1. If*

$$\mathbf{E}_{\rho_i}\left(\sum_{x \in X} \bar{\pi}_M^i(x) \, \mathbf{D}\left(\bar{\pi}_M^i(x, \cdot) \,\|\, \mu(x, \cdot) \right) \right)$$

$$\text{and} \quad \mathbf{E}_{\rho_i}\left(\sum_{y \in Y} \bar{\pi}_M^i(y) \, \mathbf{D}\left(\bar{\pi}_M^i(\cdot, y) \,\|\, \mu(\cdot, y) \right) \right) \tag{6.4}$$

are "small" then $\mathbf{E}_{\rho_i}(w(\bar{\pi}_M^i))$ is upper bounded by some well behaved function of $w(\mu)$.

We will neither give the proof for Lemma 6.12 nor the proof for the existence of such a good coordinate (see [Raz95, Section 5]). Instead we give a more intuitive formulation for the condition in Lemma 6.12 similar to the one given by Feige.

Consider a fixed $M \in \mathcal{M}^i$ with corresponding partition $\{1, \ldots, k\} - \{i\} = J \cup L$ and questions $a_j \in X$ for $j \in J$ and $b_\ell \in Y$ for $\ell \in L$. The term $\mathbf{D}\left(\bar{\pi}^i_M(x, \cdot) \| \mu(x, \cdot)\right)$ can be interpreted as a measure for the information that the first prover can get on the i-th question of the other prover from the knowledge of the questions $x_j = a_j$ for $j \in J$ and $y_\ell = b_\ell$ for $\ell \in L$. Using a notion of Feige we call this information the extended side information for coordinate i with regard to M. Thus the condition given in Lemma 6.12 can be reformulated as:

Property 3: On an average (with regard to ρ_i), the extended side information for coordinate i conveys almost no additional information on the question that the other prover receives on coordinate i.

As mentioned above with Properties 1 and 3 defining good coordinates, Raz showed that for large A at least one good coordinate i exists and the value of the coordinate game $\bar{G}^i_{\bar{\pi}}$ is bounded as claimed in Theorem 6.6.

Exercises

Exercise 6.1. What is the value of the game given in Example 6.10?

Exercise 6.2. What is the value $w(G)$ of the game G given in Example 6.3? Show that 2-fold parallel repetition does not reduce the error probability, i.e., $w(G^{\otimes 2}) = w(G)$.

Exercise 6.3. Multiple-prover games are a natural extension of two prover games. Can you generalize the game given in Example 6.3 to the case of k provers, for $k > 2$, such that $w(G^{\otimes k}) = w(G)$?

Exercise 6.4. Can the value of a game be increased by allowing the provers to use random strategies?

Exercise 6.5. For fixed X, Y, U, V, Q, consider the value $w(G)$ of the game G as a function $w(\mu)$ of the measure $\mu : X \times Y \to [0, 1]$. Show that this function is concave and continuous with Lipschitz constant 1.

7. Bounds for Approximating MaxLinEq3-2 and MaxEkSat

Sebastian Seibert, Thomas Wilke

7.1 Introduction

As we have seen in Chapter 3, Theorems 3.3 and 3.4, there are polynomial time approximation algorithms for MaxEkSat and MaxLinEq3-2 with performance ratio at most $1 + 1/(2^k - 1)$ and 2, respectively. In this chapter, we show that this is the best possible, provided $\mathcal{P} \neq \mathcal{NP}$. In particular, we give proofs for the following two theorems by Johan Håstad [Hås97b]:

Theorem 7.1. *Let $\varepsilon > 0$ small. There is no polynomial-time algorithm which distinguishes instances of LinEq3-2 where at least $1 - \varepsilon$ of the equations can be satisfied at the same time from instances where at most $1/2 + \varepsilon$ of the equations can be satisfied at the same time, unless $\mathcal{P} = \mathcal{NP}$.*

Observe that by contrast LinEq3-2 itself is in $\mathcal{P}$.

Theorem 7.2. *Let $k \geqslant 3$ and $\varepsilon > 0$ small. There is no polynomial-time algorithm which distinguishes satisfiable instances of EkSat from instances where at most a $1 - 2^{-k} + \varepsilon$ fraction of the clauses can be satisfied at the same time, unless $\mathcal{P} = \mathcal{NP}$.*

For showing that the approximation factor $1/(1 - 2^{-k})$ of the algorithm from Chapter 3 is optimal (unless $\mathcal{P} = \mathcal{NP}$), it would be sufficient to proof a weaker claim than Theorem 7.2. For this purpose it suffices to consider the distinction of EkSat instances where at least $1 - \varepsilon$ of the clauses can be satisfied at the same time from instances where at most a $1 - 2^{-k} + \varepsilon$ of the clauses can be satisfied at the same time. This is, in fact, much easier to prove, as we will see in Theorem 7.11.

However, the above version states (in case $k = 3$) also an optimal result for $\mathcal{NP}$-hardness of ε-RobE3Sat, the problem of distinguishing satisfiable E3Sat instances from those where at most a fraction of $1 - \varepsilon$ of the clauses can be satisfied at the same time, for a fixed $\varepsilon > 0$. For this problem, Theorem 7.2 pushes ε, i.e. the portion of unsatisfiable clauses, arbitrarily close to $1/8$. Note that the polynomial time $8/7$-approximation algorithm for MaxE3Sat implies that in case of at least 1/8 unsatisfiable clauses the corresponding decision problem is in $\mathcal{P}$.

The reader is referred to [Hås97b] for a proof that p is an optimal lower bound for the performance ratio of the version of MAxLINEQ3-2 where one works over $\mathbb{F}_p$ (p prime) instead of $\mathbb{F}_2$. The matching upper bounds can be obtained just as in the case of $\mathbb{F}_2$. Explanatory notes about the proof of Theorem 7.1 can be found in [SW97].

For the hardness of the versions of MAxLINEQ3-2 and MAXE*k*SAT where each equation respectively clause occurs at most once, see Chapter 8 and, in particular, Theorem 8.17.

7.2 Overall Structure of the Proofs

Put into one sentence, the proofs of Theorems 7.1 and 7.2 are reductions from ε-ROBE3SAT-b, a further restricted version of ε-ROBE3SAT. Here the condition is added that for each instance φ any variable from VAR(φ) occurs exactly b times in φ.

By Lemma 4.9, we know that this problem is $\mathcal{NP}$-hard for certain constants ε and b.

Note that the proof of Theorem 4.5 gives only a very small constant ε ($\varepsilon = 1/45056$ for the less restricted problem ROBE3SAT), whereas Theorem 7.2 gives $\varepsilon = 1/8 - \varepsilon'$ for arbitrarily small $\varepsilon' > 0$ (and this even for ε-ROBE3SAT). We will state this formally in Corollary 7.16.

The reductions are extracted from suitably chosen verifiers for ε-ROBE3SAT. Proving good soundness properties of these verifiers will be our main task. In fact, this will involve the weak parallel repetition theorem from Chapter 6 and a technically quite involved analysis via Fourier transforms. Conceptually, the verifiers are fairly simple as they essentially implement a parallel version of the MIP from Example 6.2. What will be new in these verifiers is an encoding of satisfying assignments using the so-called "long code" as introduced in [BGS95].

The proofs of Theorems 7.1 and 7.2 are very similar not only in structure but also in many details. We will therefore have an introductory section following this one where common groundwork is laid, including the definition of long code, a description of the basic tests the verifiers will use, a review of what Fourier transforms are about, and some technical lemmas. The connection between the soundness of the verifiers and the MIP from Example 6.2 (including the connection to the weak parallel repetition theorem) will be explained in Section 7.4.

The presentation we give is based on the proofs provided in [Hås97b].

7.3 Long Code, Basic Tests, and Fourier Transform

The verifiers to be constructed in the proofs of Theorems 7.1 and 7.2 will assume that a proof presented to them consists, roughly speaking, of a satisfying assignment for the input formula, encoded in what is called the long code. This encoding scheme will be explained first in this section. When checking whether a proof is in fact the encoding of a satisfying assignment the verifiers will make use of different tests all of which are very similar. The coarse structure of these tests is explained second in this section. The main technical tool used to show good soundness properties of the verifiers are Fourier transforms. A short review of Fourier transforms and some of their properties is the third issue in this section.

In the following, $\mathbf{B}$ denotes the set $\{-1, 1\}$. We will use the term bit string for strings over $\mathbf{B}$. This has technical reasons: when working with Fourier transforms, strings over $\mathbf{B}$ come in more handy than strings over $\mathbb{F}_2$ or strings over $\{\text{TRUE}, \text{FALSE}\}$. When passing from $\mathbb{F}_2$ to $\mathbf{B}$, 0 is mapped to 1, 1 is mapped to -1, and addition becomes multiplication. Similarly, when passing from $\{\text{TRUE}, \text{FALSE}\}$ to $\mathbf{B}$, FALSE is mapped to 1, TRUE is mapped to -1, conjunction becomes maximum, and disjunction becomes minimum.

7.3.1 Long Code

Let M be a finite set of cardinality m. The long code of a mapping $x: M \to \mathbf{B}$ spells out the value that each function with domain $\mathbf{B}^M$ and range $\mathbf{B}$ takes at x, which is formalized in the following.

The long code of a mapping $x: M \to \mathbf{B}$ is the function $A: \mathbf{B}^{\mathbf{B}^M} \to \mathbf{B}$ defined by $A(f) = f(x)$. Observe that by choosing an order on $\mathbf{B}^{\mathbf{B}^M}$, each function $\mathbf{B}^{\mathbf{B}^M} \to \mathbf{B}$ can be identified with a bit string of length 2^{2^m}.

Let $f_1, \ldots, f_s$ be arbitrary functions $\mathbf{B}^M \to \mathbf{B}$ and b an arbitrary function $\mathbf{B}^s \to \mathbf{B}$. If A is the long code of $x: M \to \mathbf{B}$, then, by associativity,

$$A(b(f_1(x), \ldots, f_s(x))) = b(A(f_1(x)), \ldots, A(f_s(x))) \qquad (7.1)$$

for all $x \in \mathbf{B}^M$.

Using this identity, one can actually check (probabilistically) whether a given bit string of length 2^{2^m} is in fact the long code of some function $x: M \to \mathbf{B}$, see Chapter 9 and, in particular, Lemma 9.7. In the present chapter, (7.1) is used in a much more restricted way. Here, we need to be able to view a test as a linear equation in three variables over $\mathbb{F}_2$ (for MAXLINEQ3-2) or as a disjunction of three variables (for MAXE3SAT). In the first case, we will therefore use only $(x_1, x_2, x_3) \mapsto x_1 \cdot x_2 \cdot x_3$ for b (recall that we are working over $\mathbf{B}$ instead of $\mathbb{F}_2$, thus $\cdot$ corresponds to $+$), and in the second case, we will use only

$(x_1, x_2, x_3) \mapsto \min(x_1, x_2, x_3)$ for b (where the minimum in $\mathbf{B}$ corresponds to the Boolean expression $x_1 \vee x_2 \vee x_3$).

When we take $x \mapsto -x$ for b, then (7.1) assumes

$$A(-f) = -A(f). \tag{7.2}$$

So given either one of the values $A(f)$ or $A(-f)$, we know what the other value should be. Let us rephrase this. Let $\mathbf{1} : M \to \mathbf{B}$ map each argument to 1. Then, by (7.2), A is completely determined by the restriction of A to the set of all functions f mapping $\mathbf{1}$ to 1.

There is another property of the long code that will play a role in the next subsection too. Suppose ψ is a formula and B is the long code of some satisfying assignment $y_0 : \mathbf{Var}(\psi) \to \mathbf{B}$ (where we adopt the above convention about how to interpret the values of $\mathbf{B}$ as logical constants). Let $h_\psi : \mathbf{B}^{\mathbf{Var}(\psi)} \to \mathbf{B}$ be defined by

$$h_\psi(y) = -1 \quad \text{iff} \quad y \models \psi \tag{7.3}$$

for every $y : \mathbf{Var}(\psi) \to \mathbf{B}$. Then

$$B(g) = g(y_0) = \max(g, h_\psi)(y_0) = B(\max(g, h_\psi)) \tag{7.4}$$

for every $g : \mathbf{B}^{\mathbf{Var}(\psi)} \to \mathbf{B}$. Thus B is completely determined by the restriction of B to the set of all functions g where $\max(g, h_\psi) = g$.

7.3.2 Proof Format

According to its definition in chapter 4, a proof (to be presented to a verifier) is just a bit string or, to be more precise, the inscription of a designated proof tape with access via an oracle. As we have seen, it is, however, often enough very convenient and intuitive to think of a proof as (an encoding of) a mathematical object such as a satisfying assignment. Access to this mathematical object is then interpreted as retrieving bits from the proof tape according to a certain "access convention." In this chapter, proofs will be viewed as specific collections of functions; this is explained first in this subsection. A description of the appropriate access convention then follows.

Fix a parameter v and assume $\varphi = C_1 \wedge C_2 \wedge \ldots \wedge C_m$ is an ε-RoBE3SAT instance.

We write $\binom{\mathbf{Var}(\varphi)}{v}$ for the set of all subsets of $\mathbf{Var}(\varphi)$ of cardinality v, and, similarly, $\binom{\varphi}{v}$ for the set of all formulae of the form $C_{i_1} \wedge C_{i_2} \wedge \ldots \wedge C_{i_v}$ where no variable occurs twice and $1 \leqslant i_1 < i_2 < \ldots < i_v \leqslant m$. The first condition is important; we want to ensure that the sets of variables of the individual clauses are disjoint.

A *proof* π is a collection

$$\left((A^V)_{V \in \binom{\mathbf{Var}(\varphi)}{v}}, (B^\psi)_{\psi \in \binom{\varphi}{v}} \right)$$

where each A^V is a function $\mathbf{B}^{\mathbf{B}^V} \to \mathbf{B}$ and each B^ψ is a function $\mathbf{B}^{\mathbf{B}^{\mathbf{Var}(\psi)}} \to \mathbf{B}$. Moreover, we require

$$
\begin{aligned}
A^V(-f) &= -A^V(f), & (7.5) \\
B^\psi(-g) &= -B^\psi(g), & (7.6) \\
B^\psi(g) &= B^\psi(\max(g, h_\psi)), & (7.7)
\end{aligned}
$$

for all $V \in \binom{\mathbf{Var}(\varphi)}{v}$, $\psi \in \binom{\varphi}{v}$, $f : \mathbf{B}^V \to \mathbf{B}$, and $g : \mathbf{B}^{\mathbf{Var}(\psi)} \to \mathbf{B}$.

Clearly, an arbitrary bit string – if interpreted in a straightforward way as a collection of functions as specified above – does not necessarily meet these requirements. Therefore, the verifier accesses the proof tape according to the following convention.

Let $p_1, p_2, \ldots$ be an efficient (i.e., polynomial-time) enumeration of all tuples

– (V, f) with $V \in \binom{\mathbf{Var}(\varphi)}{v}$ and $f : \mathbf{B}^V \to \mathbf{B}$ where $f(1) = 1$, and

– (ψ, g) with $\psi \in \binom{\varphi}{v}$ and $g : \mathbf{B}^{\mathbf{Var}(\psi)} \to \mathbf{B}$ where $g(1) = 1$ and $g = \max(g, h_\psi)$.

When a verifier wants to read bit $A^V(f)$ from the proof, it proceeds as follows. If $f(1) = 1$, it computes i such that $p_i = (V, f)$, asks the oracle for the i-th bit of the proof tape, and uses this bit as the value for $A^V(f)$. If $f(1) = -1$, it computes i such that $p_i = (V, -f)$, asks the oracle for the i-th bit of the proof tape, and uses the negated value as the value for $A^V(f)$.

Similarly, when a verifier wants to read bit $B^\psi(g)$ from the proof, it proceeds as follows. If $\max(g, h_\psi)(1) = 1$, it computes i such that $p_i = (\psi, \max(g, h_\psi))$, asks the oracle for the i-th bit of the proof tape, and uses this bit as the value for $B^\psi(g)$. If $\max(g, h_\psi)(1) = -1$, it computes i such that $p_i = (\psi, -\max(g, h_\psi))$, asks the oracle for the i-th bit of the proof tape, and uses the negated value as the value for $B^\psi(g)$.

This makes sure (7.5) – (7.7) are satisfied.

7.3.3 Basic Tests

The verifiers will, presented with a formula φ as above and a proof π, perform certain tests. Each of these tests follows the same pattern and has two parameters: a positive integer v and a small constant $\delta > 0$. The parameter v will not be explicit in our notation, as no confusion can arise. In each of the tests the verifier first chooses

– a formula $\psi \in \binom{\varphi}{v}$,

- a set $V \in \binom{\mathbf{Var}(\psi)}{v}$, and

- functions $f: \mathbf{B}^V \to \mathbf{B}$ and $g_1, g_2: \mathbf{B}^W \to \mathbf{B}$.

It then reads three bits from π: $A^V(f)$, $B^\psi(g_1)$, and $B^\psi(g_2)$, and decides on the basis of these bits whether or not the test has been successful. The values ψ, V, f, g_1, and g_2 are chosen according to one of two distributions, which are introduced in the following.

Write T for the set of all pairs (ψ, V) where $\psi \in \binom{\varphi}{u}$ and $V \in \binom{\mathbf{Var}(\psi)}{v}$, and write Ω for the set of all tuples (ψ, V, f, g_1, g_2) where $(\psi, V) \in T$, $f: \mathbf{B}^V \to \mathbf{B}$, and $g_1, g_2: \mathbf{B}^{\mathbf{Var}(\psi)} \to \mathbf{B}$.

For a set W, a set $V \subseteq W$, and a function $y: W \to \mathbf{B}$, let $y{\uparrow}V$ denote the restriction of y to V.

Both distributions are distributions over Ω. The first distribution, denoted D_δ^0, is determined by the following rules how to pick an element (ψ, V, f, g_1, g_2) from Ω:

- an element (ψ, V) from T is chosen randomly, and W is set to $\mathbf{Var}(\psi)$,

- for every $x: V \to \mathbf{B}$, $f(x)$ is chosen randomly,

- for every $y: W \to \mathbf{B}$, $g_1(y)$ is chosen randomly, and

- for every $y: W \to \mathbf{B}$, $g_2(y)$ is set to $f(y{\uparrow}V) \cdot g_1(y)$ with probability $1 - \delta$ and to $-f(y{\uparrow}V) \cdot g_1(y)$ with probability δ.

The other distribution, denoted D_δ, is determined in a similar way:

- an element (ψ, V) from T is chosen randomly, and W is set to $\mathbf{Var}(\psi)$,

- for every $x: V \to \mathbf{B}$, $f(x)$ is chosen randomly,

- for every $y: W \to \mathbf{B}$, $g_1(y)$ is chosen randomly,

- for every $y: W \to \mathbf{B}$,

  - if $f(y{\uparrow}V) = 1$, then $g_2(y)$ is set to $-g_1(y)$, and

  - if $f(y{\uparrow}V) = -1$, then $g_2(y)$ is set to $-g_1(y)$ with probability δ and to $g_1(y)$ with probability $1 - \delta$.

Observe that in both definitions we could have expressed $g_1(y)$ in terms of $g_2(y)$ just in the same way we expressed $g_2(y)$ in terms of $g_1(y)$, i.e., the distributions are symmetric with respect to g_1 and g_2.

We will write Prob_δ^0 respectively Prob_δ for the corresponding probability functions.

7.3.4 Basic Properties of the Distributions

In this subsection, we prove two basic properties of the distributions D_δ^0 and D_δ.

As before, we fix a formula φ and parameters v and δ. Furthermore, we fix a pair $(\psi, V) \in T$ and write W for $\mathbf{Var}(\varphi)$. Moreover, we take expectation w.r.t. D_δ^0, and D_δ respectively, on the condition that the fixed ψ and V are chosen. However, the claims extend immediately to the unconditional expectation.

Lemma 7.3. *Let* $\alpha \subseteq \mathbf{B}^V$, $\beta_1, \beta_2 \subseteq \mathbf{B}^W$ *such that* $\beta_1 \neq \beta_2$. *Then*

$$\mathbf{E}_D \left(\prod_{x \in \alpha} f(x) \prod_{y_1 \in \beta_1} g_1(y_1) \prod_{y_2 \in \beta_2} g_2(y_2) \right) = 0, \qquad (7.8)$$

$$\mathbf{E}_D \left(\prod_{y_1 \in \beta_1} g_1(y_1) \prod_{y_2 \in \beta_2} g_2(y_2) \right) = 0. \qquad (7.9)$$

Proof. Assume $\beta_1 \neq \beta_2$, say $y_0 \in \beta_1 \setminus \beta_2$. (The other case follows by symmetry.) Then the random variable $g_1(y_0)$ is independent of the random variable

$$\prod_{x \in \alpha} f(x) \prod_{y_1 \in \beta_1 - y_0} g_1(y) \prod_{y \in \beta_2} g_2(y).$$

Thus, the expected value of the product on the right-hand side of (7.8) evaluates to

$$\mathbf{E}_D \left(g_1(y_0) \right) \cdot \mathbf{E}_D \left(\prod_{x \in \alpha} f(x) \prod_{y_1 \in \beta_1 - y_0} g_1(y) \prod_{y \in \beta_2} g_2(y) \right).$$

This is 0, as $\mathbf{E}_D(g_1(y)) = 0$ regardless of the particular distribution, and proves (7.8).

(7.9) is proved in exactly the same way with the only (obvious) modification that the factor $\prod_{x \in \alpha} f(x)$ is discarded.   ∎

The second lemma we need is the following. For $\beta \subseteq \mathbf{B}^W$, we write $\beta{\uparrow}V$ for the set $\{y{\uparrow}V \mid y \in \beta\}$.

Lemma 7.4. *Let* $\alpha \subseteq \mathbf{B}^V$, $\beta \subseteq \mathbf{B}^W$ *such that* $\alpha \not\subseteq \beta{\uparrow}V$. *Then*

$$\mathbf{E}_D \left(\prod_{x \in \alpha} f(x) \prod_{y \in \beta} g_1(y) g_2(y) \right) = 0. \qquad (7.10)$$

Proof. The proof has the same pattern as the previous one with the only difference that here $f(x_0)$ and

$$\prod_{x \in \alpha - x_0} f(x) \prod_{y \in \beta} g_1(y) g_2(y)$$

are independent random variables for an element $x_0 \in \alpha \setminus \beta \uparrow V$. ∎

7.3.5 Fourier Transforms

From an algebraic point of view (see, e.g., [LMN93]), Fourier transforms are particular linear transformations on certain vector spaces. We will use Fourier transforms as the major mathematical (combinatorial) tool for the analysis of the verifiers to be constructed in the proofs of Theorems 7.1 and 7.2. In this subsection, we first develop the notion of a Fourier transform (which the reader already familiar with Fourier transforms can skip) and then give some lemmas about Fourier transforms we will use later.

Let M be a finite set of cardinality m. We are interested in functions defined on $\mathbf{B}^M$ with values in $\mathbf{B}$.

Assume f_1 and f_2 are functions $M \to \mathbf{B}$. If $f_1 = f_2$, then

$$\sum_{\alpha \subseteq M} \prod_{x \in \alpha} f_1(x) f_2(x) = 2^m.$$

On the other hand, if $f_1 \neq f_2$, say $f_1(x_0) \neq f_2(x_0)$, then

$$\sum_{\alpha \subseteq M} \prod_{x \in \alpha} f_1(x) f_2(x)$$

$$= \sum_{\alpha \subseteq M - x_0} \prod_{x \in \alpha} f_1(x) f_2(x) + \sum_{\alpha \subseteq M - x_0} f_1(x_0) f_2(x_0) \prod_{x \in \alpha} f_1(x) f_2(x)$$

$$= \sum_{\alpha \subseteq M - x_0} \left(\prod_{x \in \alpha} f_1(x) f_2(x) - \prod_{x \in \alpha} f_1(x) f_2(x) \right) = 0.$$

That is,

$$\sum_{\alpha \subseteq M} \prod_{x \in \alpha} f_1(x) f_2(x) = \begin{cases} 2^m, & \text{if } f_1 = f_2, \\ 0, & \text{if } f_1 \neq f_2. \end{cases} \tag{7.11}$$

So, for each function $A: \mathbf{B}^M \to \mathbf{B}$ and $f: M \to \mathbf{B}$,

$$A(f) = \sum_{f': M \to \mathbf{B}} 2^{-m} A(f') \sum_{\alpha \subseteq M} \prod_{x \in \alpha} f(x) f'(x)$$

$$= \sum_{\alpha \subseteq M} \left(2^{-m} \sum_{f': M \to \mathbf{B}} A(f') \prod_{x \in \alpha} f'(x) \right) \prod_{x \in \alpha} f(x).$$

Setting

$$\hat{A}_\alpha = 2^{-m} \sum_{f:M\to B} A(f) \prod_{x\in\alpha} f(x), \tag{7.12}$$

we can thus write:

$$A(f) = \sum_{\alpha\subseteq M} \hat{A}_\alpha \prod_{x\in\alpha} f(x). \tag{7.13}$$

Observe the duality between (7.12) and (7.13). The real numbers $\hat{A}_\alpha$ are called the Fourier coefficients of A. They can as well be viewed as expectations (where we assume a uniform distribution on the set of all $f: M \to \mathbf{B}$):

$$\hat{A}_\alpha = \mathbf{E}\left(A(f) \prod_{x\in\alpha} f(x)\right).$$

Lemma 7.5 (Parseval's identity). *Let $A: \mathbf{B}^M \to \mathbf{B}$ be an arbitrary function. Then*

$$\sum_{\alpha\subseteq M} \hat{A}_\alpha^2 = 1.$$

Proof. By definition,

$$\sum_{\alpha\subseteq M} \hat{A}_\alpha^2 = \sum_{\alpha\subseteq M} \left(2^{-2m} \sum_{f,f':M\to B} A(f)A(f') \prod_{x\in\alpha} f(x)f'(x)\right),$$

which, by (7.11), evaluates to 1. ∎

Lemma 7.6. *Let $A: \mathbf{B}^M \to \mathbf{B}$ be a function such that $A(-f) = -A(f)$. Assume that $\alpha \subseteq M$ has even cardinality. Then $\hat{A}_\alpha = 0$.*

Proof. By definition,

$$\hat{A}_\alpha = 2^{-m} \sum_{f:M\to B} A(f) \prod_{x\in\alpha} f(x).$$

Fix an arbitrary element from M, say x_0. Splitting the right-hand side of the above equation according to whether $f(x_0) = 1$ and $f(x_0) = -1$ leads to

$$2^{-m}\left(\sum_{f:M\to B, f(x_0)=1} A(f) \prod_{x\in\alpha} f(x) + \sum_{f:M\to B, f(x_0)=-1} A(f) \prod_{x\in\alpha} f(x)\right),$$

which is the same as

$$2^{-m} \sum_{f:M\to B, f(x_0)=1} \left(A(f) \prod_{x\in\alpha} f(x) + A(-f) \prod_{x\in\alpha} -f(x)\right),$$

which, in turn, by assumption that α has an even number of elements, is identical with

$$2^{-m} \sum_{f:M\to\mathbf{B},f(x_0)=1} \left(A(f) \prod_{x\in\alpha} f(x) - A(f) \prod_{x\in\alpha} f(x) \right).$$

Clearly, the last expression is 0.   ∎

Lemma 7.7. *Let B be a function $\mathbf{B}^M \to \mathbf{B}$ and assume $h: M \to \mathbf{B}$ is such that $B(g) = B(\max(g,h))$ for all $g: M \to \mathbf{B}$. Furthermore, let $\beta \subseteq M$ be such that $\hat{B}_\beta \neq 0$. Then $h(y) = -1$ for all $y \in \beta$.*

Proof. The proof goes by way of contradiction. Assume $h(y_0) = 1$ for some $y_0 \in \beta$. By definition and assumption,

$$\hat{B}_\beta = 2^{-m} \sum_{g:M\to\mathbf{B}} B(\max(g,h)) \prod_{y\in\beta} g(x).$$

Just as in the proof of the previous lemma, we split the sum on the right-hand side and obtain for $\hat{B}_\beta$:

$$2^{-m} \left(\sum_{g:M\to\mathbf{B},g(y_0)=1} B(\max(g,h)) \prod_{y\in\beta} g(y) + \right.$$

$$\left. \sum_{g:M\to\mathbf{B},g(y_0)=-1} B(\max(g,h)) \prod_{y\in\beta} g(y) \right).$$

For every function g, let g' denote the function obtained from g by modifying g to take the value $-g(y_0)$ at y_0. The previous expression can then be written as

$$2^{-m} \sum_{g:M\to\mathbf{B},g(y_0)=1} \left(B(\max(g,h)) \prod_{y\in\beta} g(y) + B(\max(g',h)) \prod_{y\in\beta} g'(y) \right).$$

Since $h(y_0) = 1$, we know $\max(g,h) = \max(g',h)$. Thus the previous expression is identical with

$$2^{-m} \sum_{g:M\to\mathbf{B},g(y_0)=1} B(\max(g,h)) \left(\prod_{y\in\beta} g(y) + \prod_{y\in\beta} g'(y) \right).$$

Now observe that by definition of g', the last term can also be written as

$$2^{-m} \sum_{g:M\to\mathbf{B},g(y_0)=1} B(\max(g,h)) \left(\prod_{y\in\beta} g(y) - \prod_{y\in\beta} g(y) \right),$$

which is 0.   ∎

7.4 Using the Parallel Repetition Theorem for Showing Satisfiability

The application of the weak version of the Parallel Repetition Theorem 6.4 in the proofs of the main theorems of this chapter will be via the following lemma.

There, we use the feature of ε-RoBE3SAT-b that each variable in the given formula φ occurs at most b times for some constant $b \in \mathbb{N}$. This implies that for large $|\varphi|$ the probability of choosing non-disjoint clauses tends to zero. Let $m(v)$ be s.t. $|\varphi| \geqslant m(v)$ implies that this probability is at most $1/2$.

Lemma 7.8. *For any small $\delta, \delta', \gamma > 0$, and for any $k \in \mathbb{N}$, there exists a constant $v(\delta, \delta', \gamma, k) \in \mathbb{N}$ such that for all $v \geqslant v(\delta, \delta', \gamma, k)$, and for all instances φ of ε-RoBE3SAT-b of length $|\varphi| \geqslant m(v)$ the following holds.*

Assume that there exists a proof $\pi = \left((A^V)_{V \in \binom{\text{Var}(\varphi)}{v}}, (B^\psi)_{\psi \in \binom{\varphi}{v}} \right)$ for the formula φ such that

$$\left| \mathbf{E} \left(\sum_{\beta : |\beta| \leqslant k} \sum_{\alpha : |\alpha| \leqslant k, \alpha \sqcap \beta, |\hat{A}^V_\alpha| \geqslant \gamma} \left| \hat{A}^V_\alpha \right| (\hat{B}^\psi_\beta)^2 \right) \right| \geqslant \frac{\delta}{2},$$

where $\alpha \sqcap \beta$ may be any condition that implies the existence of some $y \in \beta$ such that $(y{\uparrow}V) \in \alpha$, and the expectation is taken w.r.t. the distribution of (ψ, V) according to $D_{\delta'}$ or $D^0_{\delta'}$.

Then φ is satisfiable.

Strictly speaking, the Weak Parallel Repetition Theorem is used only as a technical tool for proving the above lemma, and one can at least in principle imagine that the lemma can also be proven without using the theorem. But one should note that the probabilistically checkable proofs as introduced in the last section are just made in such a way that they are in a close relation to the parallel repetition of a certain MIP-protocol.

The pivotal point in the claimed correspondence is the choice of v clauses ψ and v formulae V performed by the verifiers. Under both distributions D_δ and D^0_δ, this random choice corresponds essentially to the v-fold parallel repetition of the MIP$(2,1)$-protocol 6.2 where ψ is given to the first prover and V to the second. That this correspondence is an only "essential", not "exact" one is due to the case of non-disjoint clauses ψ, which is unavoidable in the parallel repetition of Protocol 6.2 but excluded in the Distributions D_δ and D^0_δ.

To overcome this difficulty we made the assumption that $|\varphi| \geqslant m(v)$ which implies that the case of non-disjoint clauses occurs with probability at most $1/2$. This bound will be used in the proof of the lemma below in that we may restrict to the case of disjoint clauses only, without reducing the success rate of the provers by more than a factor of 2.

In case of disjoint clauses, we may see a probabilistically checkable proof $\pi = \left((A^V)_{V \in \binom{\mathbf{Var}(\varphi)}{v}}, (B^\psi)_{\psi \in \binom{\varphi}{v}} \right)$ as a (deterministic) strategy for the provers if each A^V, B^ψ is in fact the long code of an assignment to V, respectively. $\mathbf{Var}(\psi)$. The provers may then just return that assignment to the MIP-verifier. Conversely each deterministic strategy of giving back certain assignments can be converted into long codes. Let us point out that this correspondence serves only as an illustration of what is going on. How to get formally prover strategies from probabilistically checkable proofs π will be described in the course of the proof of Lemma 7.8 below. The construction given there applies to any proof π certifying a certain expectation as stated in the assumption of the lemma. We will not need to restrict to case π being a long code.

Proof of Lemma 7.8.   We describe a strategy for the provers of Protocol 6.2 to convince the verifier of the satisfiability of the given formula. We will show that under the given assumption this strategy will have a success rate greater than $\frac{\gamma\delta}{4k^2}$, a constant *independent of v*.

Then we can use the Weak Parallel Repetition Theorem stating that the success rate in the case of inputs to be rejected will tend to zero with increasing repetition factor. Hence there exists a constant $v_0 \in \mathbb{N}$, which we use as the claimed $v(\delta, \delta', \gamma, k)$, such that for all $v \geqslant v_0$:

> If the v-fold parallel repetition of Protocol 6.2 is performed on an unsatisfiable RObE3SAT instance φ, then there do not exist any strategies for the provers having a success rate greater than $\frac{\gamma\delta}{4k^2}$.

Consequently φ has to be satisfiable.

To complete the proof it remains to give prover strategies which have a success rate greater than $\frac{\gamma\delta}{4k^2}$ under the given assumption of the lemma. In advance, we remind of the fact (Lemma 7.7), that only those Fourier coefficients $\hat{B}^\psi_\beta$ are non-zero, where β contains only assignments for $\mathbf{Var}(\psi)$ which satisfy ψ. Since the strategy of prover P_1 below returns only assignments which are contained in some β with $\hat{B}^\psi_\beta \neq 0$, this guarantees the satisfaction of ψ. Thus, for convincing the verifier of Protocol 6.2, the prover strategies need only return with sufficient probability a pair of assignments (x, y) s.t. $x = y{\uparrow}V$. We will call a pair (x, y) *compatible* iff $x = y{\uparrow}V$.

In the following, we consider the situation that some fixed ψ, V are given to the provers. We use as abbreviation $W = \mathbf{Var}(\psi)$, and we omit superscripts ψ, V of B and A, respectively.

PROVERSTRATEGY If the chosen clauses have non-disjoint sets of variables, P_1 will give up, and only in this case it may happen that $|V| < v$, in which case also P_2 has to give up.

Otherwise, both provers will base their behavior on the same proof π as given by the assumption.

P_1: Choose $\beta \subseteq \mathbf{B}^W, |\beta| \leq k$, with probability proportional $\hat{B}_\beta^2$. Return randomly some $y \in \beta$ to the verifier.

P_2: Choose $\alpha \subseteq \mathbf{B}^V, |\alpha| \leq k, \left|\hat{A}_\alpha\right| \geq \gamma$, with probability proportional $\left|\hat{A}_\alpha\right|$. Now choose randomly some $x \in \alpha$ and return it to the verifier.

Note that not in any case appropriate α and β need to exist. If for some V and ψ the provers cannot find α respectively. β which meet the above conditions they will give up. In the following we will call α and β *admissible* if they meet the above conditions. Since the sum in the assumption of the lemma ranges just over admissible α and β, there exist admissible α and β with a certain probability.

In the following, we consider the case that some ψ is chosen which has clauses with disjoint sets of variables i.e., we calculate the probability under the condition of the disjointness of ψ. From this the unconditional probability will be estimated at the end of the proof.

Let us see what happens if admissible sets α and β are chosen. We observe that they have to be non-empty since by Lemma 7.6 we have $\hat{A}_\emptyset = \hat{B}_\emptyset = 0$.

By $\sum_\beta \hat{B}_\beta^2 = 1$, for each single $\beta, |\beta| \leq k$, the probability to be chosen is at least

$$\frac{\hat{B}_\beta^2}{\sum_{\beta : |\beta| \leq k} \hat{B}_\beta^2} \geq \hat{B}_\beta^2.$$

Calculating the probability of a set α to be chosen is a bit more complicated. First we will bound the sum over the absolute values of the coefficients of all admissible sets α.

$$\sum_{\alpha : |\alpha| \leq k, |\hat{A}_\alpha| \geq \gamma} \left|\hat{A}_\alpha\right| \underset{(*)}{\leq} \sum_{\alpha : |\alpha| \leq k, |\hat{A}_\alpha| \geq \gamma} \hat{A}_\alpha^2 \frac{1}{\gamma} \underset{\text{Lemma 7.5}}{\leq} \frac{1}{\gamma}. \tag{7.14}$$

Here, inequality $(*)$ is obtained by multiplying each single term $\left|\hat{A}_\alpha\right|$ with $\left|\hat{A}_\alpha\right| \frac{1}{\gamma} \geq 1$.

By (7.14), each admissible α is chosen with probability at least $\left|\hat{A}_\alpha\right| \gamma$.

We will now calculate the probability that the provers return a *compatible* pair (x, y) to the verifier, i.e. a pair with $x = y \uparrow V$.

Since the provers choose independently, each possible pair of sets (α, β) is chosen with probability at least $\gamma \left|\hat{A}_\alpha\right| \hat{B}_\beta^2$. Once the provers have chosen a pair (α, β)

containing some $x \in \alpha, y \in \beta$ with $x = y \uparrow V$, they will return such a compatible pair (x, y) to the verifier with probability at least $\frac{1}{k^2}$ since $|\alpha|, |\beta| \leqslant k$.

Overall, the probability of returning to the verifier some (x, y) such that $x = y \uparrow V$ is at least

$$\frac{1}{k^2} \gamma \; \mathbf{E} \left(\sum_{\beta \,:\, |\beta| \leqslant k} \; \sum_{\alpha \,:\, |\alpha| \leqslant k, \alpha \sqcap \beta, |\hat{A}_\alpha| \geqslant \gamma} \left| \hat{A}_\alpha \right| \hat{B}_\beta^2 \right) \geqslant \frac{1}{k^2} \gamma \frac{\delta}{2}$$

where the inequality holds by the assumption.

As pointed out above, β contains only assignments for $\mathbf{Var}(\psi)$ which satisfy ψ. Consequently, the verifier has to accept any compatible pair (x, y) it gets from the provers (see Protocol 6.2). Thus the above probability of choosing a compatible pair already gives the success probability of the prover strategies under the condition that the clauses in ψ have disjoint sets of variables.

Since by $|\varphi| \geqslant m(v)$ this condition is fulfilled with probability at least $1/2$, the unconditional success rate of the prover strategies is at least $\frac{\gamma \delta}{4k^2}$. ∎

7.5 An Optimal Lower Bound for Approximating MAXLINEQ3-2

As explained earlier, the proof of Theorem 7.1 is a reduction from ε_0-ROBE3SAT-b, and this reduction is based on a certain verifier for ε_0-ROBE3SAT-b. We first present this verifier. We then explain the completeness and soundness properties it enjoys. A description of the reduction from ε_0-ROBE3SAT-b follows.

7.5.1 The Verifier

The verifier essentially implements a basic test as described in Subsection 7.3.3. For technical reasons, the verifier rejects short formulae in Step 1.

Verifier Vlin$[\delta, v]$.

    Input: An instance $\varphi = C_1 \wedge C_2 \wedge \ldots \wedge C_m$ of ε_0-ROBE3SAT-b.

    Proof: A pair $\left((A^V)_{V \in \binom{\mathbf{Var}(\varphi)}{v}}, (B^\psi)_{\psi \in \binom{\varphi}{v}} \right)$ such that (7.5) – (7.7) hold.

1. If $\binom{\varphi}{v}$ is empty, then reject.

2. Choose $(\psi, V, f, g_1, g_2) \in \Omega$ according to the distribution D_δ^0 (see Subsection 7.3.3).

3. If
$$A^V(f) \cdot B^\psi(g_1) \cdot B^\psi(g_2) = 1, \tag{7.15}$$
then accept, else reject. (See also (7.2) and (7.4), and Subsection 7.3.2.)

7.5.2 Completeness

We first show that verifier $\mathbf{Vlin}[\delta, v]$ has the following completeness property.

Lemma 7.9 (completeness of the verifier). *Let $\delta, v > 0$. Then there exists a constant $m_1(v)$ such that for every satisfiable input with at least $m_1(v)$ clauses, the verifier $\mathbf{Vlin}[\delta, v]$ accepts some π as a proof for φ with probability at least $1 - \delta$.*

Proof. First of all, if φ is short, then $\mathbf{Vlin}[\delta, v]$ will reject in Step 1, no matter what the proof looks like.

On the other hand, if φ is long enough, i.e., if the number of clauses is greater than $3lv$ (l being the constant which bounds the number of occurrences of each variable in an ε_0-ROBE3SAT-b instance), then $\binom{\varphi}{v}$ will not be empty and $\mathbf{Vlin}[\delta, v]$ will always pass Step 1 and choose (ψ, V, f, g_1, g_2) as determined by D_δ^0. We show that there is a proof such that regardless of which ψ and V the verifier chooses, it will accept with probability $1 - \delta$ after Step 1.

Let z be a satisfying assignment for φ and consider the proof π defined by

$$
\begin{aligned}
A^V(f) &= f(z{\uparrow}V), \\
B^\psi(g) &= g(z{\uparrow}\mathbf{Var}(\psi)),
\end{aligned}
$$

for $V \in \binom{\mathbf{Var}(\varphi)}{v}$, $f: \mathbf{B}^V \to \mathbf{B}$, $\psi \in \binom{\varphi}{v}$, and $g: \mathbf{B}^{\mathbf{Var}(\psi)} \to \mathbf{B}$.

By construction, $A^V(f) = f(z{\uparrow}V)$, $B^\psi(g_1) = g_1(z{\uparrow}\mathbf{Var}(\psi))$, and $B^\psi(g_2) = g_2(z{\uparrow}\mathbf{Var}(\psi))$. This is true as A^V and B^ψ are in fact long codes and moreover, B^ψ is the long code of a satisfying assignment for ψ (see (7.2) and (7.4)).

From the definition of D_δ^0 follows that $f(z{\uparrow}V) \cdot g_1(z{\uparrow}\mathbf{Var}(\psi)) \cdot g_2(z{\uparrow}\mathbf{Var}(\psi)) = 1$ with probability $1 - \delta$. So π is accepted as a proof for φ with probability at least $1 - \delta$.   ∎

Observe that unlike usually we do not have perfect completeness for the verifier; the distribution D_δ^0 introduces an error with probability δ. The following consideration gives a motivation for this.

Since the test in Step 3 is linear, it will not only let pass any legal long code but also any linear combination of long codes. The error introduced by D_δ^0 ensures that linear combinations with less summands get more attention (see details in the proof of Lemma 7.10).

7.5.3 Soundness

The following soundness properties of the verifier are much more difficult to prove than the previous lemma.

Lemma 7.10 (soundness of the verifier). *Let $\delta > 0$. Then there exists a constant $v_1(\delta)$ such that the following holds for every input φ: if for some $v \geqslant v_1(\delta)$ verifier $\mathbf{Vlin}[\delta, v]$ accepts some π as a proof for φ with probability at least $(1 + \delta)/2$ and if φ has at least $m(v)$ clauses, then φ is satisfiable.*

For an explanation of $m(v)$, see Lemma 7.8.

Proof. We need some more notation. When β is a set of assignments $W \to \mathbf{B}$ and when $V \subseteq W$, then we write $\beta \uparrow_2 V$ for the set of all assignments $x \colon V \to \mathbf{B}$ there exists an odd number of elements $y \in \beta$ with $y \uparrow V = x$ for.

Let φ be a ε_0-RobE3Sat-b instance and π a proof that is accepted with probability at least $(1 + \delta)/2$ by $\mathbf{Vlin}[\delta, v]$. Define k by $k = \lceil -(\log \delta)/\delta \rceil$, and let T denote the set of all pairs (ψ, V) where $\psi \in \binom{\varphi}{v}$ and V is a set of variables with one variable from each clause in ψ.

If we show

$$\mathbf{E}\left(\sum_{|\beta| \leqslant k, \, |\hat{A}^V_{\beta \uparrow_2 V}| \geqslant \delta/4} \hat{A}^V_{\beta \uparrow_2 V} (\hat{B}^\psi_\beta)^2 \right) \geqslant \delta/2, \tag{7.16}$$

then, by Lemma 7.8, φ is satisfiable, provided φ has at least $m(v)$ clauses and $v \geqslant v(\delta, \delta, \delta/4, k)$. (Recall that $\left| \hat{A}^V_{\beta \uparrow_2 V} \right| \geqslant \delta/4$ implies that $\beta \uparrow_2 V \neq \emptyset$. Clearly, $|\beta| \leqslant k$ implies $|\beta \uparrow_2 V| \leqslant k$.) In other words, setting $v_1(\delta) = v(\delta, \delta, \delta/4, k)$ makes Lemma 7.10 true, provided (7.16) holds under the given assumptions. So in the rest of this proof, we will show that (7.16) holds.

The test in Step 3 is passed iff $A^V(f) \cdot B^\psi(g_1) \cdot B^\psi(g_2) = 1$. If the test is not passed, then $A^V(f) \cdot B^\psi(g_1) \cdot B^\psi(g_2) = -1$. Thus the probability that the test is passed is given by

$$\frac{1}{2} \left(1 + \mathbf{E}(A^V(f) \cdot B^\psi(g_1) \cdot B^\psi(g_2)) \right). \tag{7.17}$$

Hence, by assumption,

$$\mathbf{E}\left(A^V(f) \cdot B^\psi(g_1) \cdot B^\psi(g_2) \right) \geqslant \delta. \tag{7.18}$$

The left-hand side of (7.18) can be written as

$$\frac{1}{|T|} \sum_{(\psi,V)\in T} \mathbf{E}\left(A^V(f) \cdot B^\psi(g_1) \cdot B^\psi(g_2)\right), \tag{7.19}$$

where each expectation is taken on the condition that ψ and V are chosen.

Fix (ψ, V) from T, and consider one of the summands in the above expression. For notational convenience, we will drop the superscripts with A and B and not spell out multiplication any more.

By Fourier expansion, we obtain

$$\mathbf{E}\left(A(f)B(g_1)B(g_2)\right) =$$

$$\sum_{\alpha,\beta_1,\beta_2} \hat{A}_\alpha \hat{B}_{\beta_1} \hat{B}_{\beta_2}\, \mathbf{E}\left(\prod_{x\in\alpha} f(x) \prod_{y\in\beta_1} g_1(y) \prod_{y\in\beta_2} g_2(y)\right).$$

From Lemma 7.8 we can conclude that the right-hand side of the previous equation is identical with:

$$\sum_{\alpha,\beta} \hat{A}_\alpha \hat{B}_\beta^2\, \mathbf{E}\left(\prod_{x\in\alpha} f(x) \prod_{y\in\beta} g_1(y)g_2(y)\right). \tag{7.20}$$

Consider one of the inner expectations, namely

$$\mathbf{E}\left(\prod_{x\in\alpha} f(x) \prod_{y\in\beta} g_1(y)g_2(y)\right) \tag{7.21}$$

for fixed α and β. By independence, this expectation can be split into a product of several expectations:

$$\prod_{\substack{x\in\beta\uparrow V \\ x\in\alpha}} \mathbf{E}\left(f(x) \prod_{\substack{y\in\beta \\ y\uparrow V = x}} g_1(y)g_2(y)\right) \prod_{\substack{x\in\beta\uparrow V \\ x\notin\alpha}} \mathbf{E}\left(\prod_{\substack{y\in\beta \\ y\uparrow V = x}} g_1(y)g_2(y)\right).$$

We first calculate the expectations in the left product, where we split according to whether $f(x) = 1$ or $f(x) = -1$ and use s_x to denote the number of elements $y \in \beta$ such that $y\uparrow V = x$:

$$\mathbf{E}\left(f(x) \prod_{\substack{y\in\beta \\ y\uparrow V = x}} g_1(y)g_2(y)\right)$$

$$
= \frac{1}{2} \left(\prod_{\substack{y \in \beta \\ y\uparrow V = x \\ f(x) = 1}} \mathbf{E}(g_1(y)g_2(y)) - \prod_{\substack{y \in \beta \\ y\uparrow V = x \\ f(x) = -1}} \mathbf{E}(g_1(y)g_2(y)) \right)
$$

$$
= \frac{1}{2}\left((1-2\delta)^{s_x} - (2\delta-1)^{s_x}\right) = \begin{cases} (1-2\delta)^{s_x}, & \text{if } s_x \text{ is odd}, \\ 0, & \text{otherwise}. \end{cases}
$$

Similarly, we get

$$
\mathbf{E}\left(\prod_{y \in \beta, y\uparrow V = x} g_1(y)g_2(y) \right) = \begin{cases} (1-2\delta)^{s_x}, & \text{if } s_x \text{ is even}, \\ 0, & \text{otherwise}. \end{cases}
$$

So (7.21) is non-zero only if

- for all $x \in \beta\uparrow V$ where $x \in \alpha$, there is an odd number of elements $y \in \beta$ such that $y\uparrow V = x$, and

- for all $x \in \beta\uparrow V$ where $x \notin \alpha$, there is an even number of elements $y \in \beta$ such that $y\uparrow V = x$.

This is true if and only if $\alpha = \beta\uparrow_2 V$. In this case, (7.21) evaluates to

$$
\prod_{x \in \beta\uparrow V} (1-2\delta)^{s_x}.
$$

Now, observe that by definition, $\sum_{x \in \alpha} s_x = |\beta|$. So (7.21) evaluates to $(1-2\delta)^{|\beta|}$, and we can write (7.20) as

$$
\sum_{\beta} \hat{A}_{\beta\uparrow_2 V} \hat{B}_{\beta}^2 (1-2\delta)^{|\beta|}.
$$

We try to identify small summands in the above sum.

First, observe that because of Lemma 7.5

$$
\sum_{\beta : |\hat{A}_{\beta\uparrow_2 V}| \leqslant \delta/4} \hat{A}_{\beta\uparrow_2 V} \hat{B}_{\beta}^2 (1-2\delta)^{|\beta|} \leqslant \delta/4. \tag{7.22}
$$

Using the fact that $(1-x)^{1/x} < 1/e$ for small x, we obtain $(1-2\delta)^{-(\log \delta)/\delta} \leqslant \delta^2$. Assuming $\delta < 1/4$, this implies $(1-2\delta)^{-(\log \delta)/\delta} \leqslant \delta/4$, which gives

$$
\sum_{|\beta| \geqslant k} \hat{A}_{\beta\uparrow_2 V} \hat{B}_{\beta}^2 (1-2\delta)^{|\beta|} \leqslant \delta/4. \tag{7.23}
$$

From (7.22) and (7.23) finally follows (7.16).   ∎

7.5.4 Reduction from ε_0-RobE3Sat-b

The proof of Theorem 7.1 can now be completed by a description of a reduction from ε_0-RobE3Sat-b.

Let $\delta > 0$ be a rational number such that $1 - \delta \geqslant 1 - \varepsilon$ and $(1 + \delta)/2 \leqslant 1/2 + \varepsilon$, i.e. $\delta \leqslant \varepsilon$. Let v be greater than $v_1(\delta)$ from Lemma 7.10.

The reduction distinguishes two cases.

First case, φ has at most $m(v)$ or at most $m_1(v)$ clauses. The reduction produces a single equation if φ is satisfiable and two inconsistent equations otherwise. In the first case, all equations can be satisfied at the same time; in the second case, at most $1/2$ of the equations can be satisfied at the same time.

Second case, φ has more than $m(v)$ and more than $m_1(v)$ clauses. The reduction produces a set of equations in variables denoted $X^V(f)$ and $Y^\psi(g)$ corresponding to the positions of a proof for the verifier **Vlin**$[\delta, v]$.

A proof π itself corresponds to an $\mathbb{F}_2$-variable assignment ν_π for these variables (via the correspondence given by $-1 \mapsto 1$ and $1 \mapsto 0$). Given a proof π, the test in Step 3 is passed iff

$$X^V(f) + Y^\psi(g_1) + Y^\psi(g_2) = 0 \tag{7.24}$$

holds under the assignment ν_π.

There is a minor point we have to take care of here. We have to rewrite (7.24) in certain situations according to the convention for accessing a proof described in Subsection 7.3.2. For instance, if $f(1) = -1, g_1(1) = 1, g_2(1) = 1, \max(g_1, h_\psi) = g_1$, and $\max(g_2, h_\psi) = g_2$, then (7.24) takes the form $X^V(-f) + Y^\psi(g_1) + Y^\psi(g_2) = 1$.

Let S_π be the set of all tuples (ψ, V, f, g_1, g_2) for which (7.24) is satisfied under ν_π. The probability that the verifier accepts π as a proof for φ can then be written as

$$\frac{1}{|T|} \sum_{(\psi, V, f, g_1, g_2) \in S_\pi} \mathrm{Prob}_\delta^0(\psi, V, f, g_1, g_2).$$

As δ was chosen rational, there exists a positive integer N such that for every $(\psi, V, f, g_1, g_2) \in \Omega$ the number $N \cdot \mathrm{Prob}_\delta^0(\psi, V, f, g_1, g_2)$ is an integer.

The reduction produces for each tuple (ψ, V, f, g_1, g_2) occurring in S_π exactly $|T| \cdot N \cdot \mathrm{Prob}_\delta^0(\psi, V, f, g_1, g_2)$ copies of (7.24) respectively an appropriately modified version of (7.24).

Let E denote the resulting system of equations. Then the probability a proof π is accepted is the ratio of the maximum number of equations in E satisfiable at the same time to the overall number of equations in E.

On the one hand, if φ is satisfiable, then, by Lemma 7.9, at least $1 - \delta \geqslant 1 - \varepsilon$ of the equations can be satisfied at the same time. If, on the other hand, φ is not satisfiable, then, by Lemma 7.10, at most $(1 + \delta)/2 \leqslant 1/2 + \varepsilon$ of the equations can be satisfied at the same time.

7.6 Optimal Lower Bounds for Approximating MaxEkSat

In this section we will present the proof that MaxEkSat cannot be approximated in polynomial time within a factor of $2^k/(2^k - 1) - \varepsilon$, for any constant $\varepsilon > 0$, unless $\mathcal{P} = \mathcal{NP}$.

Grounding on the results of the previous section, it needs only a small additional effort to reach this claim in the broader sense suggested by the above formulation. But as we will see this only shows, in case $k = 3$, $\mathcal{NP}$-hardness of separating formulae where fractions of $1 - \varepsilon$ and $7/8 + \varepsilon$, respectively, of the clauses can be satisfied.

The main part of the section will be devoted to the sharpened result stating that it is $\mathcal{NP}$-hard to separate satisfiable E3Sat-instances from those where only a fraction of $7/8 + \varepsilon$ of the clauses can be satisfied. From this we can easily generalize to case $k \geqslant 3$.

Remember that, when dealing with maximization problems, we consider a version of E3Sat where formulae are sequences of clauses rather than sets. This allows multiple occurrences of clauses and can equivalently be seen as having weighted clauses, where for a fixed ε there exists a fixed (and finite) set of possible weights for the clauses. MaxE3Sat is then the problem of satisfying a subset of clauses having a maximal weight. The weights will result from probabilities in the following.

Theorem 7.11. *For MaxE3Sat, there exists no polynomial time $(8/7 - \varepsilon)$-approximation algorithm for any small constant $\varepsilon > 0$, unless $\mathcal{P}=\mathcal{NP}$.*

Proof. We give a reduction from MaxLinEq3-2 to MaxE3Sat, mapping the $(1/2 + \varepsilon', 1 - \varepsilon')$-gap of Theorem 7.1 to a $(7/8 + \varepsilon, 1 - \varepsilon)$-gap here.

An equation $x + y + z = 0$ is replaced by the clauses $(\overline{x} \vee y \vee z), (x \vee \overline{y} \vee z),$ $(x \vee y \vee \overline{z}), (\overline{x} \vee \overline{y} \vee \overline{z})$, whereas $x + y + z = 1$ is replaced by the remaining four possible clauses over the three variables.

By any assignment to the three variables exactly one of the eight possible clauses is not satisfied. This one belongs always to the equation not fulfilled by the "same" assignment (identifying the elements of $\mathbb{F}_2$ with Boolean values by "true=1" and "false=0"). This means that for each equation fulfilled by an assignment all four clauses are satisfied, and for any equation not fulfilled we have three out of four clauses satisfied.

This translates fractions of $1/2 + \varepsilon'$ and $1 - \varepsilon'$ fulfilled equations to fractions of $7/8 + \varepsilon$ and $1 - \varepsilon$ satisfied clauses, respectively, when letting $\varepsilon' = 4\varepsilon$.   ∎

7.6.1 Separating Satisfiable from at most $(7/8 + \varepsilon)$-Satisfiable Formulae

We now turn to the sharpened result.

Theorem 7.12. *For any small $\varepsilon > 0$ it is $\mathcal{NP}$-hard to distinguish satisfiable E3SAT-instances from those where at most a fraction of $7/8 + \varepsilon$ of their clauses can be satisfied.*

The strategy of the proof will be the following (roughly explained in Section 7.1, and similar to the proof of Theorem 7.1). Overall we will give for any fixed ε $(0 < \varepsilon < 1/8)$ a polynomial time reduction from ε_0-RobE3Sat-b (shown to be $\mathcal{NP}$-hard for some $\varepsilon_0 < 10^{-4}$ and some fixed b in Lemma 4.9) to $(1/8 - \varepsilon)$-RobE3Sat-b'.

Since polynomially bounded weights can be converted into multiple occurrences of clauses as described in case of equations in Section 7.5.4, it will be sufficient to give a reduction to a polynomially weighted variant of $(1/8 - \varepsilon)$-RobE3Sat-b'.

As in the proof of Theorem 7.1 the reduction will be obtained via a verifier for ε_0-RobE3Sat-b. The crucial properties of this verifier will be the following.

1. Any satisfiable formula φ will be accepted with probability 1.

2. Any unsatisfiable formula φ will be accepted with probability less than $7/8 + \varepsilon$.

3. Acceptance of the verifier on a given formula φ can be expressed by a (polynomially) weighted formula φ' where the maximal weighted fraction of satisfiable clauses of φ' will be the probability of the verifier to accept φ.

The verifier will be very similar to the one defined for MaxLinEq3-2. Again we let the verifier choose a set ψ of clauses with variables W and a subset V of one variable per clause. The "proof of satisfiability" for a formula φ will again contain for an assignment to the variables of φ the long codes A^V and B^ψ for each such ψ and V. The difference to the previous construction lies in the test the verifier performs on φ and the collection of long codes, which will be a bit more complicated this time.

Before giving the proof of Theorem 7.12 we describe the verifier and state the main lemmas for showing the claimed probability in the outcome of it.

The basic test performed by the verifier will use again functions from V and W into $\mathbf{B} = \{-1, 1\}$ with the correspondence to Boolean functions as explained in Section 7.3.

Test 7.13. (Parameters: $0 < \delta' < 1$, $v \in \mathbb{N}$.)

> **Input:** an instance φ of ε_0-RobE3Sat-b with $|\varphi| \geqslant m(v)$.
>
> **Proof:** $\pi = \left((A^V)_{V \in \binom{\mathbf{Var}(\varphi)}{v}}, (B^\psi)_{\psi \in \binom{\varphi}{v}} \right)$.

1. Randomly choose a set of v clauses $\psi \in \binom{\varphi}{v}$ with variables $W = \mathbf{Var}(\psi)$ and a set $V \in \binom{\mathbf{Var}(\varphi)}{v}$ consisting of one variable per chosen clause.

2. Pick f and g_1 randomly as functions on V and W, respectively.

3. The function g_2 on W is taken randomly within the following constraints (where $x = y{\uparrow}V$):

   a) if $f(x) = 1$ then $g_1(y)g_2(y) = -1$ must hold;

   b) if $f(x) = -1$ then with probability δ' we choose g_2 such that $g_1(y)g_2(y) = -1$, and $g_1(y)g_2(y) = 1$ otherwise.

4. Check if $\min(A^V(f), B^\psi(g_1), B^\psi(g_2)) = -1$.

The above distribution of chosen tuples (ψ, V, f, g_1, g_2) is just the distribution $D_{\delta'}$ as described in a slightly different manner in Subsection 7.3.3.

Now we remind of the access conventions of the proofs as introduced in Section 7.3.2. The conventions about negation, i.e. conditions (7.5) and (7.6) will only be used later in the main proof. Here we concentrate on condition (7.7) under which the above test can be restated as checking whether

$$\min(A^V(f), B^\psi(\max(g_1, h_\psi)), B^\psi(\max(g_2, h_\psi))) = -1$$

holds. In the corresponding logical notation this means the satisfaction of

$$A^V(f) \vee B^\psi(g_1 \wedge \psi) \vee B^\psi(g_2 \wedge \psi).$$

Note that the possibility of rejecting unsatisfiable formulae for Test 7.13 results only from this "masking" by h_ψ: Without it any fixed assignment (satisfying or not) would guarantee success. If π contains only long codes which are all obtained from the same assignment for the entire formula we would always have $A^V(f) = f(x)$, $B^\psi(g_1) = g_1(y)$, and $B^\psi(g_2) = g_2(y)$ for the restrictions y and x of this assignment to $\mathbf{Var}(\psi)$ and V, respectively. But this would imply

$$\min(A^V(f), B^\psi(g_1), B^\psi(g_2)) = \min(f(x), g_1(y), g_2(y)) = -1 \qquad (7.25)$$

according to distribution D_δ.

Now by conditions $B^\psi(g_1) = B^\psi(\max(g_1, h_\psi))$ and $B^\psi(g_2) = B^\psi(max(g_2, h_\psi))$, we can have $B^\psi(g_1) = g_1(y)$ and $B^\psi(g_2) = g_2(y)$ for every ψ only if the fixed assignment is *satisfying* for the *entire* formula (in which case always $h_\psi(y) = -1$ and hence $\max(g, h_\psi)(y) = g(y)$). Clearly this cannot be the case for an unsatisfiable formula.

By the above argumentation, we have already convinced ourselves that (7.25) holds if π consists of long codes generated from an assignments which satisfies the whole formula. Hence by using such a proof, Test 7.13 will accept any satisfiable formula with probability 1.

Based on the elementary Test 7.13, the actual test done by the verifier will consist of choosing an instance of Test 7.13 on a random distribution over several possible parameters $\delta_1, \ldots, \delta_t$. The existence of an appropriate rational constant c and positive integer $v_2(\delta, \delta_i)$ will be part of the proofs of the respective lemmas. Constant c is needed, and determined, in the proof of Lemma 7.17, and the existence of $v_2(\delta, \delta_i)$ will be shown in the proof of Lemma 7.15. For giving the reader an impression how the constants δ_i are chosen below, we remark that c may be fixed to be $\frac{1}{2208}$ (see also the comments after Lemma 7.17).

Verifier Vsat[δ].

(It uses the constants $t = \lceil \delta^{-1} \rceil$, $\delta_1 = \delta^2$, $\delta_{i+1} = \delta_i^{8/c}$ for $i = 1, 2, \ldots, t - 1$, and $v := \max\{v_2(\delta, \delta_i) : i = 1, \ldots, t\}$).

   Input: an instance φ of ε_0-ROBE3SAT-b s.t. $|\varphi| \geq m(v)$.

   Proof: $\pi = \left((A^V)_{V \in \binom{\text{Var}(\varphi)}{v}}, (B^\psi)_{\psi \in \binom{\varphi}{v}} \right)$.

   1. Randomly choose $j \in \{1, \ldots, t\}$ with uniform distribution.

   2. Perform Test 7.13 using parameters δ_j and v and accept iff the answer is positive.

The resulting general distribution over tuples (ψ, V, f, g_1, g_2), obtained as a sum over the distributions $D_{\delta_j}, j = 1, \ldots, t$, divided by t, will be denoted D_{gen}.

Besides these definitions, we will need in the proof of Theorem 7.12 the following two lemmas which themselves will be proved in Section 7.6.2. They give estimates on the important parts in the calculation of the acceptance probability of Verifier **Vsat[δ]**, where especially the second one bounds this probability for unsatisfiable formulae.

Lemma 7.14. *Let φ be an instance of ε_0-RobE3Sat-b. Then for any proof $\pi = \left((A^V)_{V \in \binom{\mathrm{Var}(\varphi)}{v}}, (B^\psi)_{\psi \in \binom{\varphi}{v}} \right)$ for φ, and for any small constant $\delta' > 0$ holds*

$$\left| \mathbf{E} \left(B^\psi(g_1) B^\psi(g_2) \right) \right| \leqslant 3\delta'^{1/2} + \sum_{\beta \,:\, \delta'^{-1/2} < |\beta| \leqslant \delta'^{-4/c}} (\hat{B}^\psi_\beta)^2 \qquad (7.26)$$

where the expectation is taken w.r.t. distribution $D_{\delta'}$.

Lemma 7.15. *For any small constants $\delta, \delta' > 0$, there exists a constant $v_2(\delta, \delta') \in \mathbb{N}$ such that for all $v \geqslant v_2(\delta, \delta')$, and for any formula φ with $|\varphi| \geqslant m(v)$, the following holds.*

If there exists some proof $\pi = \left((A^V)_{V \in \binom{\mathrm{Var}(\varphi)}{v}}, (B^\psi)_{\psi \in \binom{\varphi}{v}} \right)$ for the formula φ such that w.r.t. distribution $D_{\delta'}$

$$\left| \mathbf{E} \left(A^V(f) B^\psi(g_1) B^\psi(g_2) \right) \right| \geqslant \delta,$$

then φ is satisfiable.

In the following, we will usually omit superscripts V and ψ for simplicity.

Proof of Theorem 7.12. We will give a reduction from ε_0-RobE3Sat-b to a weighted version where in case of unsatisfiability at most a fraction of $7/8 + \varepsilon$ can be satisfied. We choose δ as a rational number such that $(7 + 5\delta)/8 \leqslant 7/8 + \varepsilon$, i.e. $\delta \leqslant 8\varepsilon/5$. We use the constants $t, \delta_1, \ldots, \delta_t$, and v from the definition of verifier $\mathbf{Vsat}[\delta]$, and we let $m(v)$ be as defined in Section 7.4.

For formulae of length less than $m(v)$, we have to do the reduction by hand. That means mapping each satisfiable formula to a single clause and each unsatisfiable formula to a formula consisting of all eight possible clauses over the same three variables. By any assignment exactly seven out of these eight clauses are satisfied.

In case $|\varphi| \geqslant m(v)$ we will first convince ourselves that it suffices to show: if Verifier $\mathbf{Vsat}[\delta]$ accepts a formula φ with probability $(7 + 5\delta)/8$ then φ is satisfiable, and for each satisfiable formula, there is a proof such that the verifier accepts with probability 1.

Since the success criterion of each single elementary test is a disjunction of the form $A(f) \vee B(g_1) \vee B(g_2)$, it can be translated immediately into a clause with three literals, when using a separate variable for each position in the long codes A and B. (The use of long code values $-A(f)$ instead of $A(-f)$, and analogously for B, as described in Subsection 7.3.2, causes here the use of negated variables where appropriate.)

Together with the weights resulting from the probability that Verifier $\mathbf{Vsat}[\delta]$ performs each single test, this gives already the claimed formula with the maximal acceptance probability of the verifier as maximal fraction of satisfiability.

Since for each fixed δ the other constants will be fixed too, we have as desired only finitely many weights.

Also, this assures that the occurrence of each variable in the resulting formula can be bounded by some constant b'.

Now we want to calculate the acceptance probability of $\mathbf{Vsat}[\delta]$. As we have noted above, Test 7.13 will always succeed if π is generated from a satisfying assignment. This transfers immediately to the whole verifier, which means that for any satisfiable formula there is a proof such that $\mathbf{Vsat}[\delta]$ will accept with probability 1. It remains to show the upper bound of $(7+5\delta)/8$ for the acceptance probability of unsatisfiable formulae.

We observe that the result of Test 7.13 can be calculated as

$$1 - \frac{1}{8}\Big(1 + A(f))(1 + B(g_1))(1 + B(g_2)\Big) \tag{7.27}$$

if we identify "accept" with 1 and "reject" with 0.

Therefore the acceptance probability of Verifier $\mathbf{Vsat}[\delta]$ is just the expected value of (7.27) under the distribution D_{gen}.

Expanding the expectation of the above product results in the term

$$1 - \frac{1}{8} - \frac{1}{8}\mathbf{E}(A(f)) - \frac{1}{8}\mathbf{E}(B(g_1)) - \frac{1}{8}\mathbf{E}(B(g_2)) - \frac{1}{8}\mathbf{E}(A(f)B(g_1))$$

$$- \frac{1}{8}\mathbf{E}(A(f)B(g_2)) - \frac{1}{8}\mathbf{E}(B(g_1)B(g_2)) - \frac{1}{8}\mathbf{E}(A(f)B(g_1)B(g_2))$$

This is equal to

$$\frac{7}{8} - \frac{1}{8}\mathbf{E}(B(g_1)B(g_2)) - \frac{1}{8}A(f)\,\mathbf{E}(B(g_1)B(g_2))$$

since all others terms have expected value 0 as being products of independent functions only, which are chosen equally distributed from $\mathbf{B}^V \to \mathbf{B}$ and $\mathbf{B}^W \to \mathbf{B}$, respectively.

For any unsatisfiable formula φ, Lemma 7.15 bounds the absolute value of the last term by $\delta/8$.

For $\mathbf{E}(B(g_1)B(g_2))$ Lemma 7.14 gives us

$$|\mathbf{E}\left(B(g_1)B(g_2)\right)| \leqslant \frac{1}{t}\sum_{i=1}^{t}\left(3\delta_i^{1/2} + \sum_{\beta:\,\delta_i^{-1/2}<|\beta|\leqslant\delta_i^{-4/c}} \hat{B}_\beta^2\right) \leqslant 3\delta_1^{1/2} + \frac{1}{t} \leqslant 4\delta.$$

For obtaining the second inequality above we use the fact that the choice of the δ_i was made such that the intervals of the inner sum are consecutive, i.e.

$$\sum_{i=1}^{t} \sum_{\beta \,:\, \delta_i^{-1/2} < |\beta| \leqslant \delta_i^{-4/c}} \hat{B}_\beta^2 = \sum_{\beta \,:\, \delta_t^{-1/2} < |\beta| \leqslant \delta_1^{-4/c}} \hat{B}_\beta^2 \leqslant 1.$$

Overall the expected value of (7.27), being the probability of success for Verifier **Vsat**$[\delta]$, is shown to be less than $(7 + 5\delta)/8$ for any unsatisfiable formula φ. ∎

As mentioned in the introduction of this chapter we get immediately

Corollary 7.16. *For any $\varepsilon > 0$, there exists some constant $b \in \mathbb{N}$ such that $(1/8 - \varepsilon)$-RobE3Sat-b is $\mathcal{NP}$-complete.*

It remains to give the generalization from E3Sat to EkSat for $k \geqslant 3$.

Proof of Theorem 7.2.   We proceed by induction, starting from case $k = 3$.

In the induction step, we map a $(1 - 1/2^k, 1)$-gap of satisfiable clauses in case of EkSat to a $(1 - 1/2^{k+1}, 1)$-gap for E$(k+1)$Sat.

This is done by introducing for each EkSat-instance φ a new variable x and substituting each clause C by the two clauses $C \vee x$ and $C \vee \overline{x}$. This produces an E$(k+1)$Sat-instance φ'.

If we look at an arbitrary assignment to the variables of φ', half of the clauses are satisfied just by the value of x, and in the rest x can be omitted, yielding a copy of φ. Thus any assignment to the variables of φ satisfying a fraction of ρ clauses yields for both possible assignments to x an assignment satisfying $\frac{1+\rho}{2}$ clauses of φ' and vice versa. This gives immediately the desired mapping of gaps. ∎

7.6.2 Estimating Expected Values of Success

It remains to give proof sketches of the two technical lemmas estimating the important parts of the acceptance probability.

For that purpose we need another lemma stating roughly speaking that in the choice of V there is only a small probability that large sets β of assignments for W are projected onto small sets α of assignments for V.

Lemma 7.17. *If V is chosen out of W according to Test 7.13, there exists a constant $c, 0 < c < 1$ such that for any fixed $\beta \subseteq 2^W$*

$$\mathbf{E}\left(\frac{1}{|\beta \!\uparrow\! V|}\right) \leqslant \frac{1}{|\beta|^c}.$$

The proof of this lemma is omitted here. It is based mainly on the following observation. One can think of the choice of V as a sequential process of choosing subsequently for the clauses of ψ one variable per clause. Assume that, before choosing a variable from $C = x \vee y \vee z$, two assignments in β can differ on their restriction to V only on x or y (depending on the choices made before). Then the choice of z will force the two assignments to have identical restrictions to V. Starting from a large set β of different assignments, one can estimate the expectation on the number of different restrictions according to the part of V chosen so far, changing during this process from $|\beta|$ to $|\beta{\uparrow}V|$.

In the course of this calculation, one gets some inequalities c has to fulfill for making the claim of the lemma true. As stated above, $c = \frac{1}{2208}$ turns out to be one possible value. For the details see [Hås97b].

Proof of Lemma 7.14.   We have to estimate the expected value of $B(g_1)B(g_2)$ under the chosen distribution $D_{\delta'}$.

As stated in Lemma 7.3, the only non-zero contributions in the expected value of the Fourier transform are those where $\beta_1 = \beta_2$, and here factors with different projections onto V are independent of each other.

Consequently we have

$$
\mathbf{E}\left(B(g_1)B(g_2)\right) = \mathbf{E}\left(\sum_\beta \hat{B}_\beta^2 \prod_{y \in \beta} g_1(y)g_2(y)\right)
$$

$$
= \mathbf{E}\left(\sum_\beta \hat{B}_\beta^2 \prod_{x \in \beta{\uparrow}V} \prod_{y \in \beta: y{\uparrow}V = x} g_1(y)g_2(y)\right),
$$

which under distribution $D_{\delta'}$ can be calculated to be

$$
\sum_\beta \hat{B}_\beta^2 \; \mathbf{E}\left(\prod_{x \in \beta{\uparrow}V} \frac{1}{2}\left((-1)^{s_x} + (1 - 2\delta')^{s_x}\right)\right) \tag{7.28}
$$

where s_x is the number of $y \in \beta$ with $x = y{\uparrow}V$. To check this we consider the two possible values for $f(x)$ which occur each with probability $1/2$. If $f(x) = 1$, we have $g_1(y)g_2(y) = -1$ for each of the s_x many assignments y being restricted to x. Otherwise we have for each y independently $\mathbf{E}(g_1(y)g_2(y)) = \delta'(-1) + (1 - \delta')1 = 1 - 2\delta'$.

Now we make a case distinction in accordance to the summation boundaries stated in the claim of the lemma, i.e. we distinguish the cases

1. $|\beta| \leqslant \delta'^{-1/2}$,

2. $\delta'^{-1/2} < |\beta| \leqslant \delta'^{-4/c}$, and

3. $\delta'^{-4/c} < |\beta|$.

We want to estimate the corresponding three partial sums of (7.28) such that they all together meet the right hand side of inequality (7.26). In the second case there is nothing to do since the absolute value of the large product is obviously at most one, and the term

$$\sum_{\beta:\,\delta'^{-1/2}<|\beta|\leqslant\delta'^{-4/c}} \hat{B}_\beta^2$$

occurs on the right hand side of (7.26).

In case 1, independently of V, we will show the corresponding sum to be at most $\delta'^{1/2}$ in absolute size. By Lemma 7.6, for all β of even size, we have $\hat{B}_\beta = 0$ and hence nothing to prove. For β of odd size, there must be some $x \in \beta{\uparrow}V$ having an odd s_x. The corresponding factor in (7.28) evaluates to

$$\frac{1}{2}((-1)^{s_x} + (1 - 2\delta')^{s_x}) = \frac{1}{2}(-1 + (1 - 2\delta')^{s_x}). \tag{7.29}$$

By $s_x \leqslant |\beta|$ we have $s_x \leqslant 1/\delta'$ (we are considering case 1), and thus obtain as an estimate for odd s_x:

$$(1 - 2\delta')^{s_x} \geqslant 1 - 2\delta's_x.$$

Consequently, the factor (7.29) can be bounded by

$$0 \geqslant \frac{1}{2}(-1 + (1 - 2\delta')^{s_x}) \geqslant \frac{1}{2}(-1 + (1 - 2\delta's_x)) = -\delta's_x \geqslant -\delta'^{1/2},$$

where the last inequality comes again from $s_x \leqslant |\beta| \leqslant \delta'^{1/2}$.

All other factors in (7.28) can be bounded by 1 in absolute value, which bounds the absolute value of the large product by $\delta'^{1/2}$. In total, we obtain the following bound for the absolute value of the contribution of case 1 to the sum (7.28).

$$\left| \sum_{\beta:\,|\beta|\leqslant\delta'^{-1/2}} \hat{B}_\beta^2\, \mathbf{E}\left(\prod_{x\in\beta{\uparrow}V} (\frac{1}{2}((-1)^{s_x} + (1 - 2\delta')^{s_x})) \right) \right|$$

$$\leqslant \sum_{\beta:\,|\beta|\leqslant\delta'^{-1/2}} \hat{B}_\beta^2\, \delta'^{1/2} \underset{\text{Lemma 7.5}}{\leqslant} \delta'^{1/2}.$$

To finish the proof we have to bound the contribution of case 3 by $2\delta'^{1/2}$. Here, we need to consider the distribution over the choices of V out of $W = \mathbf{Var}(\psi)$, using Lemma 7.17. We observe that $0 < (1 - 2\delta')^{s_x} \leqslant 1 - 2\delta'$, bounding the absolute value of each factor in (7.28) by $1 - \delta'$.

We now estimate

$$\sum_{\beta\,:\,\delta'-4/c<|\beta|} \hat{B}_\beta^2 \ \mathbf{E}\left(\prod_{x\in\beta\uparrow V}\frac{1}{2}\left((-1)^{s_x}+(1-2\delta')^{s_x}\right)\right)$$

$$\leqslant \sum_{\beta\,:\,\delta'-4/c<|\beta|} \hat{B}_\beta^2 \ \mathbf{E}\left((1-\delta')^{|\beta\uparrow V|}\right).$$

By Lemma 7.17, case $|\beta\uparrow V| < |\beta|^{c/2}$ occurs with probability at most $|\beta|^{-c/2} \leqslant \delta'^2$. Hence this case contributes in absolute value at most δ'^2 to the expectation.

In case $|\beta\uparrow V| \geqslant |\beta|^{c/2} \geqslant \delta'^{-2}$ we obtain $(1-\delta')^{|\beta\uparrow V|} \leqslant (1-\delta')^{\delta'^{-2}} \leqslant \delta'$.

Overall the expected value of the product is bounded by $\delta'^2 + \delta' \leqslant 2\delta'^{1/2}$ which also bounds the whole sum, when applying again $\sum_\beta \hat{B}_\beta^2 \leqslant 1$ (Lemma 7.5). $\blacksquare$

Lemma 7.15 is in a certain correspondence to Lemma 7.9. In fact the difference in the proof lies mainly in the use of a different distribution over the functions f, g_1, g_2, and after calculating the impact of this we will reach a point where again Lemma 7.8 will be applied, using implicitly the Weak Parallel Repetition Theorem.

Proof of Lemma 7.15. Here we have to show that for any sufficiently large formula φ the satisfiability follows from

$$|\mathbf{E}\left(A(f)B(g_1)B(g_2)\right)| \geqslant \delta. \qquad (7.30)$$

Remember that the functions f, g_1, g_2 are assumed to be chosen according to $D_{\delta'}$, and that the number of clauses the verifier chooses for the test is assumed to be $v \geqslant v_2(\delta, \delta')$. The appropriate value for $v_2(\delta, \delta')$ will be determined at the end of the proof.

As in the proof of Lemma 7.9 it is again our aim to reach from the assumption (7.30) a point where we can set up the MIP$(2,1)$-protocol.

For this we use again the Fourier transform, representing $\mathbf{E}(A(f)B(g_1)B(g_2))$ as

$$\mathbf{E}\left(\sum_\alpha \sum_{\beta_1,\beta_2} \hat{A}_\alpha \hat{B}_{\beta_1} \hat{B}_{\beta_2} \prod_{x\in\alpha} f(x) \prod_{y\in\beta_1} g_1(y) \prod_{y\in\beta_2} g_2(y)\right). \qquad (7.31)$$

Again by Lemmas 7.3 and 7.4, only the terms where $\beta_1 = \beta_2 =: \beta$ and $\alpha \subseteq \beta\uparrow V$ contribute to the expected value.

This transforms the sum (7.31) to

$$\mathbf{E}\left(\sum_\beta \hat{B}_\beta^2 \sum_{\alpha\subseteq\beta\uparrow V} \hat{A}_\alpha \ \mathbf{E}\left(\prod_{x\in\alpha} f(x) \prod_{y\in\beta} g_1(y)g_2(y)\right)\right). \qquad (7.32)$$

Note that we have decomposed the calculation of the expectation under distribution $D_{\delta'}$ into two aspects. The outer expectation is taken w.r.t. the distribution of clauses ψ and variable sets V. Here we will obtain the same result for all ψ but the distribution of sets V for fixed ψ will play an important role. Consequently, the inner expectation is (for fixed ψ, V) taken w.r.t. the distribution of the functions f, g_1, g_2.

Now we use the shorthands

$$\mathbf{E}_{\alpha,\beta} := \mathbf{E}\left(\prod_{x \in \alpha} f(x) \prod_{y \in \beta} g_1(y) g_2(y) \right),$$

and

$$\mathbf{E}_{\beta,V} := \sum_{\alpha \subseteq \beta \uparrow V} \hat{A}_\alpha \, \mathbf{E}_{\alpha,\beta},$$

which convert (7.32) to

$$\mathbf{E}\left(\sum_\beta \hat{B}_\beta^2 \, \mathbf{E}_{\beta,V} \right).$$

For applying Lemma 7.8 we need to show that large β give only a small contribution to (7.31). Therefore, the following calculations are meant to be applied for large β only, regardless of the fact that most of them hold for all β.

As in the previous proof we can split the inner product into different components, according to which $y \in \beta$ restrict to the same $x \in \beta \uparrow V$. Since α may be a proper subset of $\beta \uparrow V$, there may be also some $x \in (\beta \uparrow V) \setminus \alpha$. For those, we have to calculate

$$\mathbf{E}_{\alpha,\beta,x} := \mathbf{E}\left(\prod_{y \in \beta : y \uparrow V = x} g_1(y) g_2(y) \right),$$

and in case $x \in \alpha$ the respective part of the product is

$$\mathbf{E}'_{\alpha,\beta,x} := \mathbf{E}\left(f(x) \prod_{y \in \beta : y \uparrow V = x} g_1(y) g_2(y) \right).$$

Then we get

$$\mathbf{E}_{\alpha,\beta} = \prod_{x \in \alpha} \mathbf{E}'_{\alpha,\beta,x} \prod_{x \in (\beta \uparrow V) \setminus \alpha} \mathbf{E}_{\alpha,\beta,x} . \tag{7.33}$$

As in the proof of Lemma 7.14, see (7.28), we obtain from the definition of distribution $D_{\delta'}$

$$\mathbf{E}_{\alpha,\beta,x} = \frac{1}{2}\left((-1)^{s_x} + (1 - 2\delta')^{s_x} \right) \tag{7.34}$$

by a simple case distinction depending on $f(x)$.

Similarly we get

$$E'_{\alpha,\beta,x} = \frac{1}{2}\left((-1)^{s_x} - (1 - 2\delta')^{s_x}\right) \tag{7.35}$$

where the only difference lies in the fact that we have to multiply by the value $f(x) = -1$ in the second case.

Now we estimate

$$E_{\beta,V} = \sum_{\alpha \subseteq \beta \uparrow V} \hat{A}_\alpha \, E_{\alpha,\beta}$$

$$\leqslant \left(\sum_{\alpha \subseteq \beta \uparrow V} \hat{A}_\alpha^2\right)^{\frac{1}{2}} \left(\sum_{\alpha \subseteq \beta \uparrow V} E_{\alpha,\beta}^2\right)^{\frac{1}{2}} \leqslant \left(\sum_{\alpha \subseteq \beta \uparrow V} E_{\alpha,\beta}^2\right)^{\frac{1}{2}}. \tag{7.36}$$

The first inequality of (7.36) is obtained by interpreting the first sum as scalar product and applying the Cauchy-Schwartz inequality. The second one is just another application of Lemma 7.5.

Next, we want to convince ourselves that

$$\sum_{\alpha \subseteq \beta \uparrow V} E_{\alpha,\beta}^2 =$$

$$\prod_{x \in \beta \uparrow V} \left(\left(\frac{1}{2}((-1)^{s_x} - (1 - 2\delta')^{s_x})\right)^2 + \left(\frac{1}{2}((-1)^{s_x} + (1 - 2\delta')^{s_x})\right)^2\right) \tag{7.37}$$

In view of (7.33), (7.34), and (7.35) this means that we have to show

$$\sum_{\alpha \subseteq \beta \uparrow V} \prod_{x \in \alpha} E'^2_{\alpha,\beta,x} \prod_{x \in (\beta \uparrow V) \setminus \alpha} E^2_{\alpha,\beta,x} = \prod_{x \in \beta \uparrow V} \left(E'^2_{\alpha,\beta,x} + E^2_{\alpha,\beta,x}\right).$$

This claim is obtained by showing by induction on the size of some helping set $N \subseteq \beta \uparrow V$ that

$$\prod_{x \in (\beta \uparrow V) \setminus N} \left(E'^2_{\alpha,\beta,x} + E^2_{\alpha,\beta,x}\right) \left(\sum_{\alpha \subseteq N} \prod_{x \in \alpha} E'^2_{\alpha,\beta,x} \prod_{x \in (\beta \uparrow V) \setminus \alpha} E^2_{\alpha,\beta,x}\right)$$

$$= \prod_{x \in \beta \uparrow V} \left(E'^2_{\alpha,\beta,x} + E^2_{\alpha,\beta,x}\right).$$

Each factor in the product in (7.37) is of the form $a^2 + b^2$ with $|a| + |b| = 1$ and $z := \max(|a|, |b|) \leqslant 1 - \delta'$. This implies $z \geqslant \frac{1}{2}$ and $1 - z \geqslant \delta'$. Hence each factor is bounded by $a^2 + b^2 = 1 - 2|a||b| = 1 - 2z(1 - z) \leqslant 1 - 2\frac{1}{2}(1 - z) \leqslant 1 - \delta'$. Consequently

$$\sum_{\alpha \subseteq \beta \uparrow V} E_{\alpha,\beta}^2 \leqslant (1 - \delta')^{|\beta \uparrow V|},$$

and by (7.36)

$$\mathbf{E}_{\beta,V} \leqslant (1 - \delta')^{\frac{|\beta \uparrow V|}{2}}.$$

Here, we may apply Lemma 7.17 analogously as in the previous proof, yielding that for $|\beta| \geqslant k := (\delta\delta')^{-2/c}$ we have the case $|\beta \uparrow V| < (\delta\delta')^{-1}$ with probability at most $\delta\delta'$. This gives us, again analogously as in the previous proof,

$$\mathbf{E}\left(\sum_{\beta:|\beta|>k} \mathbf{E}_{\beta,V}\right) \leqslant \frac{\delta}{4}. \tag{7.38}$$

Now we remind that we started from the assumption

$$\delta \leqslant |\mathbf{E}\left(A(f)B(g_1)B(g_2)\right)| = \left|\mathbf{E}\left(\sum_{\beta} \hat{B}_{\beta}^2 \mathbf{E}_{\beta,V}\right)\right|$$

which by (7.38) can be transformed to

$$\frac{3}{4}\delta \leqslant \left|\mathbf{E}\left(\sum_{\beta:|\beta|\leqslant k} \hat{B}_{\beta}^2 \mathbf{E}_{\beta,V}\right)\right| = \left|\mathbf{E}\left(\sum_{\beta:|\beta|\leqslant k} \sum_{\alpha \subseteq \beta \uparrow V} \hat{B}_{\beta}^2 \hat{A}_{\alpha} \mathbf{E}_{\alpha,\beta}\right)\right|$$

$$\leqslant \mathbf{E}\left(\sum_{\beta:|\beta|\leqslant k} \sum_{\alpha \subseteq \beta \uparrow V} \hat{B}_{\beta}^2 \left|\hat{A}_{\alpha}\right|\right)$$

where the last inequality comes from the observation $|\mathbf{E}_{\alpha,\beta}| \leqslant 1$.

Finally, we want to eliminate all α with $\left|\hat{A}_{\alpha}\right| \leqslant 2^{-k}\frac{\delta}{4}$. For each β with $|\beta| \leqslant k$ there may be at most 2^k possible $\alpha \subseteq \beta \uparrow V$. Hence

$$\sum_{\alpha \subseteq \beta \uparrow V:|\hat{A}_{\alpha}|<2^{-k}\frac{\delta}{4}} \left|\hat{A}_{\alpha}\right| \leqslant \frac{\delta}{4},$$

and by $\sum_{\beta} \hat{B}_{\beta}^2 \leqslant 1$, also

$$\mathbf{E}\left(\sum_{\beta:|\beta|\leqslant k} \sum_{\alpha \subseteq \beta \uparrow V:\ |\hat{A}_{\alpha}|<2^{-k}\frac{\delta}{4}} \hat{B}_{\beta}^2 \left|\hat{A}_{\alpha}\right|\right) \leqslant \frac{\delta}{4}.$$

Thus we have

$$\frac{1}{2}\delta \leqslant \mathbf{E}\left(\sum_{\beta:|\beta|\leqslant k} \sum_{\alpha \subseteq \beta \uparrow V:|\hat{A}_{\alpha}|\geqslant 2^{-k}\frac{\delta}{4}} \hat{B}_{\beta}^2 \left|\hat{A}_{\alpha}\right|\right).$$

Now we apply Lemma 7.8. We let $\gamma = 2^{-k}\delta/4 = 2^{-(\delta'\delta)^{-2/c}}\delta/4$, and obtain the claimed $v_2(\delta, \delta')$ as $v(\delta, \delta', \gamma, k)$ of that lemma. Please note, that condition $\alpha \subseteq \beta\!\uparrow\!V$ as well implies $|\alpha| \leqslant k$ (by $|\beta| \leqslant k$) as it serves as condition $\alpha \sqcap \beta$, guaranteeing the existence of some $y \in \beta$ with $y\!\uparrow\!V \in \alpha$ (remember that α has to be non-empty by $\hat{A}_\emptyset = 0$).

Thus Lemma 7.8 gives the desired satisfiability of φ.   ∎

8. Deriving Non-Approximability Results by Reductions

Claus Rick, Hein Röhrig

8.1 Introduction

In the last chapter we saw how optimal non-approximability results have been obtained for two canonical optimization problems. Taking these problems as a starting point, we will show how the best known non-approximability results for some other $\mathcal{APX}$-problems can be derived by means of computer-generated reductions. Most of the results surveyed in this chapter are from the work of Bellare, Goldreich, and Sudan [BGS95], Trevisan, Sorkin, Sudan, and Williamson [TSSW96], Crescenzi, Silvestri, and Trevisan [CST96] and Håstad [Hås97b].

8.1.1 The Concept of a Gadget

Since the early days of $\mathcal{NP}$-completeness theory, the technique of local replacements [GJ79] has been a common way to devise reductions. For example, in the reduction from SAT to 3SAT, each clause of a SAT formula is represented by a collection of clauses, each containing exactly three literals and built over the original variables and some new auxiliary variables. In the following we will call such finite combinatorial structures which translate constraints of one problem into a set of constraints of a second problem a "gadget". In order to prove the $\mathcal{NP}$-hardness of 3SAT we only need to make sure that *all* clauses in a gadget are satisfiable if and only if the corresponding SAT clause is satisfied (of course we also have to ensure that the gadgets can be constructed in polynomial time).

The concept is easily extended to show the $\mathcal{NP}$-hardness of optimization problems rather than decision problems. For example, consider the following reduction from 3SAT to MAX2SAT. Each clause $C_k = X_1 \vee X_2 \vee X_3$ is replaced by ten clauses

$$X_1, X_2, X_3, \neg X_1 \vee \neg X_2, \neg X_2 \vee \neg X_3, \neg X_3 \vee \neg X_1,$$

$$Y^k, X_1 \vee \neg Y^k, X_2 \vee \neg Y^k, X_3 \vee \neg Y^k.$$

Y^k is an auxiliary variable local to each gadget whereas X_1, X_2 and X_3 are primary variables which can also occur in gadgets corresponding to other clauses of the 3SAT instance. If clause C_k is satisfied, then 7 of the 10 clauses in the corresponding gadget are satisfiable by setting Y^k appropriately, and this is the maximum number of clauses satisfiable in each gadget; otherwise only 6 of the 10 clauses are satisfiable. So if we could solve MAX2SAT exactly we would also be able to decide 3SAT. This proves MAX2SAT to be $\mathcal{NP}$-hard.

8.1.2 Gap-Preserving Reductions

Gadgets conserve their importance in the field of non-approximability results of optimization problems. In order to prove non-approximability results we need an initial optimization problem which is $\mathcal{NP}$-hard to approximate to within a certain factor, i.e., which exhibits a certain gap. Gadgets then are to implement so-called "gap-preserving" reductions (for a formal definition see below).

The previous chapter showed a gap for MAXE3SAT: it is $\mathcal{NP}$-hard to distinguish satisfiable instances from instances where at most a fraction of $\frac{7}{8} + \varepsilon$ of the clauses can be satisfied at the same time. Using the same gadgets as above to replace each clause by ten 2SAT clauses we can immediately derive a non-approximability result for MAX2SAT. If the instance of MAXE3SAT is satisfiable, then exactly $\frac{7}{10}$ of the clauses of the corresponding instance of MAX2SAT are satisfiable. On the other hand, if a fraction of at most $\frac{7}{8}$ clauses of the MAXE3SAT instance can be satisfied at the same time, then a fraction of at most $\frac{6}{10}\frac{1}{8} + \frac{7}{10}\frac{7}{8}$ clauses in the corresponding instance of MAX2SAT are satisfiable. Thus we have established an $\mathcal{NP}$-hard gap for instances of MAX2SAT and we conclude that it is hard to approximate MAX2SAT to within a factor of $\frac{56}{55} - \varepsilon$.

In general, gap problems can be specified by two thresholds $0 \leqslant s \leqslant c \leqslant 1$. Let $|I|$ denote the size of a problem instance I, e.g., the number of clauses in an unweighted MAX3SAT instance or the sum of the weights associated to the clauses in a weighted MAX3SAT instance. Then $\mathbf{Opt}(I)/|I| \leqslant 1$ is the normalized optimum, i.e., the maximum fraction of satisfiable clauses in I. We can separate instances I where $\mathbf{Opt}(I)/|I| \geqslant c$ from instances where $\mathbf{Opt}(I)/|I| < s$, i.e., where the normalized optimum is either above or below the thresholds. For an optimization problem Π the corresponding gap problem is

GAP-$\Pi_{c,s}$
Instance: Given an instance I of Π in which the normalized optimum is guaranteed to be either above c or below s
Question: Which of the two is the case for I?

We are of course interested in gap problems which are $\mathcal{NP}$-hard to decide since this implies that the underlying problem is hard to approximate to within a factor of $\frac{c}{s}$. The larger the gap, the stronger the non-approximability result.

The PCP based approach to find such gap problems is as follows [Bel96]: First show that $\mathcal{NP}$ is contained in some appropriate PCP class, then show that this class reduces to a gap problem $\text{GAP-}\Pi_{c,s}$. The second step was straightforward in the last chapter: The computations of the verifiers were in one to one correspondence to instances of MAXE3SAT and MAXLINEQ3-2, respectively. The completeness and the soundness of the verifiers serve as thresholds for the corresponding gap problems.

Once we have identified an $\mathcal{NP}$-hard gap problem we can try to design reductions to other problems which preserve the gap within certain limits. We define a *gap-preserving* reduction between optimization problems as a reduction between the corresponding gap problems:

Definition 8.1 (gap-preserving reduction [AL96]). *Let Π and Π' be two maximization problems. A* gap-preserving *reduction from Π to Π' with parameters (c, s) and (c', s') is a polynomial time algorithm f that satisfies the following properties:*

- for every instance I of Π, algorithm f produces an instance $I' = f(I)$ of Π',

- $\mathbf{Opt}(I)/|I| \geqslant c \Rightarrow \mathbf{Opt}(I')/|I'| \geqslant c'$,

- $\mathbf{Opt}(I)/|I| < s \Rightarrow \mathbf{Opt}(I')/|I'| < s'$.

Lemma 8.2. *Given a gap-preserving reduction from Π to Π' with parameters (c, s) and (c', s'), $\mathcal{NP}$-hardness of $\text{GAP-}\Pi_{c,s}$ implies that achieving an approximation ratio c'/s' is $\mathcal{NP}$-hard for Π'.*

In the previous example we saw how to construct a gap-preserving reduction using a gadget. Gadgets suitable for deriving non-approximability results must meet some structural requirements. The following sections show how for a restricted class of problems, the notion of a gadget can be formalized, how its performance can be measured, and how optimal gadgets can be constructed using a computer.

8.2 Constraint Satisfaction Problems

Usually non-approximability results have been improved by constructing new PCP verifiers optimizing the query complexity, the soundness or other parameters of the proof systems (see Chapters 4.6 and 4.7 for a discussion on the developments concerning the clique problem). Such a clear interaction between the parameters of a proof system and the approximation ratio which is hard to achieve is not available for $\mathcal{APX}$-problems in general [Bel96]. A further improvement in recent PCP verifiers, especially in [BGS95] and in [Hås97b], has been the fact that the way in which a particular verifier accepted or rejected on the basis of the answers to the queries was made explicit and could be expressed by simple Boolean functions.

Theorem 8.3. *[Hås97b] For any $\varepsilon, \delta > 0$ there is a verifier V for* ROBE3SAT, *which*

- *uses $\mathcal{O}(\log n)$ random bits*
- *has completeness $1 - \varepsilon$ and soundness $\frac{1+\delta}{2}$*
- *reads exactly three bits from the proof and accepts iff the exclusive or of the three bits matches a value, which depends only on the input and the random string.*

This led Trevisan et al. [TSSW96, Tre97] to generalize the MAXSAT problem to "constraint satisfaction" problems. In a constraint satisfaction problem, arbitrary Boolean functions, called constraints, take the rôle of the clauses in MAXSAT. An instance of a constraint satisfaction problem becomes a collection of constraints and the goal is to maximize the number of satisfied constraints, i.e., to find an assignment to the Boolean variables such that a maximum number of constraints evaluate to TRUE. Restrictions in the choice of Boolean functions define different constraint satisfaction problems.

For a verifier from Theorem 8.3 we can associate a constraint with each random string of the verifier. These constraints are parity checks. If $x \in L$ then a fraction of c (the completeness of the verifier) of the constraints is satisfiable. If $x \notin L$ then a fraction less than s (soundness of the verifier) of the constraints are satisfiable.

Definition 8.4 (constraint function).   *A (k-ary) constraint function is a Boolean function $f : \{0,1\}^k \to \{0,1\}$.*

Definition 8.5 (constraint). *A constraint C over a variable set $\{X_1, \ldots, X_n\}$ is a pair $C = (f, (i_1, \ldots, i_k))$ where $f : \{0,1\}^k \to \{0,1\}$ is a constraint function and $i_j \in [n]$ for $j \in [k]$. Variable X_j is said to occur in C if $j \in \{i_1, \ldots, i_k\}$. All variables occuring in C are different. The constraint C is satisfied by an assignment $\vec{a} = a_1, \ldots, a_n$ to $X_1, \ldots, X_n$ if $C(a_1, \ldots, a_n) := f(a_{i_1}, \ldots, a_{i_k}) = 1$.*

Definition 8.6 (constraint family). *A constraint family $\mathcal{F}$ is a finite collection of constraint functions. We say that constraint C is from $\mathcal{F}$ if $f \in \mathcal{F}$.*

Example 8.7. We give a number of constraint families that will be used subsequently.

- *Parity check* is the constraint family $PC = \{PC_0, PC_1\}$, where, for $i \in \{0,1\}$, PC_i is defined as follows:

$$PC_i(a,b,c) = \begin{cases} 1 & \text{if } a \oplus b \oplus c = i \\ 0 & \text{otherwise} \end{cases}$$

Note that PC is the constraint family for encoding the operation of the type of verifier of Theorem 8.3.

- *Exactly-k-SAT* is the constraint family $E k \text{SAT} = \{f : \{0,1\}^k \rightarrow \{0,1\} : |\{\vec{a} : f(\vec{a}) = 0\}| = 1\}$

- *k-SAT* is the constraint family $k \text{SAT} = \bigcup_{\ell \in [k]} E l \text{SAT}$

- CUT is the constraint function $CUT : \{0,1\}^2 \rightarrow \{0,1\}$ with $CUT(a,b) = a \oplus b$

- DICUT is the constraint function $DICUT : \{0,1\}^2 \rightarrow \{0,1\}$ with $CUT(a,b) = \neg a \wedge b$

A *constraint satisfaction problem* can now be stated as

MAX $\mathcal{F}$

Instance: Given a collection $[C_1, C_2, \ldots, C_m]$ of constraints from a constraint family $\mathcal{F}$ on n variables

Problem: Find an assignment to the variables which maximizes the number of satisfied constraints.

Note that we allow the same constraint to occur several times in an instance of a constraint satisfaction problem and that MAXPC is equivalent to MAXLINEQ3-2. The computation of a verifier from Theorem 8.3 can be expressed as an instance of MAXPC. If the instance of ROBE3SAT is satisfiable then a fraction of $1 - \varepsilon$ of the constraints are satisfiable. Otherwise a fraction of at most $\frac{1+\delta}{2}$ of the constraints are satisfiable. This establishes a gap-preserving reduction from ROBE3SAT to Gap-MAXPC$_{1-\varepsilon, \frac{1+\delta}{2}}$. By Lemma 8.2 it is hard to approximate MAXPC to within $2 - \varepsilon$.

8.3 The Quality of Gadgets

In order to obtain hardness results for other constraint satisfaction problems our aim is now to develop gap-preserving reductions from MAXPC to the problem of interest. This section explains how gadgets can be used to construct such reductions and how the "quality" of the gadgets affects the preserved gap. We start by defining a first type of gadget.

Definition 8.8 ((α, β)-gadget).   *For $\alpha, \beta \in \mathbb{N}^+$, a constraint function $f : \{0,1\}^k \rightarrow \{0,1\}$, and a constraint family $\mathcal{F}$:*
an (α, β)-gadget reducing f to $\mathcal{F}$ is a finite collection $\Gamma = [C_1, C_2, \ldots, C_m]$ of constraints from $\mathcal{F}$ over primary *variables $X_1, \ldots, X_k$ and* auxiliary *variables $Y_1, \ldots, Y_n$, with the property that for Boolean assignments $\vec{a}$ to $X_1, \ldots, X_k$ and $\vec{b}$ to $Y_1, \ldots, Y_n$ the following relations are satisfied:*

$$(\forall \vec{a} : f(\vec{a}) = 1)(\forall \vec{b}) : \sum_j C_j(\vec{a}, \vec{b}) \ \leqslant \ \alpha \tag{8.1}$$

$$(\forall \vec{a} : f(\vec{a}) = 1)(\exists \vec{b}) : \sum_j C_j(\vec{a}, \vec{b}) \ = \ \alpha \tag{8.2}$$

$$(\forall \vec{a} : f(\vec{a}) = 0)(\forall \vec{b}) : \sum_j C_j(\vec{a}, \vec{b}) \ \leqslant \ \alpha - \beta. \tag{8.3}$$

The gadget is called strict if, in addition,

$$(\forall \vec{a} : f(\vec{a}) = 0)(\exists \vec{b}) : \sum_j C_j(\vec{a}, \vec{b}) \ = \ \alpha - \beta. \tag{8.4}$$

Remark 8.9.

– Our introductory reduction from 3SAT to MAX2SAT is a $(7, 1)$-gadget.

– The notion of strictness is not important to prove non-approximability; however, strict gadgets have applications in the construction of approximation algorithms (see Exercise 8.3).

A gadget may be thought of as an instance of the constraint satisfaction problem which is the target of the reduction. With respect to a reduction from a constraint function f, a gadget has the property that when restricted to any satisfying assignment $\vec{a}$ to $X_1, \ldots, X_k$ its maximum is exactly α, and when restricted to any unsatisfying assignment its maximum is at most $\alpha - \beta$. This means that there is a gap in the number of constraints that are satisfiable in these two cases. The fraction $\frac{\alpha}{\beta}$ is a characteristic value for the quality of a gadget with respect to the gap-preserving reduction in which it is used. The smaller this fraction, the better the gap preserved by the reduction, i.e., the stronger the obtained non-approximability result. This interaction will be made explicit in Theorem 8.13. If our source problem of the reduction consists only of instances formed by constraints from a single constraint function f, we can immediately state the following

Lemma 8.10. Let c and $s \in (0, 1)$. An (α, β)-gadget reducing $\mathcal{F}_0 = \{f\}$ to a constraint family $\mathcal{F}$ yields a gap-preserving reduction from MAX $\mathcal{F}_0$ to MAX $\mathcal{F}$ with parameters (c, s) and (c', s'), where c' and s' are constants satisfying

$$\frac{c'}{s'} \ \geqslant \ \frac{c\,\alpha}{\alpha - (1 - s)\beta}.$$

Proof. Assume that the instance from $\mathcal{F}_0 = \{f\}$ consists of m constraints (our source). The gadget produces an instance from $\mathcal{F}$ (our target). Let ℓ be the total number of constraints in a gadget used to replace each constraints of the source. In order to prove the claimed gap, we first need a lower bound on the number of constraints that is satisfiable in the target when more than cm constraints are satisfied in the source. From (8.2) we can conclude that

$$c' \geqslant \frac{cm\alpha}{lm}$$

On the other hand we need an upper bound when less than sm constraints are satisfied in the source. In this case, the error introduced by the gadget contributes $m(\alpha - \beta)$ satisfiable target constraints; for each satisfiable source constraint, we get β more satisfiable target constraints, i.e.,

$$s' \leqslant \frac{m(\alpha - \beta) + sm\beta}{lm}.$$

Note that (8.1) and (8.3) are necessary here to get the desired upper bound. ∎

What happens if our "source" problem is made up from constraint families with more than one constraint function, like PC= $\{PC_0, PC_1\}$? For $\mathcal{F} = \bigcup_i f_i$ and weights $w_i \in \mathbb{Q}^+, \sum_i w_i = 1$, we define

WMax $\mathcal{F}$

Instance: Given m constraints from a constraint family $\mathcal{F} = \bigcup_i f_i$ on n variables so that mw_i constraints are from f_i.

Problem: Find an assignment to the variables which maximizes the number of satisfied constraints.

For notational convenience, the following lemma, which is a generalization of Lemma 8.10, is only formulated for the case of $\mathcal{F}_0 = \{f_1, f_2\}$.

Lemma 8.11. *Let $\mathcal{F}_0 = \{f_1, f_2\}$ be a constraint family and let $w_1, w_2 \in \mathbb{Q}^+$, $w_1 + w_2 = 1$, $\alpha_1, \alpha_2, \beta \in \mathbb{N}$, $\alpha_1 \leqslant \alpha_2$. Let $\mathcal{F}$ be a constraint family. If there exist an (α_1, β)-gadget reducing f_1 to $\mathcal{F}$ and an (α_2, β)-gadget reducing f_2 to $\mathcal{F}$ then, for any $c \geqslant w_1$ and s, we have a gap-preserving reduction from WMax $\mathcal{F}_0$ to Max $\mathcal{F}$ with parameters (c, s), (c', s'), such that c' and s' satisfy*

$$\frac{c'}{s'} \geqslant \frac{w_1\alpha_1 + (c - w_1)\alpha_2}{w_1\alpha_1 + w_2\alpha_2 - (1 - s)\beta}.$$

Remark 8.12. We may assume that the gadgets have the same parameter β, since we can transform an (α_1, β_1)-f_1-gadget and an (α_2, β_2)-f_2-gadget into an $(\alpha_1\beta_2, \beta_1\beta_2)$-$f_1$-gadget and an $(\alpha_2\beta_1, \beta_1\beta_2)$-$f_2$-gadget, respectively. The quality of the gadgets is preserved by this transformation.

Proof of Lemma 8.11. Assume again that there are m constraints in our source problem, $w_1 m$ from f_1 and $w_2 m$ from f_2. Let ℓ_1, ℓ_2 be the number of replacement constraints. By the same reasoning as above,

$$c' \geqslant \frac{mw_1\alpha_1 + m(c - w_1)\alpha_2}{mw_1\ell_1 + mw_2\ell_2}$$

$$s' \leqslant \frac{mw_1(\alpha_1 - \beta) + mw_2(\alpha_2 - \beta) + ms\beta}{mw_1\ell_1 + mw_2\ell_2}.$$

∎

Recall that our main interest in gadgets was to use them to encode the computation of the verifier from Theorem 8.3. The following theorem shows that we should try to find (α, β)-gadgets where the ratio $\frac{\alpha}{\beta}$ is as small as possible.

Theorem 8.13 (induced gap). *For any constraint family $\mathcal{F}$, if there exists an (α_1, β)-gadget reducing PC_0 to $\mathcal{F}$ and an (α_2, β)-gadget reducing PC_1 to $\mathcal{F}$, then for any $\gamma > 0$, $\mathrm{MAX}\,\mathcal{F}$ is hard to approximate to within*

$$\left(1 - \frac{\beta}{\alpha_1 + \alpha_2}\right)^{-1} - \gamma.$$

Proof. We can simply apply the previous lemma with the parameters $c = 1 - \varepsilon$ and $s = \frac{1}{2} + \delta$ of the verifier from Theorem 8.3. For $\gamma = \gamma(\varepsilon, \delta)$, this gives us

$$\frac{c'}{s'} \geqslant \frac{1}{1 - \frac{\beta}{2(w_1\alpha_1 + w_2\alpha_2)}} - \gamma$$

and $\gamma \to 0$. We can ensure that at most half of the PC-constraints must be reduced with the larger α_i since otherwise, we can simply complement all variables in the reduction. The proof is completed by noting that under this condition, the given term is minimal for $w_1 = w_2 = 1/2$. ∎

Using Theorem 8.13 and the (4,1)-gadgets of Bellare et al. [BGS95] to reduce PC to MAXE3SAT, Håstad [Hås97b] was able to conclude that MAXE3SAT is hard to approximate to within $\frac{8}{7} - \varepsilon$ (see Theorem 7.11).

Theorem 8.13 tells us that we should use gadgets with ratio $\frac{\alpha}{\beta}$ as small as possible. It follows that an improved construction of a gadget may yield an improved non-approximability result. Therefore, it is desirable to have lower bounds on this "quality" of a certain gadget to understand if it is possible to improve the known construction or if it is the best possible one. This question was raised in [BGS95].

Despite its importance, the construction of gadgets remained an ad hoc task and no uniform methods were known; for each verifier using a different kind of test and each target problem new gadgets had to be found. Finally, Trevisan et al. [TSSW96] developed an approach to resolve this situation. They showed how the search for a gadget can be expressed as a linear program (LP). Solving the LP with the help of a computer allows us to find provably optimum gadgets.

The main idea is to limit the search space of possible gadgets to a finite and reasonably sized one. Note that this is not always possible (see Exercise 8.1). But we can achieve this goal for some important constraint families by addressing the following two questions.

1. Recall that a gadget is a collection of constraints and that constraints are formed from some constraint family and a set of variables. Can we limit the number of different possible constraints? The key will be a limitation of the number of auxiliary variables.

2. A gadget may contain the same constraint more than once. How do we cope with multiple occurrences of constraints? Note that in general there is no a priori bound on the number of occurrences of a constraint in an optimal gadget.

8.4 Weighted vs. Unweighted Problems

A simple way to deal with the second question is to assign a weight to each possible constraint. Multiple occurrences of the same constraint can then be represented by giving it a higher weight. So we are lead to define weighted constraint satisfaction problems.

<u>WEIGHTEDMAX $\mathcal{F}$</u>

Instance: Given m constraints $\{C_1, C_2, \ldots, C_m\}$ from a constraint family $\mathcal{F}$ on n variables and associated positive rational weights $w_1, w_2, \ldots, w_m$.

Problem: Find an assignment to the variables which maximizes the weighted sum $\sum_{i=1}^{m} w_i C_i$.

A gadget may again be thought of as an instance of a weighted constraint satisfaction problem. Note however, that in this way we would only get non-approximability results for weighted versions of optimization problems rather than for unweighted ones. But fortunately, there is a result due to Crescenzi et al. [CST96] which shows that for a large class of optimization problems, including constraint satisfaction problems, the weighted versions are exactly as hard to approximate as the unweighted ones.

In an instance of WEIGHTEDMAX $\mathcal{F}$ the weights can be arbitrary rationals. Let POLYMAX $\mathcal{F}$ denote those instances of WEIGHTEDMAX $\mathcal{F}$ where the weights are polynomially bounded integers and let SINGLEMAX $\mathcal{F}$ denote those instances of MAX $\mathcal{F}$ where a constraint occurs at most once. The key step in showing that unweighted constraint satisfaction problems are exactly as hard to approximate as weighted versions of constraint satisfaction problems is to prove that the polynomially bounded weighted version is $\alpha - AP$ reducible (see Definition 1.32) to the unweighted problem, i.e., POLYMAX $\mathcal{F} \leqslant_{AP}^{\alpha}$ SINGLEMAX $\mathcal{F}$, for any $\alpha > 1$. This implies that the approximation threshold of POLYMAX $\mathcal{F}$ is at most equal to the approximation threshold of SINGLEMAX $\mathcal{F}$. So the lower bounds obtained for the weighted versions are also valid for the unweighted versions and we can conclude that both versions have the same approximation threshold. The proof of the $\alpha - AP$ reducibility relies on the notion of a *mixing set* of tuples which is defined as follows.

Definition 8.14. *Let $n > 0, k > 0$ be integers. A set $S \subseteq [n]^k$ is called k-ary $(n, w, \varepsilon, \delta)$-mixing if for any k sets $A_1, \ldots, A_k \subseteq [n]$, each with at least δn elements, the following holds:*

$$\left| |S \cap A_1 \times \cdots \times A_k| - w \frac{|A_1| \cdots |A_k|}{n^k} \right| \leqslant \varepsilon w \frac{|A_1| \cdots |A_k|}{n^k}.$$

Note that by setting each $A_i = [n]$ we can immediately derive some bounds on the size of S from the definition, namely $(1 - \varepsilon)w \leqslant |S| \leqslant (1 + \varepsilon)w$. The existence of mixing sets can be shown by an easy probabilistic construction.

Lemma 8.15. *For any integer $k \geqslant 2$ and for any two rationals $\varepsilon, \delta > 0$, two constants n_0 and c exist such that for any $n \geqslant n_0$ and for any w with $cn \leqslant w \leqslant n^k$, a k-ary $(n, w, \varepsilon, \delta)$-mixing set exists.*

Proof. Construct a set $S \subseteq [n]^k$ by randomly selecting each element from $[n]^k$ with probability w/n^k. Then for any k sets $A_1, \ldots, A_k \subseteq [n]$ we have

$$m := \mathbf{E}(|S \cap A_1 \times \ldots \times A_k|) = w \frac{|A_1| \cdot \ldots \cdot |A_k|}{n^k}$$

and therefore using Chernoff bounds (see Theorem 3.5) we get under the assumption that all sets A_i satisfy $|A_i| \geqslant \delta n$ for $i = 1, \ldots, k$

$$\begin{aligned}
\text{Prob}\left[||S \cap A_1 \times \ldots \times A_k| - m| \geqslant \varepsilon m\right] &\leqslant \left(\frac{e^\varepsilon}{(1 + \varepsilon)^{1+\varepsilon}} \right)^m \\
&\leqslant \left(\frac{e^\varepsilon}{(1 + \varepsilon)^{1+\varepsilon}} \right)^{cn\delta^k}.
\end{aligned}$$

Since $\left(\frac{e^\varepsilon}{(1+\varepsilon)^{1+\varepsilon}} \right)^{\delta^k} < 1$ for all $\varepsilon > 0$ and all $0 < \delta < 1$ we have shown that there exists a set S that is $(n, w, \varepsilon, \delta)$-mixing for $c = 1$ and $n_0 = 1$. ∎

In [Tre97] Trevisan proved that in fact mixing sets can be constructed in polynomial time.

Theorem 8.16. *For any integer $k \geqslant 2$ and for any two rationals $\varepsilon, \delta > 0$, two constants n_0 and c exist such that, for any $n \geqslant n_0$ and for any w with $cn \leqslant w \leqslant n^k$, a k-ary $(n, w, \varepsilon, \delta)$-mixing set can be constructed in time polynomial in n.*

Theorem 8.17. *Let $\mathcal{F}$ be a constraint family where each constraint function has arity $\geqslant 2$. Then, for any $\alpha > 1$, it is the case that POLYMAX $\mathcal{F} \leqslant_{AP}^{\alpha}$ SINGLEMAX $\mathcal{F}$.*

Proof. Let $r > 1$ be an arbitrary but fixed rational (a supposed approximation factor of an approximation algorithm for SINGLEMAX $\mathcal{F}$). Our goal is to develop a reduction which, given any $\alpha > 1$, allows us to construct an approximation algorithm for POLYMAX $\mathcal{F}$ with approximation factor $1 + \alpha(r - 1)$.

Let $\Phi = (C_1, \ldots, C_m, w_1, \ldots, w_m)$ be an instance of POLYMAX $\mathcal{F}$ over the variable set $X = \{x_1, \ldots, x_n\}$. Let h be the largest arity of a constraint in $\mathcal{F}$. The idea to represent the weights in an unweighted instance is to duplicate the variables and constraints in a suitable way by making use of the properties of mixing sets. We can choose $\varepsilon, \delta \in (0, 1/2)$ such that

$$\alpha \geqslant \frac{1}{r-1} \left(\frac{1 + \varepsilon + 2^h(1 + \varepsilon - (1 - \varepsilon)(1 - \delta)^h)}{1 - \varepsilon} r - 1 \right).$$

Let n_0 and c be constants such that Theorem 8.16 holds for every $k = 2, \ldots, h$ and let $w_{max} = \max_i\{w_i\}$. Now we rescale the weights by multiplying each w_i with a factor of $c^2 n_0^2 w_{max}$ and choose $N = c n_0 w_{max}$. Then $N > n_0$ and $cN \leqslant w_i \leqslant N^2$ holds for all $i \in [m]$. Theorem 8.16 ensures that, for every $cN \leqslant w \leqslant N^2$ and every $k = 2, \ldots, h$, a k-ary $(N, w, \varepsilon, \delta)$-mixing set S_k^w can be efficiently constructed. We are now ready to define an instance $\hat{\Phi}$ of SINGLEMAX $\mathcal{F}$ as follows. The variable set $\hat{X}$ of $\hat{\Phi}$ has N copies of every variable of X:

$$\hat{X} = \{x_i^j : i \in [n], j \in [N]\}.$$

For a k-ary constraint $C_j = f(x_{i_1}, \ldots, x_{i_k})$ of weight w_j we define a set of constraints, $\hat{C}_j$, as follows:

$$\hat{C}_j = \{f(x_{i_1}^{a_1}, \ldots, x_{i_k}^{a_k}) : (a_1, \ldots, a_k) \in S_k^{w_j}\}.$$

Finally $\hat{\Phi}$ is defined as the union of the $\hat{C}_j$ sets

$$\hat{\Phi} = \bigcup_{j \in [m]} \hat{C}_j.$$

First of all we can show that the values of optimal solutions are similar. We claim that for any assignment τ for Φ there exists an assignment $\hat{\tau}$ for $\hat{\Phi}$ such that $v(\hat{\Phi}, \hat{\tau}) \geqslant (1 - \varepsilon)v(\Phi, \tau)$. To see this, simply define $\hat{\tau}$ as follows: for any variable $x_i^j \in \hat{X}$ set $\hat{\tau}(x_i^j) = \tau(x_i)$. We then have

$$v(\hat{\Phi}, \hat{\tau}) = \sum_j |\{C \in \hat{C}_j : \hat{\tau} \text{ satisfies } C\}| = \sum_{j:\tau \text{ satisfies } C_j} |\hat{C}_j|$$

$$= \sum_{j:\tau \text{ satisfies } C_j} |S_k^{w_j}| \geqslant (1 - \varepsilon)v(\Phi, \tau).$$

As a corollary we get $\mathbf{Opt}(\hat{\Phi}) \geqslant (1 - \varepsilon)\mathbf{Opt}(\Phi)$. On the other hand, we need a way to derive an assignment τ for Φ given any assignment $\hat{\tau}$ for $\hat{\Phi}$. We claim that this can be done in polynomial time in such a way that

$$v(\Phi, \tau) \geqslant \frac{1}{(1+\varepsilon)(1+2^h(1-(1-\delta)^h))} v(\hat{\Phi}, \hat{\tau}).$$

In a first step, we will transform $\hat{\tau}$ into an assignment $\hat{\tau}'$ such that for every variable $x_i \in X$ at least δN and at most $(1-\delta)N$ copies of x_i are set to TRUE. If this condition is not fulfilled for, say $x_i \in X$, we can switch the value of some of its copies. Note that it is always possible to achieve the condition by switching at most δN copies. Unfortunately, this might reduce the number of satisfied constraints. The difference between the measure of $\hat{\tau}$ and that of $\hat{\tau}'$ is bounded by the number of constraints where a switched variable occurs. We claim that in $\hat{C}_j$ this number is at most $(1+\varepsilon)w_j - (1-\varepsilon)(1-\delta)^k w_j$ where k is the arity of C_j. By construction $|\hat{C}_j| \leqslant (1+\varepsilon)w_j$. Since at most δN copies of a variable are switched, we know that at least $(1-\delta)N$ copies remain unchanged. Since $\hat{C}_j$ was constructed by a mixing set $S_k^{w_j}$, we can conclude that in at least $(1-\varepsilon)(1-\delta)^k w_j \leqslant (1-\varepsilon)(1-\delta)^h w_j$ constraints of $\hat{C}_j$ no variable was switched, and this proves the claim. Summing over all $j \in [m]$ we get

$$v(\hat{\Phi}, \hat{\tau}) - v(\hat{\Phi}, \hat{\tau}') \leqslant w_{tot} \cdot (1 + \varepsilon - (1-\varepsilon)(1-\delta)^h)$$

where $w_{tot} = \sum_j w_j$. Now we will show how to derive an assignment for τ from $\hat{\tau}'$. For any variable $x_i \in X$, let

$$p_i = \frac{|\{j : \hat{\tau}'(x_i^j) = \text{TRUE}\}|}{N}$$

be the fraction of copies of x_i whose value is set to TRUE by $\hat{\tau}'$. We can derive a random assignment τ_R for the variables in X by setting variable x_i to TRUE with probability p_i, i.e., $\text{Prob}[\tau_R(x_i) = \text{TRUE}] = p_i$. We are now interested in the probability that such a random assignment satisfies a k-ary constraint $C_j = f(x_{i_1}, \ldots, x_{i_k})$ of Φ of weight w_j. Let $SAT_j \subseteq \{\text{TRUE}, \text{FALSE}\}^k$ be the set of satisfying assignments to the variables of C_j. We then have

$$\text{Prob}[\tau_R \text{ satisfies } C_j]$$

$$= \sum_{(b_1, \ldots, b_k) \in SAT_j} \text{Prob}[\tau_R(x_{i_1}) = b_1, \ldots, \tau_R(x_{i_k}) = b_k]$$

$$= \sum_{(b_1, \ldots, b_k) \in SAT_j} \frac{\prod_{i=1}^k |\{j \in [N] : \hat{\tau}'(x_i^j) = b_i\}|}{N^k}$$

Note that each set of indices $A_i = \{j \in [N] : \hat{\tau}'(x_i^j) = b_i\}$ has size at least δN. Thus, we can use the mixing property of the set $S_k^{w_j}$ and continue

$$\geqslant \sum_{(b_1, \ldots, b_k) \in SAT_j} \frac{|S_k^{w_j} \cap A_1 \times \ldots \times A_k|}{(1+\varepsilon)w_j}$$

$$\geqslant \frac{|\{C \in \hat{C}_j : \hat{\tau}' \text{ satisfies } C\}|}{(1+\varepsilon)w_j}.$$

So the expectation of the measure of our random assignment is $\mathbf{E}[v(\Phi, \tau_R)] \geqslant v(\hat{\Phi}, \hat{\tau}')/(1 + \varepsilon)$. Using the method of conditional probabilities introduced in Chapter 3.2, we can turn this random assignment into a deterministic assignment τ of measure at least the expectation in polynomial time. Also note that setting each variable to TRUE or FALSE with probability $1/2$ has average measure at least $2^{-h}w_{tot}$. Taking the better of these two possibilities, the measure $v(\Phi, \tau)$ can be estimated as follows:

$$v(\Phi, \tau) \;\geqslant\; \frac{1}{1 + \varepsilon}(v(\hat{\Phi}, \hat{\tau}) - w_{tot}(1 + \varepsilon - (1 - \varepsilon)(1 - \delta)^h))$$

$$\geqslant\; \frac{1}{1 + \varepsilon}(v(\hat{\Phi}, \hat{\tau}) - 2^h v(\Phi, \tau)(1 + \varepsilon - (1 - \varepsilon)(1 - \delta)^h))$$

$$v(\Phi, \tau) \geqslant \frac{1}{(1 + \varepsilon + 2^h(1 + \varepsilon - (1 - \varepsilon)(1 - \delta)^h))}v(\hat{\Phi}, \hat{\tau}).$$

Assuming that $\hat{\tau}$ is a solution for $\hat{\Phi}$ with performance ratio at most r we have that

$$\frac{\mathbf{Opt}(\Phi)}{v(\Phi, \tau)} \;\leqslant\; \frac{\mathbf{Opt}(\hat{\Phi})/(1 - \varepsilon)}{v(\hat{\Phi}, \hat{\tau})/(1 + \varepsilon + 2^h(1 + \varepsilon - (1 - \varepsilon)(1 - \delta)^h))}$$

$$\leqslant\; \frac{(1 + \varepsilon + 2^h(1 + \varepsilon - (1 - \varepsilon)(1 - \delta)^h))}{1 - \varepsilon}r$$

$$\leqslant\; 1 + \alpha(r - 1)$$

by the choice of ε and δ.   ■

It remains to show that instances of WEIGHTEDMAX $\mathcal{F}$ are exactly as hard to approximate as instances of POLYMAX $\mathcal{F}$. This is in fact true for a wide range of problems in which a feasible solution is a subset of objects from the input, the measure is the sum of weights associated with the objects, and the feasibility of the solution is independent of these weights. WEIGHTEDMAX $\mathcal{F}$ and WEIGHTEDMAXCUT are typical examples of such problems. The proof, which uses a scaling argument, can be found in [Tre97].

8.5 Gadget Construction

From now on we will look at weighted sums in (8.1) – (8.4) and the search for gadgets reduces to the problem of selecting a weight for each possible constraint (a weight of zero would mean that the corresponding constraint is not present in the gadget). This would be no progress, however, if we allowed only integer weights. Allowing real weights makes the problem even harder at the first glance. However, this allows us to formulate a linear program instead of an integer program. Thus, choosing the weights will be no problem if the LP is reasonably sized.

A further advantage of allowing real weights is the following rescaling property which enables us to introduce normalized gadgets. The absolute values of α and β are not important to us as long as the gap $\frac{\alpha}{\alpha-\beta}$ remains the same.

Lemma 8.18. *For every (α, β)-gadget there exists an $(\alpha', 1)$-gadget satisfying $\frac{\alpha'}{\alpha'-1} = \frac{\alpha}{\alpha-\beta}$.*

Proof. Choose $\alpha' = \frac{\alpha}{\beta}$ and rescale all the weights by $\frac{1}{\beta}$. ∎

As a consequence we can drop the second parameter β and this gives a simplified definition of a gadget.

Definition 8.19 (α-gadget). *For $\alpha \in \mathbb{R}^+$, a constraint function f, and a constraint family $\mathcal{F}$:*
an α-gadget reducing f to $\mathcal{F}$ is a finite collection of constraints C_j from $\mathcal{F}$ and associated real weights $w_j \geqslant 0$ such that the following relations are satisfied:

$$(\forall \vec{a} : f(\vec{a}) = 1)(\forall \vec{b}) : \sum_j w_j C_j(\vec{a}, \vec{b}) \leqslant \alpha \tag{8.5}$$

$$(\forall \vec{a} : f(\vec{a}) = 1)(\exists \vec{b}) : \sum_j w_j C_j(\vec{a}, \vec{b}) = \alpha \tag{8.6}$$

$$(\forall \vec{a} : f(\vec{a}) = 0)(\forall \vec{b}) : \sum_j w_j C_j(\vec{a}, \vec{b}) \leqslant \alpha - 1. \tag{8.7}$$

The gadget is called strict *if, in addition,*

$$(\forall \vec{a} : f(\vec{a}) = 0)(\exists \vec{b}) : \sum_j w_j C_j(\vec{a}, \vec{b}) = \alpha - 1. \tag{8.8}$$

So our aim will be the construction of an α-gadget for a particular reduction with α as small as possible. Now we address the other question raised at the end of Section 8.3, namely how to limit the number of different constraints that can be formulated over a given constraint family and a set of variables. In some cases, depending on the properties of the constraint family which is the target of the reduction, the gadget search space can be made finite. This is done by showing how auxiliary variables which are "duplicates" of others can be eliminated by means of proper substitutions. Our goal is to show how an α-gadget Γ can be transformed into an α'-gadget Γ' which uses fewer auxiliary variables and where $\alpha' \leqslant \alpha$. In order to keep the gap of Γ during the transformation we have to pay special attention to equation (8.6) since it ensures the maximum of the gadget to be α when $f(\vec{a}) = 1$.

Definition 8.20 (witness). *We say that function $b : \{0,1\}^k \to \{0,1\}^n$ is a witness for the gadget if equation (8.6) (and, for a strict gadget, also equation (8.8)) is satisfied when setting $\vec{b} = b(\vec{a})$. Let $b_j(\vec{a}) := b(\vec{a})_j$.*

For our introductory example, which is a strict 7-gadget, one witness (it is not unique) is the function $b(X_1, X_2, X_3) = X_1 \wedge X_2 \wedge X_3$.

Definition 8.21 (witness matrix). *Let f be a k-ary function with s satisfying assignments $\vec{a}^{(1)}, \ldots, \vec{a}^{(s)}$. For a gadget Γ with n auxiliary variables reducing f to a family $\mathcal{F}$, a witness matrix of Γ is an $s \times (k + n)$ matrix (c_{ij}) such that*

$$
c_{ij} = \begin{cases} \vec{a}_j^{(i)} & j \leqslant k \\ b_{j-k}(\vec{a}^{(i)}) & k < j \leqslant k + n \end{cases}
$$

where b is the witness function for the gadget. For strict gadgets the matrix is augmented appropriately.

Example 8.22. Assume the following are the constraints of an α-gadget reducing PC_0 to 3SAT: $[x_1, x_2, x_3, x_1 \vee y_1, \bar{x}_3 \vee y_2, x_2 \vee \bar{y}_1, \ldots, x_3 \vee y_1 \vee y_4]$ (see Figure 8.1). A witness matrix might look like this

$$
\begin{pmatrix}
0 & 0 & 0 & 0 & 1 & 1 & 0 & \cdots & 1 \\
0 & 1 & 1 & 1 & 1 & 0 & 1 & \cdots & 1 \\
1 & 0 & 1 & 0 & 0 & 1 & 0 & \cdots & 1 \\
1 & 1 & 0 & 1 & 0 & 1 & 1 & \cdots & 0
\end{pmatrix}.
$$

Note that the first three columns correspond to the four assignments to variables x_1, x_2 and x_3 which make $PC_0(x_1, x_2, x_3)$ TRUE. Given the assignment to the variables prescribed by one row, the weights of those constraints which evaluate to TRUE sum up to exactly α.

The first k columns of a witness matrix correspond to primary variables and the other columns correspond to auxiliary variables. Suppose that two columns j and j' in the witness matrix are the same and that at least one of these columns, say j', corresponds to an auxiliary variable $Y_{j'}$.

Now consider replacing each occurrence of $Y_{j'}$ in the gadget by the variable corresponding to column j thereby eliminating $Y_{j'}$. This new gadget Γ' satisfies (8.6) (and (8.8)) since the two columns in the witness matrix of Γ were identical. Moreover, since the range of the universal quantification for Γ' is smaller than that for Γ the new gadget will also satisfy (8.5) and (8.7).

If none of the constraints of Γ which contained $Y_{j'}$ contained the variable corresponding to column j prior to the replacement, everything is fine and we have constructed an equivalent gadget using one auxiliary variable less. In general, however, a variable may occur twice in a constraint after the substitution. So the new gadget would no longer be a gadget over the family $\mathcal{F}$. Fortunately a lot of constraint families satisfy the following definition which resolves this inconvenience.

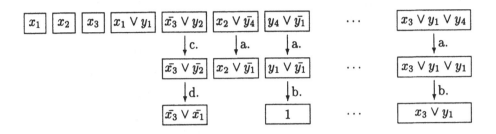

Fig. 8.1. Eliminating auxiliary variables y_4 and y_2 from a gadget (the corresponding witness matrix is given in Example 8.22): a. replacing y_4 by y_1; b. using the hereditary property; c. exploiting that 3SAT is complementation-closed; d. replacing y_2 by x_1.

Definition 8.23 (hereditary). *A constraint family $\mathcal{F}$ is hereditary if for any $f_i(X_1, \ldots, X_{n_i}) \in \mathcal{F}$, and two indices $j, j' \in [n_i]$, the function f_i when restricted to $X_j \equiv X_{j'}$ and considered as a function of $n_i - 1$ variables, is identical (up to the order of arguments) to some other function $f_{i'} \in \mathcal{F} \cup \{0, 1\}$, where $n_{i'} = n_i - 1$ (and 0 and 1 denote the constant functions).*

If, for example, we have an illegal constraint $(X_1 \vee X_1 \vee X_3)$ after the substitution, we may replace this constraint by $(X_1 \vee X_3)$ if the constraint family is 3SAT. This is not possible for E3-SAT. CUT is hereditary, too. We thus have

Lemma 8.24. *Let Γ be an α-gadget reducing f to a hereditary family $\mathcal{F}$ such that it has a witness function for which two auxiliary variables are identical, or an auxiliary variable is identical to a primary variable. Then there is an α'-gadget Γ' using one auxiliary variable less, and with $\alpha' \leqslant \alpha$. If Γ is strict, so is Γ'.*

Proof. The proof is immediate from the discussion above. To see that $\alpha' \leqslant \alpha$ we note that if constants are produced after the substitutions, subtracting them from both sides of the gadget-defining inequalities produces an $(\alpha - w)$-gadget, where w is the total weight of constraints replaced by 1's. ∎

Definition 8.25. *For a constraint f, call two primary variables $X_{j'}$ and X_j distinct if there exists an assignment $\vec{a}$, satisfying f, for which $a_{j'} \neq a_j$.*

Corollary 8.26 (upper bound on auxiliary variables). *Suppose f is a constraint on k variables, with s satisfying assignments and $k' \leqslant k$ distinct variables. If there is an α-gadget reducing f to a hereditary constraint family $\mathcal{F}$, then there is an α-gadget with at most $2^s - k'$ auxiliary variables.*

Proof. The domain of the witness function b is $\{a : f(a) = 1\}$, a set of cardinality s (note that for strict gadgets the domain is of size 2^k), so the witness matrix has s rows, and the number of distinct columns is at most 2^s. Taking into account that k' primary variables are distinct, there can be no more than $2^s - k'$ distinct columns corresponding to auxiliary variables. Exceeding this limit would allow us to eliminate one auxiliary variable via Lemma 8.24. ∎

Another additional property allows us to reduce the number of auxiliary variables even further.

Definition 8.27 (complementation-closed). *A family $\mathcal{F}$ is complementation-closed if it is hereditary and, for any $f_i(X_1, \ldots, X_{n_i}) \in \mathcal{F}$, and any index $j \in [n_i]$, the function f_i' given by $f_i'(X_1, \ldots, X_{n_i}) = f_i(X_1, \ldots, X_{j-1}, \neg X_j, X_{j+1}, \ldots, X_{n_i})$ is contained in $\mathcal{F}$.*

Notice that for a complementation-closed family $\mathcal{F}$, the hereditary property implies that if $f_i(X_1, \ldots, X_{n_i})$ is contained in $\mathcal{F}$ then so is the function f_i restricted to $X_j \equiv \neg X_{j'}$ for any two distinct indices $j, j' \in [n_i]$. This guarantees that we can make substitutions even if two columns in the witness matrix are complements of each other. This gives us an improved upper bound on the number of auxiliary variables in Corollary 8.26, namely $2^{s-1} - k'$. 2SAT is a complementation-closed family but not CUT or DICUT.

In some cases (e.g. 2SAT but not CUT), there is also no need to consider columns in the witness matrix which are identically 0 or identically 1 (clauses in which the corresponding auxiliary variables appear can be replaced by shorter clauses eliminating those variables).

We now show how to find optimal gadgets via an appropriate formulation of a linear program (LP). As an example, we are looking for an α-gadget reducing PC_0 to 2SAT with minimum value of α. According to our previous discussion we can restrict our search to gadgets over at most 7 variables: 3 primary variables and $4 = 2^{4-1} - 4$ auxiliary variables. 2SAT is complementation-closed, there are 4 satisfying assignments to PC_0, and all primary variables are distinct. Over 7 variables, $2 \cdot 7 + 4 \cdot \binom{7}{2}$ 2SAT constraints can be defined; call them $C_1, \ldots, C_{98}$. A gadget over 7 variables can thus be identified with the vector $(w_1, \ldots, w_{98})$ of the weights of the constraints. We can now formulate the following linear program with $4 \cdot 16 + 4 + 4 \cdot 16 = 132$ constraints over 99 variables:

$$
\begin{aligned}
&\text{minimize } \alpha \\
&\text{subject to} \\
&(\forall \vec{a} : f(\vec{a}) = 1) \quad (\forall \vec{b}) : \sum_j w_j C_j(\vec{a}, \vec{b}) \leq \alpha \\
&(\forall \vec{a} : f(\vec{a}) = 1) : \quad \sum_j w_j C_j(\vec{a}, b(\vec{a})) = \alpha \\
&(\forall \vec{a} : f(\vec{a}) = 0) \quad (\forall \vec{b}) : \sum_j w_j C_j(\vec{a}, \vec{b}) \leq \alpha - 1 \\
&\hspace{5.5cm} \alpha \geq 0 \\
&(\forall j \in [98]) : \hspace{3.7cm} w_j \geq 0.
\end{aligned}
$$

The attentive reader might wonder which witness function has to be used in the above LP since it is not known a priori. Recall however, that we limited the number of auxiliary variables in such a way that, apart from permutations, there is only one possibility to form the witness matrix without one column being the same or the complement of another column.

Theorem 8.28 (optimal gadget). *An optimal LP solution yields an optimal α-gadget (one where α is as small as possible).*

Proof. Of course any feasible solution to the LP is a weight vector which describes an α-gadget. Recall from the discussion in the previous section that any "duplicated" variables can be eliminated from an α-gadget to give a "simplified" α'-gadget ($\alpha' \leqslant \alpha$). The α'-gadget can be associated with its (fixed-length) vector of weights and provides a feasible solution to the associated LP. ∎

The LP given above has optimal solution $\alpha = 11$, proving the optimality of the gadget given in [BGS95]. A similar gadget can be used to reduce PC_1 to 2SAT. Applying Theorem 8.13 this shows that MAX2SAT is hard to approximate to within $\frac{22}{21} - \varepsilon$.

8.6 Improved Results

As an application of the presented techniques, we give an improved non-approximability result for the weighted variant of MAXCUT:

WEIGHTEDMAXCUT
Instance: A pair (G, ω) where $G = (V, E)$ is an undirected graph and $\omega : E \to \mathbb{Q}$.
Problem: Find a partition of V into disjoint sets V_1 and V_2, so that the sum of the weights of the edges with one endpoint in V_1 and one endpoint in V_2 is maximal.

The idea is to find gadgets reducing PC_0 and PC_1 to the CUT-family and to apply Theorem 8.13. Unfortunately, no gadget as in Definition 8.19 can reduce any member of PC to CUT: For any setting of the variables which satisfies line (8.6), the complementary setting of the primary variables has the opposite parity, i.e., is a non-satisfying assignment. However, complementing the auxiliary variables too, the CUT-constraints are unchanged, therefore the gadget's value is still α, violating relation (8.7) in Definition 8.19. As a way out, [BGS95] generalized the definition of a gadget by introducing a "special" auxiliary variable, which decides on what side of the cut we reside:

Definition 8.29. *A gadget with auxiliary constant 0 is a gadget like in Definition 8.19, except that relations (8.5-8.8) are only required to hold when $Y_1 = 0$.*

All previous results about gadgets still hold for gadgets with auxiliary constant 0 (reductions remain valid as long as solutions remain optimal when all variables are complemented).

There are $s = 4$ satisfying assignments for PC_0. The CUT constraint family is hereditary and all $k = 3$ variables in PC_0 are distinct, therefore (by Corollary 8.26) at most $2^s - k = 13$ auxiliary variables are needed. Setting up the linear program, we note that only $\binom{16}{2} = 120$ CUT constraints are possible on 3+13 variables, so there will be 120 weight variables in the LP. Since we only need to consider the case $Y_1 = 0$, there will be $2^{16-1} + 4$ LP-constraints. Solving this LP and the corresponding LP for PC_1 on the computer, we get

Lemma 8.30. *There exists an 8-gadget with constant 0 reducing PC_0 to CUT, which is optimal and strict. There exists a 9-gadget reducing PC_1 to CUT, which is optimal and strict.*

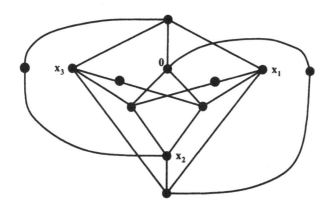

Fig. 8.2. 8-gadget reducing PC_0 to CUT. Every edge has weight 0.5. Four nodes (variables) are missing since all constraints (edges) involving these variables were assigned zero weight.

In Figure 8.2 the PC_0 gadget constructed in [TSSW96] is shown, vertices correspond to variables and edges to constraints. As an example, consider setting x_1, x_2, and x_3 to 0, which is a satisfying assignment for PC_0. Setting all auxiliary variables to 1, i.e., placing them on one side, we see that exactly four edges crossing the cut leave each of the four nodes on the other side. Plugging the gadgets into Theorem 8.13, we get

Theorem 8.31. *For every $\gamma > 0$, WEIGHTEDMAXCUT is hard to approximate to within $17/16 - \gamma$.*

Corollary 8.32. *For every* $\gamma > 0$, MAXCUT *is hard to approximate to within* $17/16 - \gamma$.

Proof. See [CST96].   ∎

Table 8.1 gives some additional non-approximability results obtained by constructing optimal gadgets which reduce MAXLINEQ3-2 (PC) to the problem of interest.

problem	α-gadget	lower bound		upper bound
MAXE2SAT	11	$\frac{22}{21} - \varepsilon$	1.0476	1.0741
MAXLINEQ2-2	6	$\frac{12}{11} - \varepsilon$	1.0909	1.1383
MAXCUT	8, 9	$\frac{17}{16} - \varepsilon$	1.0624	1.1383
MAXDICUT	6.5	$\frac{13}{12} - \varepsilon$	1.0833	1.164

Table 8.1. Results obtained by reductions using gadgets.

Exercises

Exercise 8.1. Consider reducing $f(x) = x$ to $g(x_1, x_2, x_3) = x_1 \oplus x_2 \oplus x_3$. For any fixed k we can build the following gadget: $x \oplus y_i \oplus y_j \; \forall 1 \leqslant i < j \leqslant k$.

– Show that, as k tends to infinity, the gap of the gadget approaches 2.

– Show that no finite gadget can achieve this.

Exercise 8.2. Let $\mathcal{F}$ be a constraint family where each constraint function has at most r arguments. Assume $\mathcal{F}$ is not hereditary. Consider reducing f to $\mathcal{F}$ by means of an α-gadget Γ. Assume the witness matrix has $r+1$ equal columns and at least one column corresponds to an auxiliary variable. Can we construct an α-gadget Γ' reducing f to $\mathcal{F}$ using one auxiliary variable less? This would prove that there is an α-gadget with at most $r2^s - k'$ auxiliary variables.

Exercise 8.3. There is an application of strict gadgets to approximation problems: The best known approximation algorithms for WEIGHTEDMAX3SAT take advantage of the fact that we have much better approximation algorithms for WEIGHTEDMAX2SAT than for WEIGHTEDMAX3SAT.

Assume that for each $f \in$ E3SAT, we have a strict α-gadget reducing f to 2SAT. Let I be an instance of WEIGHTEDMAX3SAT with clauses of length three of total weight m. Let I' be the WEIGHTEDMAX2SAT-instance obtained by applying the gadgets (the weight of the new constraints is the product of the source constraint's weight and the weights associated to the gadget). Show that

1. $\mathbf{Opt}(I') = \mathbf{Opt}(I) + m(\alpha - 1)$

2. Given a feasible solution of I' of weight w, the corresponding solution of I has weight at least $w - m(\alpha - 1)$

3. This gives an L-reduction [Pap94].

Thus, approximation algorithms for WEIGHTEDMAX2SAT can be used to obtain solutions for WEIGHTEDMAX3SAT. Based on this construction and a WEIGHTEDMAX2SAT algorithm from [FG95], Trevisan [Tre97] derives an improved approximation algorithm for WEIGHTEDMAX3SAT.

9. Optimal Non-Approximability of MaxClique

Martin Mundhenk, Anna Slobodová

9.1 Introduction

In Chapter 4 we have seen how parameters of PCP-systems – like query complexity, free-bit complexity, and amortized free bit complexity – for $\mathcal{NP}$-complete problems imply non-approximability results for MaxClique. For example, Bellare, Goldreich, and Sudan [BGS95], constructed a verifier with amortized free bit complexity $2 + \varepsilon_0$, for any $\varepsilon_0 > 0$. The resulting non-approximability factor for was $n^{1/3-\varepsilon_0}$. The novelty of the verifier in [BGS95] with respect to previous works was the use of the so-called long code as an encoding scheme for the proofs. Using a new test for the long code, Håstad decreased the amortized free bit complexity to $1 + \varepsilon_0$ [Hås96b], which improved the non-approximability factor of MaxClique towards $n^{1/2-\varepsilon_0}$. The history of the non-approximability results for MaxClique ended-up with an optimal factor of $n^{1-\varepsilon_0}$ by Håstad, obtained by the construction of a verifier that has logarithmic randomness, perfect completeness and arbitrarily small amortized free bit complexity.

Theorem 9.1 (Main Theorem). [Hås96a]
    Let A be a set in $\mathcal{NP}$. Then $A \in \overline{\mathrm{FPCP}}[\log n, \varepsilon_0]$, for every $\varepsilon_0 > 0$.

Before we go into details of the proof, we want to review how it implies the non-approximability result for MaxClique from Theorem 4.19. A standard technique to show hardness of approximation is to prove the $\mathcal{NP}$-hardness of the corresponding promise problem with a specified gap. Particularly, for MaxClique we define GAP-MaxClique$_{c,s}$, for any functions $0 < s(n) \leqslant c(n) \leqslant 1$, as a promise problem (A, B) where

1. A is the set of all graphs G with $\omega(G)/|G| \geqslant c(|G|)$, and
2. B is the set of all graphs G with $\omega(G)/|G| < s(|G|)$,

where $\omega(G)$ denotes the size of the largest clique in graph G.

It is straightforward to see that if GAP-MaxClique$_{c,s}$ is $\mathcal{NP}$-hard under polynomial reduction (cf. Definition 1.6), then there is no polynomial-time approximation algorithm for MaxClique within the factor $g(n) = c(n)/s(n)$, unless $\mathcal{P} = \mathcal{NP}$. Similarly, from $\mathcal{NP}$-hardness under randomized polynomial reduction follows the same non-approximability result for MaxClique, unless

$\mathcal{NP} = \mathcal{ZPP}$. In [FGL+91] it was shown that the amortized free bit complexity of $\mathcal{NP}$-hard sets and the hardness of GAP-MAXCLIQUE are strongly related.

Theorem 9.2. [FGL+91] *Let $\delta > 0$ be a constant, and let c and s be functions such that $c(n)/s(n) = n^{1-\frac{4}{1+\delta}}$. If $\mathcal{NP}$ is contained in $\overline{\text{FPCP}}[\log, \delta]$, then GAP-MAXCLIQUE$_{c,s}$ is $\mathcal{NP}$-hard under randomized polynomial reduction.*

It was shown in [ALM+92] that ROBE3SAT is hard for $\mathcal{NP}$(see also 4.4). Therefore, if ROBE3SAT is in $\overline{\text{FPCP}}[\log, \delta]$ for every $\delta > 0$, then by Theorem 9.2 we can conclude that there is no polynomial-time approximation algorithm for MAX-CLIQUE with approximation factor $n^{1-\varepsilon_0}$, for any $\varepsilon_0 > 0$, unless $\mathcal{NP} = \mathcal{ZPP}$. A PCP-system with these parameters is constructed and analyzed by Håstad in [Hås97a].

The goal of this chapter is to present the proof of Theorem 9.1 in its simplified form as contained in [Hås97a]. The starting point is a verifier of a PCP-system for ROBE3SAT similar to the two-prover interactive proof system from Chapter 6, which will be discussed in Section 9.2. The verifier has access to a proof that is encoded by the long code (cf. Chapter 7.3). In Section 9.3 we discuss how the verifier checks the proof for being encoded correctly. The entire verifier for ROBE3SAT and the analysis of its parameters are discussed in Section 9.4.

9.2 A PCP-System for ROBE3SAT and Its Parallel Repetition

In Example 6.2 a verifier of a two-prover interactive proof system for ROBE3SAT was presented. Here, we define a similar verifier of a PCP-system, which will be used in Section 9.4 in the proof of the Main Theorem of this chapter. In order to simplify the analysis of its soundness, we only consider instances φ of ROBE3SAT which fulfill additional syntactic requirements: each clause of φ contains exactly 3 literals, and each variable appears exactly 6 times in φ (in other words, we consider instances of ROBE3SAT-6 as discussed in Section 4.5). The respective subset of ROBE3SAT can be shown to be $\mathcal{NP}$-complete using an extension of a technique from Papadimitriou and Yannakakis [PY91]. In the rest of this chapter we only consider Boolean formulae which meet these requirements.

The following verifier T^1 for ROBE3SAT gets as input a formula φ. The set $\mathbf{Var}(\varphi)$ consists of the variables $x_1, \ldots, x_n$ of φ, and φ is considered as a set of clauses $C_1, \ldots, C_m$. The verifier T^1 assumes that a proof π is a collection of assignments to the variables and to the clauses of φ, i.e.,

$$\pi = \left((\alpha^{\{x\}})_{x \in \mathbf{Var}(\varphi)} , (\alpha^{\mathbf{Var}(C)})_{C \in \varphi} \right),$$

where α^M is an assignment to the variables in the set M. If π is considered as a bit string, it has length $n + 3m$, where the first n bits are assignments to each of the n variables of the input formula, and the following $3m$ bits are assignments to the variables of each of the m clauses of the input formula.

For each set C, the assignment (resp. function) $\alpha^B \uparrow C$ is the restriction of α^B on variables in C.

VERIFIER T^1 FOR RobE3Sat

input : φ (* an instance $\varphi = C_1 \wedge \cdots \wedge C_m$ of RobE3Sat *)
proof : $((\alpha^{\{x\}})_{x \in \mathsf{Var}(\varphi)}, (\alpha^{\mathsf{Var}(C)})_{C \in \varphi})$
choose randomly $i \in \{1, \ldots, n\}$ (* a variable x_i of φ *)
choose randomly $j \in \{1, \ldots, m\}$ s.t. x_i appears in C_j (* a clause C_j of φ *)
if $\alpha^{\mathsf{Var}(C_j)}$ satisfies C_j, and $(\alpha^{\mathsf{Var}(C_j)} \uparrow \{x_i\}) = \alpha^{\{x_i\}}$
    then accept
    else reject
end

Note that compared to the two-prover interactive proof system in Example 6.2, the choice of the variable and the clause are swapped. The accessed proof contains an assignment to the chosen variable and an assignment to the chosen clause. The verifier accepts, if the clause is satisfied and both assignments are consistent. We obtain similar completeness and soundness results for T^1 as with Example 6.2.

Lemma 9.3. *Verifier T^1 decides* RobE3Sat *with perfect completeness and with soundness $\frac{2+\varepsilon}{3}$, where ε is the fraction of simultaneously satisfiable clauses of unsatisfiable instances of* RobE3Sat.

Proof. If φ is satisfiable, the proof π obtained according to a satisfying assignment makes T^1 accept with probability 1. Therefore, T^1 has perfect completeness. For unsatisfiable instances φ of RobE3Sat, we estimate the acceptance probability as follows. Imagine a tree of all possible sequences (i, j) of random guesses of T^1 on input φ, where φ has m clauses and n variables. This tree has $6n$ leaves. Fix a proof π, and label the leaves either *accept* or *reject*, dependent on how T^1 on φ with proof π decides. Since φ is unsatisfiable, it has at most $\varepsilon \cdot m$ simultaneously satisfiable clauses. The proof π contains an assignment $\alpha^{\{x_1\}}, \ldots, \alpha^{\{x_n\}}$ to φ. Let C_j be a clause not satisfied by this assignment, and let $x_{j_1}, x_{j_2}, x_{j_3}$ be the variables in C_j. Then at least one of the sequences $(j_1, j), (j_2, j), (j_3, j)$ has a rejecting leaf, either because the considered assignment $\alpha^{\mathsf{Var}(C_j)}$ from π does not satisfy the clause or because an assignment $\alpha^{\{x_{j_a}\}}$ is not consistent with the assignment to the same variable in $\alpha^{\mathsf{Var}(C_j)}$. Therefore, at least $(1 - \varepsilon) \cdot m$ leaves of the tree correspond to rejecting computations of T^1 under a fixed proof π. Thus, the acceptance probability is at most $\frac{6n - (1-\varepsilon)m}{6n}$. Since every variable appears in exactly 6 clauses and each clause consists of exactly 3 literals, the number of clauses m equals $2n$. So the above fraction is equal to $\frac{2+\varepsilon}{3}$. ∎

Let us now consider the complexity parameters of verifier T^1. It has polynomial running time, uses logarithmically many random bits, and reads 4 bits from the proof. The free bit complexity equals $f_1 \leqslant 3$, because the bits read as assignment for the clause determine the bit read as assignment to the variable, to make the verifier accept. From the soundness in Lemma 9.3 we obtain amortized free bit complexity $\delta_1 = \frac{f_1}{-\log \frac{2+\varepsilon}{3}}$, which appears to be a constant independent on the size of the input.

Our goal is to get a verifier for ROBE3SAT with an arbitrarily small amortized free bit complexity $\delta > 0$. Because this complexity depends on the free bit complexity and on the soundness of the verifier, the first idea how to decrease δ_1 may be to improve the soundness of T^1. This can be achieved by a parallel repetition of T^1, as shown in Chapter 6. Let T^v denote the verifier which performs v parallel repetitions of T^1. I.e., T^v chooses a set V of v variables (instead of one) and a set ψ of v clauses, one clause for each variable in V. Note that a proof for T^v consists of assignments to all choices of v variables and v clauses, i.e., a proof π is a collection

$$\pi = \left(\left(\alpha^V \right)_{V \in \binom{\mathrm{Var}(\varphi)}{v}}, \; \left(\alpha^{\mathrm{Var}(\psi)} \right)_{\psi \in \binom{\varphi}{v}} \right)$$

where $\binom{M}{v}$ denotes the set of all v element subsets of set M. Since v is fixed, the chosen variables are different and the chosen clauses are disjoint with probability $1 - \mathcal{O}(1/n)$. For simplicity, we assume that this is always the case (since $1/n$ goes to 0 with increasing n). As with T^1, T^v has perfect completeness. For unsatisfiable instances of ROBE3SAT, by the Parallel Repetition Theorem of Raz (Theorem 6.4), the acceptance probability is bounded by c^v, where $c < 1$ is some constant, and hence, goes to 0 when v increases.

Lemma 9.4. *For every $s \in \mathbb{N}$ there exists a $v_0 \in \mathbb{N}$ such that for all $v \geqslant v_0$ it holds that T^v decides ROBE3SAT with perfect completeness and with soundness 2^{-s}.*

Unfortunately, the free bit complexity of T^v equals $v \cdot f_1$. If we calculate the amortized free bit complexity δ_v of T^v, we see that we did not make any gain at all. By the above estimations, we only obtain $\delta_v = \frac{v f_1}{-\log c^v} = \frac{f_1}{-\log c}$, which is constant and independent on v. Therefore, we need to play another card.

Following the paradigm of Arora and Safra [AS92], we let the proof consist of *encoded* assignments. This allows to decrease the soundness of the verifier by smaller increase of its free bit complexity, and hence to make the amortized free bit complexity decreasing. Then the task of the verifier consists of checking the correctness of the encodings, the consistency of the assignments with respect to the same variables and satisfiability of the chosen clauses. The used encoding and corresponding tests are described in the following section.

9.3 The Long Code and Its Complete Test

The choice of the encoding scheme has a big influence on the parameters of a proof system. While Arora et al. [ALM+92] used the Hadamard code, one of the main contributions of Bellare, Goldreich, and Sudan in [BGS95] is the introduction of the so-called *long code* (cf. Chapter 7.3). It encodes a binary string a into a binary string of length $2^{2^{|a|}}$ which consists of the values of every $|a|$-place Boolean function in point a.

From the PCP-system with verifier T^v for ROBE3SAT we will construct a PCP-system with a similar verifier T. One main difference is that the assignments checked by T are encoded by the long code. After choosing a set of variables V and a set of clauses ψ, the verifier T has to test whether the proof consists of long codes for assignments to the respective sets of variables, and whether these assignments are consistent on V and satisfy ψ. We will present now the "subprograms" (Complete Test and its extended version) used by T in order to perform these tests. A formal definition of T and the analysis of its parameters will be presented in Section 9.4.

9.3.1 Testing the Long Code

Assume that a string A of length 2^{2^n} is claimed to be a long code for some input x_0 of length n, i.e., for every Boolean function $g : \{0,1\}^n \mapsto \{0,1\}$, bit $A(g)$ at position g of A is claimed to be $g(x_0)$. Any Boolean function g over n variables can be uniquely represented by the string of its values in points $0^n, \ldots, 1^n$, as $g(0^n) \cdots g(1^n)$. This string considered as an integer from $[0, 2^{2^n} - 1]$ determines the position of $A(g)$ in the long code. The Complete Test checks A at randomly chosen positions in order to verify whether it is a long code of some string. The length n of the encoded string is the input.

COMPLETE TEST
input : n (* the length of the encoded string *)
proof : A
for $i := 1, \ldots, s$:
    choose randomly function $f_i : \{0,1\}^n \to \{0,1\}$
    ask for $A(f_i)$
for all Boolean functions $a : \{0,1\}^s \to \{0,1\}$:
    construct the function $g = a(f_1, \ldots, f_s)$
    ask for $A(g)$
    if $A(g) \neq a(A(f_1), \ldots, A(f_s))$ **then** quit rejecting
accept

Whereas the parameter s is fixed, n depends on the size of the checked proof A. The running time of the Complete Test is exponential in n.

In the first loop, the Complete Test uses s independent and therefore free queries. Each of the possible answers may lead to acceptance. In the next loop, it switches to the "checking mode". The Complete Test makes 2^{2^s} queries, but unlike in the first loop, the acceptable answers are predetermined by the previous answers. Because in the checking mode there is exactly one sequence of answers which makes the Complete Test accept, there are no further free queries. Hence, the Complete Test uses s free bits. Since the test asks every possible non-adaptive question that it knows the answer to, it is called *complete*.

If the Complete Test rejects A, then A is *not* a long code. But since not all points of A are checked, the test may accept by mistake. Nevertheless, if the test accepts A, then the bits of A which the Complete Test has seen look like an evaluation of the chosen functions f_i on some point x. Note that it is not the goal of the test to reconstruct that x.

Lemma 9.5. *If the Complete Test accepts A, then there exists an input x such that $A(f_i) = f_i(x)$ for all tested functions f_i.*

Proof. We show the contraposition. Assume that for every x there exists an f_i such that $A(f_i) \neq f_i(x)$. Let $a : \{0,1\}^s \to \{0,1\}$ be defined as $a(b_1, \ldots, b_s) = \bigwedge_i (b_i = A(f_i))$. Then $g = a(f_1, \ldots, f_s)$ is the n-place constant 0-function. Let c_0 be the s-place constant 0-function. Then also $g = c_0(f_1, \ldots, f_s)$. Because $a \neq c_0$, in the second loop of the Complete Test it is checked whether $A(g) = a(A(f_1), \ldots, A(f_s))$ and whether $A(g) = c_0(A(f_1), \ldots, A(f_s))$. Since $a(A(f_1), \ldots, A(f_s)) = 1$ and $c_0(A(f_1), \ldots, A(f_s)) = 0$, one of these checks must fail. Therefore, the Complete Test rejects. ∎

As mentioned above, the test may accept even if A is not a long code. This is illustrated by the following example.

Example 9.6. Fix some x, y, z and let $A(f) = f(x) \oplus f(y) \oplus f(z)$.

A is not a long code of any input. Apply the Complete Test on that A. With probability 2^{-s} it holds that $f_i(x) = f_i(y)$, for all $i = 1, \ldots, s$. In this case, $A(f) = f(z)$, for all queried functions f, and therefore $a(A(f_1), \ldots, A(f_s)) = a(f_1(z), \ldots, f_s(z))$. Thus, the Complete Test accepts and A looks like a long code for z. This example gives an evidence that the failure probability of the test is at least 2^{-s}. We could decrease it by taking a larger s but it would contradict our goal to obtain an arbitrarily small amortized free bit complexity as s is the number of free bits in the test.

Håstad proved that, when the Complete Test accepts, with high probability what the test has seen from A looks like a code-word from a very small set S_A. The important properties are: firstly, that S_A depends only on A and not on the functions chosen during the test, and, secondly, that the acceptance probability in the remaining cases (i.e., when the input looks like a code-word not in S_A) can be arbitrarily decreased without increase of the free bit complexity.

Lemma 9.7. [Hås96b] $\forall \gamma > 0, l \in \mathbb{N} \; \exists s_0 \; \forall s \geqslant s_0 \; \exists v_0 \; \forall v \geqslant v_0$:
for every A of length 2^{2^v} there is a set $S_A \subseteq \{0,1\}^v$ of size $|S_A| \leqslant 2^{\gamma \cdot s}$, such
that with probability $\geqslant 1 - 2^{-ls}$ the Complete Test on A results in one of the
following two cases.

1. The Complete Test rejects, or

2. there is an $x \in S_A$ such that for every function f_i checked in the Complete
   Test, $A(f_i) = f_i(x)$.

With other words, the result of the test can be analyzed as follows. Either the
test rejects and we are sure that A is not a long code, or it accepts while the
checked points look like being the part of a long code. In the latter case we have
a set of candidates – strings x for which what the Complete Test has seen of A
looks like a long code of. The lemma tells that there is a small set of candidates,
such that the cases when the checked points are consistent with a long code for
an element outside of this subset are very rare. This set is independent from the
functions randomly chosen in the test.

The proof of this lemma takes "a bulk of paper" [Hås97a] and is omitted here.
However, it is useful to know what the set S_A is. Remind from Chapter 7 that
each bit $A(f)$ of the long code can be represented using Fourier coefficients
(see 7.13) as

$$A(f) = \sum_{\alpha \subseteq \{-1,1\}^n} \hat{A}_\alpha \prod_{x \in \alpha} f(x) \; .$$

The set S_A now is defined as

$$S_A := \{x \mid \exists \alpha, |\alpha| \leqslant l' : x \in \alpha \wedge \hat{A}_\alpha^2 \geqslant l' 2^{-\gamma s}\}$$

where l' is a constant depending on γ and l only. By Parseval's identity (see
Lemma 7.5) it follows that S_A contains at most $2^{\gamma s}$ points.

9.3.2 The Complete Test with Auxiliary Functions

In the context of this chapter, the Complete Test is used by a verifier for
ROBE3SAT. Besides checking whether the proof consists of long codes, the verifier checks whether the long codes encode satisfying assignments for the clauses
of the formula under consideration, and it also checks consistency of different
assignments. For this purpose, it uses an extended version of the Complete Test
described in this subsection. The additional properties of acceptable proofs are
expressed via so-called *auxiliary functions*.

More abstractly, let h be any Boolean function. Let H_1 be the set of inputs x
with $h(x) = 1$, and let H_0 be defined respectively. The verifier wants to check,
whether the proof A is the long code of an input x such that $h(x) = 1$. The
most straightforward way to do this is to access $A(f \wedge h)$ instead of $A(f)$. Then

the value depends only on $f(x)$, for $x \in H_1$. Hence the Fourier transform has support only on sets of inputs x with $h(x) = 1$ and all inputs in the special set S_A have this property too, i.e., $S_A \subset H_1$. Unfortunately, this approach does work only if h is fixed before the test starts, what does not apply in our case. The choice of h during the test will make S_A dependent on h. This problem is solved as follows.

We define A' by

$$A'(f) = 2^{-|H_0|} \sum_{\{g|g \wedge h = f \wedge h\}} A(g) \quad .$$

The functions g appearing in the sum are those that may disagree with f on the points $x \in H_0$ only. There are exactly $2^{|H_0|}$ functions with this property.

Assuming A is the long code for some $x \in H_1$, we have $g(x) = f(x)$, for all g with $g \wedge h = f \wedge h$. Hence $A'(f) = A(f)$. For $x \in H_0$, some functions g considered in the sum are different from f in point x. This causes $A'(f)$ to take some value different from $A(f)$.

Using this observation, we slightly change the Complete Test in order to make it accept only those long codes which may encode an x with $h(x) = 1$. Instead of asking for $A(f)$ in the Complete Test, we will ask for $A(g)$ for all functions $g \in \{g \mid g \wedge h = f \wedge h\}$. One of them is f, which makes it easy to check whether $A'(f) = A(f)$ – all respective queries must be answered by the same value. If this is not the case, the modified test rejects. Consequently, the free bit cost of these $2^{|H_0|}$ queries is the free bit cost of the first query, which determines the answers on all later queries. Note that instead of one function h we also can use any fixed number of functions in the same way. The resulting new test is called *Complete Test with Auxiliary Functions*.

Besides that, the Fourier coefficients of A' for the sets that contain some x from H_0 are 0. The Fourier coefficients of A' for all other sets, i.e. for any α consisting of the elements from H_1, are equal to respective coefficients of A, i.e. $\hat{A}'_\alpha = \hat{A}_\alpha$. This implies that $S_{A'} = S_A \cap H_1$. Now we are able to modify Lemma 9.7.

Lemma 9.8. [Hås96b] $\forall \gamma > 0, l \in \mathbb{N} \; \exists s_0 \; \forall s \geqslant s_0 \; \exists v_0 \; \forall v \geqslant v_0 :$
for every A of length 2^{2^v} there is a set $S_A \subseteq \{x \in \{0,1\}^v \mid \bigwedge_{i=1}^r h_i(x) = 1\}$ of size $|S_A| \leqslant 2^{\gamma \cdot s}$, such that with probability $\geqslant 1 - 2^{-ls}$ the Complete Test with Auxiliary Functions $h_1, \ldots, h_r$ on A results in one of the following two cases.

1. *The Complete Test with Auxiliary Functions rejects, or*

2. *there is an $x \in S_A$ such that for every function f_i checked in the test, $A(f_i) = f_i(x)$.*

One of the applications of auxiliary functions in the proof of the Main Theorem, is the test whether a proof is a long code of an assignment satisfying a certain clause. We can take the function which evaluates to 1 on all assignments satisfying that clause as auxiliary function. By Lemma 9.8 it follows that if the test with that auxiliary function accepts, then the proof looks like an encoding of a satisfying assignment of the considered clause. Although the very assignment is not known, with high probability it is an element from the small set specified above.

Since auxiliary functions can be chosen during the test, we can use them also for testing consistency between the answers.

9.4 The Non-Approximability of MAXCLIQUE

In the previous sections we have seen how to decide ROBE3SAT with a PCP-system, and how a verifier can test whether a supposed long code encodes a satisfying assignment of a formula. We can now put the parts together in order to get a PCP-system for ROBE3SAT with arbitrarily small amortized free bit complexity.

Theorem 9.1. [Hås96a]
  Let A be a set in $\mathcal{NP}$. Then $A \in \overline{\mathrm{FPCP}}[\log n, \varepsilon_0]$, for every $\varepsilon_0 > 0$.

Proof. As already discussed, it is sufficient to construct a verifier T for a PCP-system with the required parameters which decides ROBE3SAT. Our starting point is the verifier T^1 defined in Section 9.2. Like in its parallel repetition T^v, on input formula φ, the verifier T randomly chooses a set V of v variables that appear in the input formula φ. Then, for each variable in V, it chooses a clause containing the variable. This latter step will be repeated k times. So we come out with a set V of variables and sets $\psi_1, \ldots, \psi_k$ of clauses. By the arguments as in the case of T^v, we can assume that the clauses in ψ_i are pairwise disjoint. Since v and k are fixed, the verifier needs logarithmically many random bits to choose the sets. After these random choices, the proof is checked. A proof π is assumed to contain the long codes for assignments to each choice of V and ψ_i. I.e., π is a collection

$$\pi = \left(\left(A^V \right)_{V \in \binom{\mathrm{Var}(\varphi)}{v}} \ , \ \left(A^{\mathrm{Var}(\psi)} \right)_{\psi \in \binom{\varphi}{v}} \right),$$

where A^M is a string of length $2^{2^{|M|}}$. At first, the Complete Test is performed on A^V, in order to check whether it is a long code of any assignment. At second, the Complete Test with Auxiliary Functions is performed on each of $A^{\mathrm{Var}(\psi_1)}, \ldots, A^{\mathrm{Var}(\psi_k)}$. It checks whether each $A^{\mathrm{Var}(\psi_i)}$ is a long code for an assignment which satisfies all clauses in ψ_i, and whether it is consistent with the assignment to V encoded in A^V. The verifier accepts when all tests succeed.

Before we present verifier T more formally, we need to introduce a notation for extensions of Boolean functions. For a function f over variables X and a set Y, the function $f \downarrow Y$ on Y is the function with $(f \downarrow Y)(a) = f(a \uparrow X)$ for every a, i.e., $f \downarrow Y$ depends on variables in X only.

VERIFIER T FOR ROBE3SAT

input : φ with n variables and m clauses

proof : $((A^V)_{V \in \binom{\mathbf{Var}(\varphi)}{v}}, (A^{\mathbf{Var}(\psi)})_{\psi \in \binom{\varphi}{v}})$

choose randomly $V \in \binom{\mathbf{Var}(\varphi)}{v}$

perform Complete Test on A^V

    during which δs functions $f_1, \dots, f_{\delta s}$ on V are randomly chosen

for $i = 1, 2, \dots, k$, construct $\psi_i \in \binom{\varphi}{v}$ as follows:

    for each $x \in V$:

        include a randomly chosen clause from φ that contains x into ψ_i

for each i:

    perform Complete Test with Auxiliary Functions on $A^{\mathbf{Var}(\psi_i)}$

        during which $\delta s / k$ functions are randomly chosen;

    the auxiliary functions stem from $f_1 \downarrow \mathbf{Var}(\psi_i), \dots, f_{\delta s} \downarrow \mathbf{Var}(\psi_i)$ and ψ_i:

        the $f_j \downarrow \mathbf{Var}(\psi_i)$ must evaluate to the previously obtained values

        (i.e., $A^{\mathbf{Var}(\psi_i)}(f_j \downarrow \mathbf{Var}(\psi_i)) = A^V(f_j))$,

        and the clauses of ψ_i must evaluate to true (i.e., $A^{\mathbf{Var}(\psi_i)}(\psi_i) = 1$).

if all tests succeed

    then accept

    else reject

end

The reason why we work with the long codes for partial assignments instead of the long code for a full assignment is the complexity of the Complete Test. It is exponential in the number of considered variables. As the number of variables under consideration is constant, the running time of the Complete Test is polynomially bounded.

Therefore, verifier T is polynomial-time bounded. It uses logarithmically many random bits and $2\delta s$ free bits during the different applications of Complete Test. T has perfect completeness. The proof of the Theorem will be accomplished by proving that T has soundness 2^{-s} (Claim 9.9), what yields an upper bound $\varepsilon_0 = 2\delta$ on the amortized free bit complexity.

Without loss of generality, we can assume that $\delta \leqslant \frac{1}{4}$. We will see during the rest of the proof how k, s, and v can be chosen.

Claim 9.9. Verifier T decides ROBE3SAT with soundness 2^{-s}.

The analysis of the soundness of T is based on a use of verifier T^v from Section 9.2, for v chosen such that T^v has soundness strictly less than 2^{-3s} (by

Lemma 9.4). We show that every formula φ which is accepted by T with probability greater than 2^{-s}, is accepted by T^v with probability greater than 2^{-3s}. Hence, φ is satisfiable.

By Lemma 9.8, for every $A^{\text{Var}(\psi_i)}$, there exists a set $S_{A^{\text{Var}(\psi_i)}}$ of assignments of cardinality $|S_{A^{\text{Var}(\psi_i)}}| \leqslant 2^{\delta s/(2k)}$ such that, except with probability 2^{-2s}, either the Complete Test rejects or the string $A^{\text{Var}(\psi_i)}$ looks like the long code for an element from $S_{A^{\text{Var}(\psi_i)}}$. Consequently, the probability that, for some i, $A^{\text{Var}(\psi_i)}$ looks like a long code for a string not in $S_{A^{\text{Var}(\psi_i)}}$, is bounded by $k2^{-2s}$. Since this value can be made arbitrarily small by the choice of k and s, we exclude these cases from further considerations.

For the rest of the proof, we fix an arbitrary formula φ, an arbitrary proof

$$\pi = \left((A^V)_{V \in \binom{\text{Var}(\varphi)}{v}} \quad, \quad (A^{\text{Var}(\psi)})_{\psi \in \binom{\varphi}{v}} \right),$$

and consider the computation of T on input φ and proof π. In the first part of the computation, T guesses sets V and $\psi_1, \ldots, \psi_k$. Since $V \subseteq \text{Var}(\psi_i)$, for all $1 \leqslant i \leqslant k$, the assignments for $\text{Var}(\psi_i)$ are extensions of the assignments for V, i.e. there is a projection function which given an assignment α_i to $\text{Var}(\psi_i)$ outputs the corresponding sub-assignment $\alpha_i \uparrow V$ to V. Take some $\alpha^V \in S_{A^V}$ and $\alpha_i \in S_{A^{\text{Var}(\psi_i)}}$. If $\alpha_i \uparrow V = \alpha^V$, then the verifier T has no chance to find an inconsistency between A^V and $A^{\text{Var}(\psi_i)}$, and therefore this computation path will be accepting. The probability of the existence of such projections depends on the choice of V. Let $\binom{\varphi}{V}$ denote the set of all possible choices ψ of T after choosing V, i.e., $\binom{\varphi}{V}$ consists of all sets of clauses from φ obtained by picking for every $x \in V$ one clause from φ which contains x. There may be choices of variables V which are "inherently dangerous" for the verifier in the following sense: there is an assignment α to V such that with high probability for a randomly chosen $\psi \in \binom{\varphi}{V}$, the proof π contains $A^{\text{Var}(\psi)}$ for which $S_{A^{\text{Var}(\psi)}}$ contains an extension of α. "High probability" means here any value larger than the soundness we want to prove. Assignments for variables in V with this property form the set common(V).

Definition 9.10. *Let $V \in \binom{\text{Var}(\varphi)}{v}$. The set* common$(V)$ *is the set of assignments to V defined by*

$$\text{common}(V) := \{\alpha \mid \text{Prob}_{\psi \in \binom{\varphi}{V}}[\alpha \in (S_{A^{\text{Var}(\psi)}} \uparrow V)] \geqslant 2^{-s}\}$$

where $S_{A^{\text{Var}(\psi)}} \uparrow V = \{\alpha \uparrow V \mid \alpha \in S_{A^{\text{Var}(\psi)}}\}$.

If T accepts φ under proof π with probability at least 2^{-s}, then some choice of V must have a non-empty common(V). When T has chosen a set V with non-empty common(V), then it accepts with high probability.

We proceed as follows to prove the soundness of T. Using π and common(V), we construct a proof ρ for T^v. Assuming that T accepts φ under proof π with probability at least 2^{-s}, we show that T^v accepts φ under proof ρ with probability at least 2^{-3s}. Because v was chosen such that the soundness of T^v is strictly less than 2^{-3s}, it follows that φ is satisfiable.

In order to estimate the acceptance probability of T^v, we need to estimate the probability that common(V) is empty.

Claim 9.11. If common$(V) = \emptyset$, then T accepts after choosing V with probability less than $d_k 2^{-2s}$, for some constant d_k that depends only on k.

Proof.   Consider an accepting computation of T, on which V, $\psi_1, \ldots, \psi_k$ are chosen and common$(V) = \emptyset$. According to the discussion above, we can assume that, for each i, $A^{\mathrm{Var}(\psi_i)}$ looks like a long code for some assignment $\alpha_i \in S_{A^{\mathrm{Var}(\psi_i)}}$ and that A^V looks like a long code for some α^V. Because the functions $f_1, \ldots, f_{\delta s}$, which are randomly chosen during the Complete Test of A^V, are used as auxiliary functions in the Complete Tests for the $A^{\mathrm{Var}(\psi_i)}$, it holds that $f_j(\alpha^V) = (f_j \downarrow \mathrm{Var}(\psi_i))(\alpha_1) = \cdots = (f_j \downarrow \mathrm{Var}(\psi_i))(\alpha_k)$ for all j and some collection of α_i. Therefore, in order to estimate an upper bound of the probability that T accepts, it suffices to estimate the probability that for randomly chosen ψ_i, such a collection exists.

This probability depends on the number m of different projections $\alpha_i \uparrow V$, i.e. $m = |\{ \alpha_i \uparrow V \mid i = 1, 2, \ldots, k \}|$. Having m different projections, we know that at least $m - 1$ assignments from the collection differ from each other and are not consistent with α^V. There are $2^{(m-1)\delta s}$ possible values that take the randomly chosen functions $f_1, \ldots, f_{\delta s}$ in these points, but for only one of those the collection survives the Complete Test. Therefore, the probability of surviving is at most $2^{-(m-1)\delta s}$.

Our further estimation considers collections with $m > k/2$ and collections with $m \leqslant k/2$ separately. It turns out that the probability of surviving for the collections with $m > k/2$ is small and collections with $m \leqslant k/2$ appear rarely.

Claim 9.12. Choose randomly $\psi_1, \ldots, \psi_k \in \binom{\varphi}{V}$. The probability, that any collection of $\alpha_i \in S_{A^{\mathrm{Var}(\psi_i)}}$ has at most $k/2$ projections on V, is at most $c_k 2^{-2s}$, where c_k is a constant dependent on k only.

Proof.   Let us analyze the probability that the set of assignments $S_{A^{\mathrm{Var}(\psi_i)}} \uparrow V$ intersects $\bigcup_{j=1}^{i-1}(S_{A^{\mathrm{Var}(\psi_j)}} \uparrow V)$. Using Lemma 9.7 with $\gamma = \delta/(2k)$, the latter contains at most $k 2^{\delta s/(2k)}$ elements. Since common$(V) = \emptyset$, the probability of any fixed single element occurring in $S_{A^{\mathrm{Var}(\psi_i)}} \uparrow V$ is less than 2^{-s}. Hence, the probability of a non-empty intersection is at most

$$2^{\delta s/(2k)} \cdot 2^{-s} \cdot k 2^{\delta s/(2k)}$$

what equals $k 2^{[2\delta s/(2k)]-s}$. Now, let k be chosen such that $k > 10$. Since $\delta/k < \frac{1}{4}$, we get that $k 2^{[2\delta s/(2k)]-s} \leqslant k 2^{-s/2}$. In order to prove the Claim, at least $k/2$

times a non-empty intersection must be obtained. Thus we have to estimate $(k2^{-s/2})^{k/2}$. The latter equals $c_k \cdot 2^{-sk/4}$ for a constant $c_k = k^{k/2}$ which depends only on k. Because of the choice of k, we finally obtain that the latter product is at most $c_k 2^{-2s}$. ∎

Now we continue in the proof of Claim 9.11. The probability that a collection of α_i passes the test is at most the probability that a collection with $m > k/2$ passes the test plus the probability that a collection has $m \leqslant k/2$, i.e.

$$\begin{aligned}
&\leqslant \quad (2^{-\frac{k}{2}\delta s}) + c_k 2^{-2s} \\
&= \quad 2^{-2s} \cdot (2^{-s((\delta k/2)+2)} + c_k) \\
&\leqslant \quad d_k 2^{-2s}
\end{aligned}$$

for $d_k = (1 + c_k)$. Since $2^{-s((\delta k/2)+2)} \leqslant 1$, the last inequality holds, and since c_k depends only on k, d_k does so too. This completes the proof of Claim 9.11. ∎

Now we analyze the probability that T^v accepts φ under a proof ρ of the following form. Let

$$\rho := \left((\alpha^V)_{V \in \binom{\mathbf{Var}(\varphi)}{v}} \quad , \quad (\alpha^{\mathbf{Var}(\psi)})_{\psi \in \binom{\varphi}{v}} \right)$$

have the property that

$$\alpha^V = \begin{cases} y, & \text{for a } y \in \text{common}(V) \\ a, & \text{for fixed } a, \text{ if common}(V) = \emptyset \end{cases}$$

$$\alpha^{\mathbf{Var}(\psi)} = z, \text{ for a } z \in S_{A^{\mathbf{Var}(\psi)}}$$

Assume that T accepts with probability at least 2^{-s}. Note that every V is chosen with equal probability. Therefore it follows from Claim 9.11 that common(V) is not empty with probability at least $2^{-s} - d_k 2^{-2s}$. Note that s can be chosen such that $2^{-s} - d_k 2^{-2s} \geqslant 2^{-s} - 2^{-1.5s}$. Therefore

$$\text{Prob}_{V \in \binom{\mathbf{Var}(\varphi)}{v}}[\text{common}(V) \neq \emptyset] \geqslant 2^{-s} - 2^{-1.5s}.$$

For V with non-empty common(V), ρ contains some $\alpha^V \in \text{common}(V)$. Given such an α^V, the probability that there is an extension of α^V in $S_{A^{\mathbf{Var}(\psi)}}$ for a randomly chosen ψ, is

$$\text{Prob}_{\psi \in \binom{\varphi}{v}}[\alpha^V \in (S_{A^{\mathbf{Var}(\psi)}} \uparrow V) \mid \alpha^V \in \text{common}(V)] \geqslant 2^{-s}$$

by the definition of common(V). The probability that z is such an element – given that it exists – is

$$\text{Prob}_{z \in S_{A^{\mathbf{Var}(\psi)}}}[\alpha^V = (z \uparrow V) \mid \alpha^V \in \text{common}(V), \psi \in \binom{\varphi}{v}] \geqslant |S_{A^{\mathbf{Var}(\psi)}}|^{-1}.$$

Consequently, the assignments $\alpha^{\mathbf{Var}(\psi)}$ for ρ can be chosen in such a way that the probability that T^v accepts φ under ρ is at least the product of the latter three probabilities. Using Lemma 9.8 to estimate $|S_{A\text{Var}(\psi)}|$, this probability is at least

$$(2^{-s} - 2^{-1.5s}) \cdot 2^{-s} \cdot 2^{-\delta s/(2k)} \geqslant 2^{-3s}.$$

By the soundness of T^v, it follows that φ is satisfiable.

Since φ and π were chosen arbitrarily, it follows that every unsatisfiable φ is accepted by T with probability at most 2^{-s} under any proof. Therefore, T has soundness 2^{-s}. ∎

As a consequence of Theorem 9.1, we get the optimal non-approximability results for MaxClique previously discussed in Chapter 4.

Theorem 9.13. [Hås96a] *Let $\varepsilon_0 > 0$ be any constant.*

1. *No polynomial time algorithm approximates MaxClique within a factor $n^{1-\varepsilon_0}$, unless $\mathcal{NP} = \mathcal{ZPP}$.*

2. *No polynomial time algorithm approximates MaxClique within a factor $n^{1/2-\varepsilon_0}$, unless $\mathcal{NP} = \mathcal{P}$.*

10. The Hardness of Approximating Set Cover

Alexander Wolff

10.1 Introduction

Imagine the personnel manager of a large company that wants to start a devision from scratch and therefore has to hire a lot of new employees. The company naturally wants to keep labour costs as low as possible. The manager knows that 100 different skills are needed to operate successfully in the new field. After placing some job ads, he gets a list of 1000 applicants each of who posesses a few of the desired skills. Whom should he hire?

This problem has long been known as SETCOVER, the problem of covering a base set (the required skills) with as few elements (the applicants' skills) of a given subset system (the list of applications) as possible. More formally:

SETCOVER

Instance: A base set S, $|S| = N$, and a collection $\mathcal{F} = \{S_1, \ldots, S_M\}$ of M subsets of S, where M is bounded by some polynomial in N.

Problem: Find a subset $\mathcal{C} \subseteq \mathcal{F}$ with minimum cardinality, such that every element of S is contained in at least one of the sets in $\mathcal{C}$.

Karp showed that the decision version of SETCOVER is NP-complete [Kar72], but the following greedy algorithm approximates an optimal solution.

GREEDY_SET_COVER$(S, \mathcal{F})$

$\mathcal{C} := \emptyset$
while $\bigcup \mathcal{C} \neq S$ **do**
    find $S_i \in \mathcal{F}$ such that the gain $|S_i \cap (S \setminus \bigcup \mathcal{C})|$ is maximal
    $\mathcal{C} := \mathcal{C} \cup \{S_i\}$
return $\mathcal{C}$

This simple algorithm always finds a cover $\mathcal{C}$ of at most $H(N) \cdot |\mathcal{C}_{opt}|$ subsets, where $\mathcal{C}_{opt}$ is a minimum cover and $H(n) = \sum_{i=1}^{n} 1/i$ is the nth harmonic number. $H(n) \in \ln n + \mathcal{O}(1)$. For the proof of the greedy algorithm's approximation ratio ([Joh74], [Lov75]) of $H(N)$, see Exercise 10.1.

In this chapter I present the paper "A threshold of ln n for approximating set cover" [Fei96] by Uriel Feige. It shows under some assumption about complexity classes that there is no efficient approximation algorithm with a quality guarantee better than that of the simple greedy algorithm above, namely $\ln N$. This improves a previous result of Lund and Yannakakis [LY93] who showed under slightly stricter assumptions that there is no polynomial $(1/2 - \varepsilon) \log_2 N$-approximation algorithm for SETCOVER. Note that $\log_2 N = \log_2 e \cdot \ln N \approx 1.44 \cdot \ln N$. I will use $\log N$ as shorthand notation for $\log_2 N$ from now on.

There are few NP-hard optimization problems with non-trivial approximation ratio for which an approximation algorithm matching the problem's approximation threshold is known. One example is the MIN-P-CENTER problem for which a 2-approximation algorithm exists ([HS85], [DF85]) but it is NP-hard to obtain anything better [HN79]. For the MAPLABELING problem where one is interested in placing rectilinear labels of maximum common size next to n sites, the situation is similar: There is a theoretically optimal approximation algorithm which finds a placement with labels of half the optimal size, but it is NP-hard to get any closer [FW91].

10.1.1 Previous Work

Lund and Yannakakis showed that one cannot expect to find a polynomial-time $(1/2 - \varepsilon) \log N$-approximation algorithm for SETCOVER for any $\varepsilon > 0$ [LY93]. In their proof they construct a subset system whose central property is sketched as follows. Let N be the size of the base set S of this system. First chose subsets of S of size $N/2$ independently at random. Put these sets and their complements into the collection $\mathcal{F}$ of subsets of S. Then obviously there are two ways to cover S. The *good* way is to take a set and its complement, thus covering S by only two sets. The *bad* way is not to do so. In this case, S is covered by random and independent subsets. Since each set which is added to the cover is expected to contain only half of the uncovered elements of S, $\log N$ sets are needed to cover S the bad way. Thus the ratio between the good and the bad case is $\frac{1}{2} \log N$.

Lund and Yannakakis use an interactive proof system as a link between the satisfiability of a Boolean formula and the size of a cover. Such systems consist of a verifier which is connected to the input, to a random source, and to two (or more) provers. Based on the input, the verifier uses random bits to pose questions to the deterministic and all-powerful provers. Depending on their answers, the verifier accepts or rejects the input.

Interactive proof systems can be seen as maximization problems. The aim is to find strategies for the provers which maximize the probability with which the verifier accepts the input. Lund and Yannakakis use a two-prover proof system for satisfiability to reduce a formula φ to a collection of sets as described above. They do this by indexing the subsets of this collection with all possible question-answer pairs which can be exchanged between verifier and provers. Their reduction has the property that if φ is satisfiable, the base set can be covered the good way, and if φ is not satisfiable, it must be covered the bad way. Knowing the sizes of a good and a bad cover, we could use a $(1/2 - \varepsilon) \log N$-approximation algorithm for SETCOVER to decide the satisfiability of φ. On the one hand their reduction is in ZTIME($N^{\mathcal{O}(\text{poly}\log N)}$) (see Definition 1.15), on the other hand every problem in $\mathcal{NP}$ can be reduced in polynomial time to that of testing the satisfiability of a Boolean formula [Coo71]. Therefore we would have $\mathcal{NP} \subseteq$ ZTIME($N^{\mathcal{O}(\text{poly}\log N)}$), which is only a little weaker than $\mathcal{P} = \mathcal{NP}$.

10.1.2 Idea of the Proof

In his proof of the $\ln N$-threshold, Feige constructs a partition system where each partition of the base set consists of k (instead of 2) random subsets of size N/k. k is a large constant. A *good* cover now consists of all k elements of one partition, while in the *bad* case d subsets of pairwise different partitions are needed. These d subsets are added successively to the cover. Each of them contains with high probability only a $1/k$-th of the yet uncovered elements. In other words, we are looking for the smallest d such that the number of uncovered points $N(1-1/k)^d$ is less than 1. This is equivalent to $d \ln \frac{k}{k-1} \approx \ln N$. $\ln n$ grows monotonically, and its slope is $1/n$, which gives us $\frac{1}{k} < \ln \frac{k}{k-1} = \ln k - \ln(k-1) < \frac{1}{k-1}$. For large k this means that $d \approx k \ln N$. Thus the ratio between a bad and a good cover approaches $\ln N$ as desired.

To take advantage of the idea above, the author uses a multi-prover proof system with k provers for the following problem.

ε-ROBE3SAT-5

Instance: A Boolean formula φ of n variables in conjunctive normal form in which every clause contains exactly three literals, every variable appears in exactly five clauses, and a variable does not appear in a clause more than once, i.e. φ is in *3CNF-5 form*. Furthermore, φ is either satisfiable or at most a $(1 - \varepsilon)$-fraction of its clauses is simultaneously satisfiable.

Question: Is φ satisfiable?

To prove that there is an $\varepsilon \in (0,1)$ such that ε-ROBE3SAT-5 is NP-hard is left to the reader, see Exercise 10.2 and Lemma 4.9 in Section 4.5.

k-prover proof systems have been used before for the special case $k = 4$ to show the hardness of approximating SETCOVER [BGLR93], but the resulting bounds were less tight than those obtained with 2-prover proof systems. Feige's main new idea is to introduce two different acceptance criteria for the verifier: The verifier of his proof system accepts "strongly" if the answers of *all* provers are mutually consistent – this corresponds to the conventional acceptance criterion. For the verifier to accept "weakly", it suffices if *two* of the k provers answer consistently, and their answers only need to be consistent in a weak sense.

The subsets of the partition system sketched above are used to build a large set system to which a Boolean formula φ can be reduced. The sets of this system are indexed by all question-answer pairs which are potentially exchanged between verifier and provers. If φ is satisfiable then it is reduced to a collection of subsets in which there is a cover of few, pairwise disjoint sets. This cover corresponds to the questions of the verifier and the mutually consistent answers of the provers which are based on a satisfying truth assignment for φ. If however only a $(1-\varepsilon)$-fraction of the clauses is satisfiable, then every cover which corresponds to a weak acceptance of the verifier has a cardinality which is greater than the cover in the satisfiable case by at least a factor of $\ln N$. Due to the complexity of the reduction we would have $\mathcal{NP} \subseteq \mathrm{DTIME}(N^{\mathcal{O}(\log \log N)})$ if for any $\varepsilon > 0$ there was a polynomial-time $(1-\varepsilon) \ln N$-approximation algorithm for SETCOVER. The complexity class $\mathrm{DTIME}(\cdot)$ is introduced in Definition 1.5.

The idea behind introducing the notion of weak acceptance is that it is a good trade-off between the following two opposing wishes. On the one hand, we want to show that the probability that the verifier accepts an unsatisfiable formula is low. On the other hand, we want to be able to design a simple random strategy for the provers which would make the verifier accept with high probability in the case where there were a cover of small size in the subset system. This will give us the necessary contradiction to show that such a cover cannot exist.

The proof of the main theorem, Theorem 10.9, is split as follows. In Section 10.2 we construct the k-prover proof system and show with which probability the verifier accepts a formula. In Section 10.3 we build a partition system with the following two properties. Firstly, the k sets of one partition suffice to cover the base set of the partition system. Secondly, a cover with sets of pairwise different partitions needs more sets by a logarithmic factor. Finally in Section 10.4 we show how to reduce an instance of ε-ROBE3SAT-5 to an instance of SETCOVER. The proof system will act as a mediator between the two. It links the satisfiability of a Boolean formula φ to the size of a cover such that the ε-gap in the number of satisfiable clauses of φ is expanded to a logarithmic gap in the cover size.

10.2 The Multi-Prover Proof System

Our aim is to build a proof system whose provers have a strategy with which they can convince the verifier to ("strongly") accept an input formula if it is satisfiable. At the same time the probability that the verifier accepts ("weakly") should be low for all stategies of the provers if an ε-fraction of the clauses is not satisfiable. This will be used in the reduction to SETCOVER (Lemma 10.6).

We first turn our attention to a less complex system, a so-called *two-prover one-round proof system*, which Feige describes as follows:

> The verifier selects a clause at random, sends it to the first prover, and sends the name of a random variable in this clause to the second prover.

Each prover is requested to reply with an assignment to the variables that it received (without seeing the question and answer of the other prover). The verifier accepts if the answer of the first prover satisfies the clause, and is consistent with the answer of the second prover.

For a more detailed and very vivid description of two-prover one-round proof systems refer to Section 6.3. Example 6.2 corresponds to the proof system used here.

Proposition 10.1. *Under the optimal strategy of the provers, the verifier of the two-prover one-round proof system accepts a 3CNF-5 formula φ with probability at most $1 - \varepsilon/3$, where ε is the fraction of unsatisfiable clauses in φ.*

Proof. The strategy of the second prover defines an assignment to the variables of φ. The verifier selects a clause which is not satisfied by this assignment with probability ε. In this case, the first prover must set at least one of the three variables differently from this assignment in order to satisfy the clause, and an inconsistency is detected with probability at least $1/3$. ∎

With the help of Raz' Parallel Repetition Theorem [Raz95] (see Chapter 6) the error in the case of unsatisfiable formulae (which initially is $1 - \varepsilon/3$) can be decreased by repeating this proof system many times in parallel, and accepting if the two provers are consistent on each of the repeated copies independently. Using the terminology of Chapter 6, our "game" has value $1 - \varepsilon/3$ and answer size $s = 2^4$. Then Theorem 6.4 yields the following:

Proposition 10.2. *If the two-prover one-round proof system is repeated l times independently in parallel, then the error is $2^{-c_1 l}$, where c_1 is a constant that depends only on ε, and $c_1 > 0$ if $\varepsilon > 0$.*

But what does the *k-prover proof system* look like? First of all, we need a binary code whose task it will be to encode the structure of the questions the verifier is going to send to the provers. We need k code words of length l, each with $l/2$ ones and $l/2$ zeros. Furthermore, we require that any two code words have a Hamming distance of at least $l/3$. Such codes exist; for instance if $k < l$ and l a power of 2, then a Hadamard matrix fulfills these requirements. Quoting Feige:

> Let l be the number of clauses the verifier picks uniformly and independently at random out of the $5n/3$ clauses of the 3CNF-5 formula. Call these clauses $C_1, \ldots, C_l$. From each clause, the verifier selects a single variable uniformly and independently at random. Call these variables $x_1, \ldots, x_l$. (We will refer to them as the *sequence of distinguished variables.*) With each prover the verifier associates a code word. Prover P_i receives C_j for those coordinates j in its code word that are one, and x_j

for those coordinates in its code word that are zero. Each prover replies
with an assignment to all variables that it received. For coordinates in
which the prover receives a variable, it replies with an assignment to that
variable. For coordinates in which the prover receives a clause, it must
reply with an assignment to the three variables that satisfies the clause
(any other reply given on such a coordinate is automatically interpreted
by the verifier as if it was some canonical reply that satisfies the clause).
Hence in this k-prover proof system, the answers of the provers are guaranteed to satisfy all clauses, and only the question of consistency among
provers arises.

The verifier *strongly* accepts if on each coordinate the answers of all provers are
mutually consistent. The verifier *weakly* accepts if there is a pair of provers which
assign consistently to the sequence of distinguished variables. Observe that they
might still assign inconsistently to those variables in a clause which were not
distinguished by the random string.

Lemma 10.3. *Consider the k-prover proof system defined above and a 3CNF-5
formula φ. If φ is satisfiable, then the provers have a strategy that causes the
verifier to always accept strongly. If at most a $(1 - \varepsilon)$ - fraction of the clauses in
φ is simultaneously satisfiable, then the verifier accepts weakly with probability
at most $k^2 \cdot 2^{-c_2 l}$, where $c_2 > 0$ is a constant which depends only on ε.*

Proof. Suppose φ is satisfiable, then the provers can base their strategy on
a fixed truth assignment for φ, say the lexicographically first. Obviously this
satisfies all clauses, and makes sure that the answers of the provers are mutually
consistent, so the verifier will accept strongly.

If at most a $(1 - \varepsilon)$-fraction of the clauses of φ is satisfiable, let γ denote the
probability with which the verifier accepts φ weakly. Due to the pigeon-hole
principle and the fact that the verifier reacts deterministically to the answers of
the provers, there must be one among the $\binom{k}{2}$ pairs of provers with respect to
which the verifier accepts at least with probability $\gamma / \binom{k}{2}$.

The code words which correspond to these two provers have a Hamming distance
of at least $l/3$. They both have weight $l/2$. Obviously they have the same weight
on the part in which they match. This forces them to also have the same weight
on those coordinates where they are complementary. Thus, for both, the number
of 0-bits equals the number of 1-bits there. It follows that the first prover has a
1-bit and therefore receives a clause on at least $l/6$ coordinates of the code word
on which the second prover has a 0-bit and receives a variable. The situation
on each of these coordinates is the same as in the two-prover one-round proof
system introduced above: the verifier gives a clause to the first, and a variable
of this clause to the second prover. It accepts if the answers are consistent.
(We assume that clauses are always satisfied.) From Proposition 10.2 we know
that the error probability, i.e. the probability with which the verifier accepts an

unsatisfiable formula, is reduced to $2^{-c_1 l/6}$ by $l/6$ parallel repetitions of the two-prover system. The answers of the two provers on the other $5l/6$ coordinates can be fixed in a way that maximizes the acceptance probability which by averaging remains at least $\gamma/\binom{k}{2}$ and is less than $2^{-c_1 l/6}$. So $\gamma \leqslant \binom{k}{2} \cdot 2^{-c_2 l} \leqslant k^2 \cdot 2^{-c_2 l}$ for some constant $c_2 = c_1/6$ which, like c_1, depends only on ε. ∎

10.3 Construction of a Partition System

We are searching for a set system whose base set is easy to cover if we take all sets belonging to one partition, and needs roughly a logarithmic factor more sets if we take sets from pairwise different partitions. For each random string which drives the verifier, we will use a distinct copy of the upcoming partition system to be able to build a set system which models the behaviour of our proof system, see Figure 10.1.

Definition 10.4. *Given a constant $c_3 > 0$, a partition system $B[m, L, k]$ has the following properties.*

1. *The base set contains m points.*

2. *There is a collection of L distinct partitions, where $L < (\ln m)^{c_3}$.*

3. *Each partition partitions the base set into k disjoint subsets, where*
   $$k < \frac{\ln m}{3 \ln \ln m}.$$

4. *Any cover of the base set by subsets that appear in pairwise different partitions requires more than $(k - 2) \ln m$ subsets.*

We want to show that the following randomized construction gives us a partition system which fulfills the properties above with high propability if m is sufficiently large:

> For each partition p decide independently at random in which of its k subsets to place each of the m points of the base set.

Let d denote the threshold $(k - 2) \ln m$. Consider a particular choice of d subsets of pairwise different partitions and show that with high probability these do not cover the base set. The probability for a single point to be covered is $1 - (\frac{k-1}{k})^d$. This can be transformed to

$$1 - (1 - 1/k)^{k \cdot (1-2/k) \ln m} < 1 - e^{-(1+1/k)(1-2/k) \ln m} < 1 - m^{-(1-1/k)}$$

by applying the inequalities $(1 - 1/k)^k > e^{-(1+1/k)}$ for $k \geqslant 2$ and

$$(1+1/k)(1-2/k) = (1-1/k-2/k^2) < (1-1/k).$$

The probability that all m points are covered is therefore $(1-m^{-1+1/k})^m$, which is less than $e^{-m^{1/k}}$.

How many ways are there for choosing the d subsets from our partition system? First we choose d out of L partitions, and then from each of these one out of the k subsets it consists of. This gives us $k^d\binom{L}{d} = \frac{k^d}{d!} \cdot \frac{L!}{(L-d)!}$ ways. Since $d \gg k$ the first fraction is less than 1, and the whole expression less than L^d. Substituting $L < (\ln m)^c$ and $d < k \ln m$ gives us that the number of choices is less than $L^d < (\ln m)^{ck \ln m}$.

But what is the probability that any of these choices covers the whole base set? Obviously it is less than $(\ln m)^{ck \ln m} \cdot e^{-m^{1/k}}$. If we can show that this converges to zero for sufficiently large m, then we have reached our goal. We can do this by applying the natural logarithm and then showing that $(ck \ln m \ln \ln m - m^{1/k})$ tends to $-\infty$. We know that $k < \frac{\ln m}{3 \ln \ln m}$, so we are certainly on the safe side if $(\frac{c}{3} \ln^2 m - m^{\frac{3 \ln \ln m}{\ln m}})$ diverges. If we apply the monotonically growing natural logarithm now to both (positive) summands, we get $\ln \frac{c}{3} + 2 \ln \ln m$ and $3 \ln \ln m$. Now we see that the second summand will dominate the first in the long run, so the sum will diverge, and the probability for a cover will converge to zero, which proves that the probabilistic construction works.

Proposition 10.5. *The construction of a partition system $B[m, L, k]$ is in* ZTIME$(m^{\mathcal{O}(\log m)})$.

Proof. The randomized construction described above requires time polynomial in m. Exhaustively checking that it has Property 4 can be done in time roughly $k^d\binom{L}{d}$ – choose d out of L partitions and in each partition one out of k subsets. As we have seen above, this is less than $(\ln m)^{ck \ln m}$. Substituting $k < \frac{\ln m}{3 \ln \ln m}$ gives us $(\ln m)^{c \ln^2 m / 3 \ln \ln m}$ which is in $m^{\mathcal{O}(\log m)}$ since (applying the natural logarithm to both expressions) $\frac{c}{3} \ln^2 m$ is in $\mathcal{O}(\log m) \cdot \ln m$. The expected number of times the randomized construction needs to be tried until it succeeds is less than $2 = \sum_{i=1}^{\infty} \frac{i}{2^i}$ since it succeeds at least with probability $1/2$ for m sufficiently large. ∎

The author states that the randomized construction can be replaced by a deterministic construction using techniques developed in [NSS95].

10.4 Reduction to Set Cover

Finally we have to show how to reduce an instance of ε-RoBE3SAT-5 to an instance of SETCOVER. The reduction must expand the ε-gap in the number of unsatisfiable clauses of a 3CNF-5 formula to a logarithmic gap in the size of a cover for the base set. By supplying us with the indices of the subsets, the proof system serves as a link between the two instances.

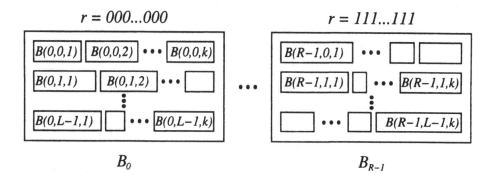

Fig. 10.1. The building blocks of the subset system.

Recall that the verifier has to select l clauses from the 3CNF-5 formula, and to distinguish a variable from each of these clauses at random. In order to do so, the verifier needs $l \cdot (\log 5n/3 + \log 3) = l \log 5n$ random bits. Let $R = 2^{l \log 5n} = (5n)^l$ be the number of 0-1-strings of this length. With each such string r, we associate a *distinct* partition system $B_r = B[m, L, k]$, where $L = 2^l$ and $m = n^{\Theta(l)}$. So the base set S of our set system will have $N = mR$ points. As we know, each random string distinguishes a sequence of l variables. Each of the $L = 2^l$ partitions in B_r is labeled with an l-bit string which corresponds to a truth assignment to the sequence of distinguished variables. Each subset in a partition is labeled by the index of a unique prover P_i. We use $B(r, j, i)$ to denote the i-th subset of partion j in partition system B_r, see Figure 10.1.

The sets of our set system will consist of a number of such partition subsets $B(r, j, i)$. A random string r and the index i of a prover tell the verifier unequivocally (via the prover's code word, see Page 253) for which of the $l/2$ coordinates encoded by r it has to ask the prover for an assignment to all variables of a clause, and for which of the $l/2$ coordinates it only has to ask for an assignment to a single variable. Thus the question q which a verifier might ask a prover is a function π of r and i. π extracts the random bits which are relevant for prover P_i from r according to code word C_i, in other words $q = \pi(r, i)$. Feige uses the notation $(q, i) \in r$ to say that "on random string r prover P_i receives question q". Now we can describe how the proof system is reduced to a family $\mathcal{F}$ of subsets of S. With each question-answer pair $[q, a]$ which is potentially exchanged between the verifier and prover P_i we associate the subset

$$S_{(q,a,i)} = \bigcup_{(q,i) \in r} B(r, a_r, i)$$

where a_r is a projection of the prover's answer a to the sequence of distinguished variables. For each clause for which the prover is supposed to answer with a truth assignment to all of its three variables, this projection simply ignores the assignments to the two variables which were not distinguished. Note that the prover's index i in $S_{(q,a,i)}$ is redundant since it can be deduced from the question q and the code words, but used nevertheless for clarity.

Let Q denote the number of possible different questions that a prover may receive. A question to a single prover includes $l/2$ variables, for which there are $n^{l/2}$ possibilities (with repetition), and $l/2$ clauses, for which there are $(5n/3)^{l/2}$ possibilities. Hence $Q = n^{l/2} \cdot (5n/3)^{l/2}$. Observe that this number is the same for all provers.

The following key lemma exploits the fact that the reduction above produces subset families which depend strongly on the answers of the provers. The idea is that if a formula φ is satisfiable, then there is a strategy for the provers to answer consistently, and the subsets corresponding to these answers fit nicely together. They will even be disjoint, hence few are needed to cover the base set. If however φ is not satisfiable, we will use a probabilistic argument to show that approximately by a logarithmic factor more sets are needed for a cover in the subset system to which φ is reduced in this case.

Lemma 10.6. *If φ is satisfiable, then the base set S of $N = mR$ points can be covered by kQ subsets. If only a $(1 - \varepsilon)$-fraction of the clauses in φ is simultaneously satisfiable, S requires more than $(k - 4)Q \ln m$ subsets in order to be covered. Thus the ratio between the two cases is at least $(1 - 4/k) \ln m$.*

Proof. Suppose there is a satisfying truth assignment for φ. Fix a strategy for the provers which is consistent with it. Let $a_1, \ldots, a_k$ be the provers' answers to the verifier's questions $q_1 = \pi(r, 1), \ldots, q_k = \pi(r, k)$ with the notation used above. Then for any random string r (i.e. no matter what the verifier asks), these answers are consistent and map to the same assignment a_r for the sequence of distinguished variables. Hence the sets $S_{(q_1,a_1,1)}, \ldots, S_{(q_k,a_k,k)}$ contain $B(r, a_r, 1), \ldots, B(r, a_r, k)$ and therefore cover B_r. Since this holds for all random strings r, the collection of all subsets $S_{(q,a,i)}$, where a is the answer given by prover P_i under the strategy above, covers the whole base set S. The number of sets used is the product of k, the number of provers, and Q, the number of possible questions to a single prover. These kQ subsets are disjoint.

Now consider the case that at most a $(1 - \varepsilon)$-fraction of the clauses in φ is simultaneously satisfiable. Lemma 10.3 tells us that the provers have no strategy which succeeds in convincing the verifier to accept with probability greater than $k^2 \cdot 2^{-c_2 l}$. Assume we had a cover $\mathcal{C} \subseteq \mathcal{F}$ of size $(k - 4)Q \ln m$ (i.e. just below the logarithmic threshold). If we can find a strategy for the provers based on this cover which succeeds with a probability that is greater than Lemma 10.3 allows, then we have the contradiction needed.

We map our cover C back into a proof system. With each question q to a prover P_i, we associate a weight $w_{q,i}$ which equals the number of answers a such that $S_{(q,a,i)} \in C$. Hence $\sum_{q,i} w_{q,i} = |C|$. Now with each random string r we associate the weight $w_r = \sum_{(q,i) \in r} w_{q,i}$. This weight equals the number of subsets that participate in covering B_r since

$$w_r = \sum_{(q,i) \in r} w_{q,i} = \sum_{(q,i) \in r} \left|\{a : S_{(q,a,i)} \in C\}\right| = \left|\{S_{(q,a,i)} \in C : (q,i) \in r\}\right|.$$

We say that a random string r is *good* if $w_r < (k-2)\ln m$.

Proposition 10.7. *Given a cover C of size $(k-4)Q\ln m$, the fraction of good r is at least $2/k$.*

Proof. Let $G = \{r : w_r \geqslant (k-2)\ln m\}$ be the set of *bad* random strings. Assume $|G| > (1-2/k)R$ where R is the number of possible random strings. Then

$$\sum_r w_r \geqslant \sum_{r \in G} w_r \geqslant (1-2/k)R \cdot (1-2/k)k\ln m > (1-4/k)kR\ln m$$

On the other hand, using the definition of w_r, we have

$$\sum_r w_r = \sum_r \sum_{(q,i) \in r} w_{q,i} = \sum_{q,i} \frac{R}{Q} w_{q,i} = \frac{R}{Q} \cdot |C|$$

Recall that Q denotes the number of possible different questions that a prover may receive. Then R/Q is the number of random strings which cause the verifier to send a particular question q to a particular prover. This explains the second equality. Putting the two lines above together, we get $|C| > (k-4)Q\ln m$ which is a contradiction. ∎

Proposition 10.8. *Given a cover C of size $(k-4)Q\ln m$. Then there is a strategy for the provers such that the verifier accepts φ with probability at least $8/(k^3 \ln^2 m)$.*

Proof. We will show that the following simple random strategy which is based on the assumed cover C succeeds with the desired probability.

> On receiving question q, P_i selects an answer a uniformly at random from the set of answers that satisfy $S_{(q,a,i)} \in C$.

For a fixed random string r there is a one-to-one correspondence between the sets $B(r,p,i)$ which participate in the cover of B_r, and sets $S_{(q,a,i)}$ that belong to C: on the one hand, it is clear that a set $S_{(q,a,i)}$ contains at most one $B(r,p,i)$ from partition system B_r, namely that with $p = a_r$. (Recall that a_r is the projection of the answer a of prover P_i to an assignment a_r for the sequence of variables distinguished by r.) On the other hand, random string r forces the verifier to pose prover P_i a unique question q.

It is enough to focus on good r, and to compute a lower bound for the probability that the verifier accepts if it receives a random string of this type. Recall Property 4 of our partition system which states that a cover of the m points of B_r by subsets that appear in pairwise different partitions requires more than $(k-2) \ln m$ subsets. Since we know that a good r is covered by fewer subsets, C must have used two (different) subsets $B(r, p, i)$ and $B(r, p, j)$ from the same partition p. Let $S_{(q_i, a_i, i)}$ and $S_{(q_j, a_j, j)}$ be the two sets in C which the above mentioned one-to-one correspondence yields.

Consider what happens when the random source supplies the verifier with a good random string r. The verifier poses question q_i to P_i and q_j to P_j. Let $A_i = \{a | S_{(q_i, a, i)} \in C\}$ be the set of answers prover P_i can choose from. Define A_j analogously. Using the random strategy described above, P_i will pick a_i with probability $1/|A_i|$, P_j analogously. $|A_i| + |A_j| = w_{q_i,i} + w_{q_j,j} \leqslant w_r < k \ln m$ since r is good. So the joint probability for the provers to answer with a_i and a_j is at least $1/(|A_i| \cdot |A_j|) \geqslant (2/(|A_i| + |A_j|))^2 > 4/(k \ln m)^2$. The first inequality holds because xy is maximized by $x = y = z/2$ under the constraint that $x + y = z$. Since these answers are weakly consistent, i.e. they map to the same a_r, the verifier will accept weakly. Given that the fraction of good r is at least $2/k$ the verifier will accept with probability at least $8/(k^3 \ln^2 m)$. ∎

To complete the proof of Lemma 10.6 we must show that the provers' random strategy succeeds with a probability that is greater than Lemma 10.3 allows, i.e. $8/(k^3 \ln^2 m) > k^2 \cdot 2^{-c_2 l}$. This is true if $8 \cdot 2^{c_2 l} > k^5 \ln^2 m$. Since Definition 10.4 requires $k < \frac{\ln m}{3 \ln \ln m}$, we are on the safe side if $8 \cdot 2^{c_2 l} > \frac{\ln^7 m}{(3 \ln \ln m)^5}$, but even more so, if $2^{c_2 l} > \ln^7 m$. Setting $l = \frac{8}{c_2} \log \log n$, and $m = (5n)^{c_4 l}$, we receive $\log^8 n > \frac{8 c_4}{c_2} \log \log n \cdot \ln^7 n$, which is certainly true for sufficiently large n. c_4 is a constant which will be fixed in the proof of the main theorem. ∎

Combining three results — the NP-hardness of ε-RoBE3SAT-5, the complexity of constructing a partition system (Proposition 10.5), and the connection between the number of satisfiable clauses in a 3CNF-5 formula and the size of a cover with sets from such a partition system (Lemma 10.6) — we obtain the main theorem.

Theorem 10.9. *[Fei96]*
If there is some $\varepsilon > 0$ such that a polynomial time algorithm can approximate SETCOVER *within $(1 - \varepsilon) \ln N$, then $\mathcal{NP} \subseteq \mathrm{DTIME}(N^{\mathcal{O}(\log \log N)})$.*

Proof. Assume that there is such an algorithm A. Reduce an instance of an arbitrary problem $X \in \mathcal{NP}$ in polynomial time to an instance of the NP-hard problem ε-RobE3Sat-5, i.e. to a 3CNF-5 formula φ with n variables which is either satisfiable or has at most a $(1 - \varepsilon)$-fraction of clauses which are simultaneously satisfiable.

According to the proof of Lemma 10.6, let $l = \frac{8}{c_2} \log \log n$ and $m = (5n)^{c_4 l}$ with $c_4 = 2/\varepsilon$. As required by Definition 10.4, let $L = 2^l < (\ln m)^{c_3}$ for some constant c_3 and choose a $k < \frac{\ln m}{3 \ln \ln m}$. Make sure that $k \geqslant 8/\varepsilon$. In order to satisfy $\frac{\ln m}{3 \ln \ln m} > 8/\varepsilon$, n must be sufficiently large, namely $\frac{2l \ln 5n}{(5n)^{l/12}} < \varepsilon$.

Now reduce φ with the help of partition systems $B[m, L, k]$ to a collection $\mathcal{C}$ of sets $S_{(q,a,i)}$ as above.

Recall that $R = (5n)^l = m^{\varepsilon/2}$ and $Q = n^{l/2} \cdot (5n/3)^{l/2}$. Therefore m, R, and Q are bounded by $n^{\mathcal{O}(\log \log n)}$ and the deterministic construction of partition systems [NSS95] is in $\mathrm{DTIME}(n^{\mathcal{O}(\log \log n)}) \subseteq \mathrm{DTIME}(N^{\mathcal{O}(\log \log N)})$ where $N = mR = m^{1+\varepsilon/2}$.

Apply algorithm A to the instance of SetCover to which φ was reduced. If φ is satisfiable, then there is a cover of size kQ according to Lemma 10.6. Therefore A must find a cover $\mathcal{C}_A$ of size at most $kQ \cdot (1 - \varepsilon) \ln N$ in the satisfiable case. With $\ln N = (1 + \varepsilon/2) \ln m$ and $\varepsilon/2 \geqslant 4/k$ the size of A's solution can be upper-bounded as follows:

$$|\mathcal{C}_A| \leqslant kQ(1 - \varepsilon)(1 + \varepsilon/2) \ln m < kQ(1 - \varepsilon/2) \ln m \leqslant kQ(1 - 4/k) \ln m$$

On the other hand Lemma 10.6 says that if at most a $(1 - \varepsilon)$-fraction of the clauses in φ is simultaneously satisfiable, S requires more than $(k - 4)Q \ln m$ subsets in order to be covered. Thus we could use A to decide deterministically an arbitrary NP-problem in time proportional to $N^{\mathcal{O}(\log \log N)}$. ∎

Exercises

Exercise 10.1. Proof that the greedy algorithm presented in Section 10.1 always finds a cover $\mathcal{C}_{gr} \subseteq \mathcal{F}$ of the base set S such that

$$|\mathcal{C}_{gr}| \leqslant (1 + \frac{1}{2} + \ldots + \frac{1}{\Delta}) \cdot |\mathcal{C}_{opt}|$$

where $|\mathcal{C}_{opt}|$ is the size of a minimum cover and Δ the size of the largest subset of S in $\mathcal{F}$ ([Joh74], [Lov75]). Show that this ratio cannot be improved.

Hint: Group the subsets which the greedy algorithm selects into phases such that all subsets of one phase cover the same number of so far uncovered elements of the base set.

Exercise 10.2. Show that there is a $\delta \in (0,1)$ such that ε-RObE3SAT-5, as defined in Section 10.1.2, is NP-hard for all $\varepsilon \leqslant \delta$.

Hint: Use that it is NP-hard to approximate MAX3SAT-b (see Lemma 4.9 or [ALM$^+$92]). The number of occurences of each variable in a formula of type 3CNF-b is bounded from above by a universal constant b.

11. Semidefinite Programming and Its Applications to Approximation Algorithms

Thomas Hofmeister*, Martin Hühne

11.1 Introduction

$\mathcal{NP}$-hard problems are often tackled using the *relaxation method* approach. The idea of this approach is to add or remove conditions to or from the original problem in such a way that the solution space is enlarged. The hope is that the new solution space will be more smooth in the sense that it allows a more efficient, i.e., polynomial-time, computation of the optimum.

One possibility of applying this method lies in computing upper or lower bounds within a branch-and-bound approach. Let us assume that we are dealing with a maximization problem. Since the modified solution space contains the original solution space as a subset, the optimal solution to the relaxed problem is an upper bound to the optimum of the original problem.

Known examples for problems which can be treated in this way include the KNAPSACK problem. Given n objects with size c_j and value t_j, we want to fill a knapsack such that its value is maximized. This is an $\mathcal{NP}$-hard problem, but if we are allowed to take fractional parts of the objects, then we can solve the problem in polynomial time, using a greedy algorithm. The greedy solution provides us with an upper bound for the original KNAPSACK problem.

One can also try to apply some transformation (often called "rounding") to a solution of the relaxed problem in order to obtain a good solution for the original problem. The quality of the so obtained solution can then be estimated by comparing it with the upper bound provided by the relaxed solution. There are cases, like "totally unimodular" linear programs, where relaxations even yield the optimal solutions, as is the case with the weighted matching problem.

The relaxation method can also be combined with *semidefinite programs*. Semidefinite programming is a generalization of linear programming. It has been studied for quite some time. For example, it is known that semidefinite programs can be solved in polynomial time. The attention of algorithm designers has turned to semidefinite programming only recently. This is due to a paper by Goemans and Williamson [GW95], where for several $\mathcal{NP}$-hard problems efficient

* This author was partially supported by the Deutsche Forschungsgemeinschaft as part of the Collaborative Research Center "Computational Intelligence" (SFB 531).

approximation algorithms based on semidefinite programming were presented. One of these problems is MAXCUT, where the set of vertices of a given graph has to be partitioned into two parts such that the number of edges between the parts is maximized. (For a definition of the other problems considered in this chapter, see the end of this introduction.) Goemans and Williamson describe a randomized polynomial-time algorithm that computes a cut of size at least 0.878 times the size of the maximum cut.

The idea is to first formulate the MAXCUT-problem as an integer quadratic program. The relaxation considers not only integer variables as possible candidates for the solution, but n-dimensional unit vectors. By this relaxation, the solution space becomes much larger, but the resulting problem can be solved in polynomial time with the help of semidefinite programming. From the solution obtained in this way, an integer solution can be constructed by randomly selecting a hyperplane and identifying the vectors with either 1 or -1, depending on which side of the plane they are. It can be shown that the integer solution so obtained has an expectation for the objective function which is at least 0.878 times the relaxed solution. Thus, Goemans and Williamson have provided a *randomized* $1/0.878 \approx 1.139$-approximation algorithm for MAXCUT, improving upon the previously best approximation ratio of 2. In the conference version of their paper, they also sketched how this method might be derandomized into a deterministic algorithm. However, their derandomization contains a flaw, which is reported in the journal version of that paper. A correct but complex derandomization procedure has been given by Mahajan and Ramesh [MR95a].

It should be noted that (unless $\mathcal{P}=\mathcal{NP}$) there can be no polynomial-time approximation algorithm with an approximation ratio of $1+\varepsilon$ for arbitrarily small ε, since the by now famous PCP-theorem has as one of its consequences that there is a constant $\varepsilon_0 > 0$ such that computing $1+\varepsilon_0$-approximations for MAXCUT is $\mathcal{NP}$-hard. In fact, this constant has recently been made more explicit by Håstad in [Hås97b] who showed that (for all $\varepsilon > 0$) already computing a $(17/16-\varepsilon)$-approximation is $\mathcal{NP}$-hard.

The article by Goemans and Williamson has started a series of research activities. Several authors have studied the question whether semidefinite programming can also be applied to other graph problems or, more general, classes of problems. The papers by Karger, Motwani and Sudan [KMS94], Klein and Lu [KL96] as well as Bacik and Mahajan [BM95] go into this direction. Karloff investigates in [Kar96] whether the estimations in [GW95] are sharp and also shows that additional linear constraints can not lead to better approximation ratios.

The essence of all these papers is that semidefinite programming is a very versatile and powerful tool. Besides MAXCUT, it has been applied to obtain better approximation algorithms for several other important $\mathcal{NP}$-hard problems. Among these are the problems MAXSAT, MAXkSAT, MAXEkSAT (especially for $k = 2$) and MAXDICUT. In the following table, the approximation ratios α obtained in [GW95] are listed. The table also lists $1/\alpha$, since that value is more commonly used in that paper and its successors.

Problem	$1/\alpha$	α
MaxDiCut	0.796	1.257
Max2Sat	0.878	1.139
MaxSat	0.755	1.324
MaxSat	0.758	1.319

(In this chapter, the approximation ratios are rounded to three decimal digits.)

The first of the two MaxSat-algorithms was given in the conference version and the second one in the final version. Before, the best known approximation ratio for MaxDiCut was 4 [PY91]. For MaxE2Sat, a 1.334-approximation algorithm was known [Yan92, GW94b].

Tuning the approach of semidefinite programming, several authors studied improvements of the approximation algorithms for these problems. At present, polynomial-time algorithms with the following approximation ratios are known:

Problem	$1/\alpha$	α	Reference
MaxDiCut	0.859	1.165	Feige/Goemans [FG95]
MaxE2Sat	0.931	1.075	Feige/Goemans [FG95]
MaxSat	0.765	1.307	Asano/Ono/Hirata [AOH96]
MaxSat	0.767	1.304	Asano/Hori/Ono/Hirata [AHOH96]
MaxSat	0.770	1.299	Asano [Asa97]

We will discuss some of the improvements in this chapter. However, our emphasis will be on *methods* rather than on results. I.e., we will not present the very last improvement on the approximation of these problems. Instead, we present the basic methods and techniques which are needed. More precisely, we will look at five basic methods used in the approximation algorithms for MaxDiCut and MaxSat:

1. The modeling of asymmetric problems such as MaxDiCut and MaxE2Sat.

2. A method for handling long clauses in MaxSat instances.

3. Combining different approximation algorithms for MaxEkSat yields better approximations for MaxSat.

4. Adding linear restrictions to the semidefinite program can improve the approximation ratio for MaxDiCut and MaxSat. However, there are no linear restrictions which improve the MaxCut approximation from [GW95].

5. Nonuniform rounding of the solution of the semidefinite program is a way to take advantage of the modeling of asymmetric problems. This improves the approximation ratio. The ratio is further improved if nonuniform rounding is combined with method 4.

Note that there are several other approaches for the approximation of MaxCut and MaxSat. For example, in Chapter 12, the "smooth integer programming" technique is described, which can be used if the instances are dense. For additional references on semidefinite programming and combinatorial optimization, we refer to [Ali95, VB96, Goe97].

Our survey will be organized as follows: First, we describe the mathematical tools necessary to understand the method of semidefinite programming (Sections 11.2–11.4). Since algorithms for solving semidefinite programs are rather involved, we only give a sketch of the so-called interior-point method. We describe a "classical" approximation of MaxCut, which is not based on semidefinite programming (Section 11.5.1). We then show how Goemans and Williamson applied semidefinite programming to obtain better approximation algorithms for MaxCut (Section 11.5.2) and analyze the quality of their algorithm. In Section 11.6 we describe the modeling of the asymmetric problems MaxDiCut and MaxE2Sat. A method for modeling long clauses in a semidefinite program is described in Section 11.7.1. We review some classical approaches for MaxSat and explain how different MaxSat algorithms can be combined in order to improve the approximation ratio (Section 11.7.2). Section 11.8 describes the effect of additional constraints and of a nonuniform rounding technique.

We give a formal definition of the problems considered in this chapter.

MaxCut

Instance: Given an undirected graph $G = (V, E)$.
Problem: Find a partition $V = V_1 \cup V_2$ such that the number of edges between V_1 and V_2 is maximized.

We will assume that $V = \{1, \ldots, n\}$.

MaxDiCut

Instance: Given a directed graph $G = (V, E)$.
Problem: Find a subset $S \subseteq V$ such that the number of edges $(i, j) \in E$ with tail i in S and head j in $\overline{S}$ is maximal.

MaxSat

Instance: Given a Boolean formula Φ in conjunctive normal form.
Problem: Find a variable assignment which satisfies the maximal number of clauses in Φ.

MaxkSat, $k \in \mathbb{N}$

Instance: Given a Boolean formula Φ in conjunctive normal form with at most k literals in each clause.
Problem: Find a variable assignment which satisfies the maximal number of clauses in Φ.

MaxEkSat, $k \in \mathbb{N}$

Instance: Given a Boolean formula Φ in conjunctive normal form with exactly k literals in each clause.

Problem: Find a variable assignment which satisfies the maximal number of clauses in Φ.

The variables of the Boolean formula are denoted by $x_1, \ldots, x_n$, the clauses are $C_1, \ldots, C_m$. Note that each algorithm for Max$k$Sat is also an algorithm for MaxEkSat. An important special case is $k = 2$.

Sometimes, weighted versions of these problems are considered.

WEIGHTED MaxCut

Instance: Given an undirected graph $G = (V, E)$ and positive edge weights.

Problem: Find a partition $V = V_1 \cup V_2$ such that the sum of the weights of edges between V_1 and V_2 is maximized.

WEIGHTED MaxDiCut

Instance: Given a directed graph $G = (V, E)$ and nonnegative edge weights.

Problem: Find a subset $S \subseteq V$ such that the total weight of the edges $(i, j) \in E$ with tail i in S and head j in $\overline{S}$ is maximal.

WEIGHTED MaxSat

Instance: Given a Boolean formula $\Phi = C_1 \wedge \ldots \wedge C_m$ in conjunctive normal form and nonnegative clause weights.

Problem: Find a variable assignment for Φ which maximizes the total weight of the satisfied clauses.

All results mentioned in this chapter hold for the unweighted and for the weighted version of the problem. The approximation ratio for the weighted version is the same as for the corresponding unweighted version. Since the modification of the arguments is straightforward, we restrict ourselves to the unweighted problems.

11.2 Basics from Matrix Theory

In this section, we will recall a few definitions and basic properties from Matrix Theory. The reader interested in more details and proofs may consult, e.g., [GvL86]. Throughout this chapter, we assume that the entries of a matrix are real numbers.

An $n \times m$ matrix is called *square* iff $n = m$. A matrix A such that $A = A^T$ is called *symmetric*. A matrix A such that $A^T A$ is a diagonal matrix is called *orthogonal*. It is *orthonormal* if in addition $A^T A = Id$, i.e., $A^T = A^{-1}$.

Definition 11.1. *An* eigenvector *of a matrix A is a vector $x \neq 0$ such that there exists a $\lambda \in \mathbb{C}$ with $A \cdot x = \lambda \cdot x$. The corresponding λ is called an* eigenvalue. *The* trace *of a matrix is the sum of the diagonal elements of the matrix.*

It is known that the trace is equal to the sum of the eigenvalues of A and the determinant is equal to the product of the eigenvalues.

Note that zero is an eigenvalue of a matrix if and only if the matrix does not have full rank.

Another characterization of eigenvalues is as follows:

Lemma 11.2. *The eigenvalues of matrix A are the roots of the polynomial $\det(A - \lambda \cdot Id)$, where λ is the free variable and Id is the identity matrix.*

It is known that if A is a symmetric matrix, then it possesses n real eigenvalues (which are not necessarily distinct). For such an $n \times n$- matrix, the canonical numbering of its n eigenvalues is $\lambda_1 \geqslant \lambda_2 \geqslant \ldots \geqslant \lambda_n$. The number of times that a value appears as an eigenvalue is also called its *multiplicity.*

Definition 11.3. *The* inner product *of two matrices is defined by*

$$A \bullet B := \sum_{i,j} A_{i,j} \cdot B_{i,j} \quad = \quad trace(A^T \cdot B).$$

Theorem 11.4 (Rayleigh-Ritz). *If A is a symmetric matrix, then*

$$\lambda_1(A) = \max_{||x||=1} x^T \cdot A \cdot x \qquad and \qquad \lambda_n(A) = \min_{||x||=1} x^T \cdot A \cdot x.$$

Semidefiniteness

Having recalled some of the basic matrix notions and lemmata, let us now turn to the definitions of definiteness.

Definition 11.5. *A square matrix A is called*

$$\left. \begin{array}{lll} \text{positive semidefinite} & \text{if} & x^T \cdot A \cdot x \geqslant 0 \\ \text{positive definite} & \text{if} & x^T \cdot A \cdot x > 0 \end{array} \right\} \textit{for all } x \in \mathbb{R}^n \setminus \{\underline{0}\}$$

In this chapter, we will abbreviate "positive semidefinite" also by "PSD."

As a corollary of Theorem 11.4, we obtain that a symmetric matrix is PSD if and only if its smallest eigenvalue $\lambda_n \geqslant 0$.

Examples of PSD matrices. It should be noted that it is not sufficient for a matrix to contain only positive entries to make it PSD. Consider the following two matrices

$$A := \begin{pmatrix} 1 & 4 \\ 4 & 1 \end{pmatrix} \qquad B := \begin{pmatrix} 1 & -1 \\ -1 & 4 \end{pmatrix}.$$

The matrix A is not PSD. This can be seen by observing that $(1, -1) \cdot A \cdot (1, -1)^T = -6$. The eigenvalues of A are 5 and -3. On the other hand, the symmetric matrix B with negative entries is PSD. The eigenvalues of B are approximately 4.30 and 0.697.

For a diagonal matrix, the eigenvalues are equal to the entries on the diagonal. Thus, a diagonal matrix is PSD if and only if all of its entries are nonnegative.

Also by the definition, it follows that whenever B and C are matrices, then the matrix

$$A = \begin{pmatrix} B & 0 \\ 0 & C \end{pmatrix}$$

is PSD if and only if B and C are PSD. This means that requiring some matrices $A_1, \ldots, A_r$ to be PSD, is equivalent to requiring *one* particular matrix to be PSD.

Also, the following property holds: If A is PSD, then the $k \times k$-submatrix obtained from A by eliminating rows and columns $k + 1$ to n is also PSD.

As a final example, consider the symmetric matrix

$$\begin{pmatrix} 1 & a & \ldots & a \\ a & 1 & \ldots & a \\ & & \ldots\ldots\ldots & \\ a & a & \ldots & 1 \end{pmatrix}$$

with ones on the diagonal and an a everywhere else. A simple calculation reveals that its eigenvalues are $1 + a \cdot (n - 1)$ (with multiplicity 1) and $1 - a$ (with multiplicity $n - 1$). Hence, the matrix is PSD for all $-1/(n - 1) \leqslant a \leqslant 1$.

Further properties. In most places in this chapter, we will only be concerned with matrices A which are symmetric. For symmetric matrices, we have the following lemma:

Lemma 11.6. *Let A be an $n \times n$ symmetric matrix. The following are equivalent:*

– *A is PSD.*

– *All eigenvalues of A are nonnegative.*

– *There is an $n \times n$-matrix B such that $A = B^T \cdot B$.*

Note that given a symmetric PSD matrix A, the decomposition $A = B^T \cdot B$ is not unique, since the scalar product of vectors is invariant under rotation, so any rotation of (the vectors given by) matrix B yields another decomposition. As a consequence, we can always choose B to be an upper triangular matrix in the decomposition.

Goemans-Williamson's approximation algorithm uses a subroutine which computes the "(incomplete) Cholesky decomposition" of a given PSD symmetric matrix A. The output of the subroutine is a matrix B such that $A = B^T \cdot B$. This decomposition can be performed in time $\mathcal{O}(n^3)$, see e.g. [GvL86].

As an example, consider the following Cholesky decomposition:

$$A = \begin{pmatrix} 1 & -1 \\ -1 & 4 \end{pmatrix} = \begin{pmatrix} 1 & 0 \\ -1 & \sqrt{3} \end{pmatrix} \cdot \begin{pmatrix} 1 & -1 \\ 0 & \sqrt{3} \end{pmatrix}$$

Remark 11.7. The problem of deciding whether a given square matrix A is PSD can be reduced to the same problem for a symmetric matrix A'. Namely, define A' to be the symmetric matrix such that $A'_{i,j} := \frac{1}{2} \cdot (A_{i,j} + A_{j,i})$. Since

$$x^T \cdot A \cdot x = \sum_{i,j} A_{i,j} \cdot x_i \cdot x_j,$$

it holds that $x^T \cdot A' \cdot x = x^T \cdot A \cdot x$ and A' is PSD if and only if A is PSD.

11.3 Semidefinite Programming

One problem for a beginner in semidefinite programming is that in many papers, semidefinite programs are defined differently. Nevertheless, as one should expect, most of those definitions turn out to be equivalent. Here is one of those possible definitions:

Definition 11.8. *A semidefinite program is of the following form. We are looking for a solution in the real variables $x_1, \ldots, x_m$. Given a vector $c \in \mathbb{R}^m$, we want to minimize (or maximize) $c \cdot x$. The feasible solution space from which we are allowed to take the x-vectors is described by a symmetric matrix SP which as its entries contains linear functions in the x_i-variables. Namely, an x-vector is feasible iff the matrix SP becomes PSD if we plug the components of x into the corresponding positions.*

Here is an example of a semidefinite program:

$$
\begin{aligned}
&\text{maximize} \quad x_1 + x_2 + x_3 \\
&\text{such that} \quad \begin{pmatrix} 1 & x_1 - 1 & 2x_1 + x_2 - 1 \\ x_1 - 1 & -x_1 & x_2 + 3 \\ 2x_1 + x_2 - 1 & x_2 + 3 & 0 \end{pmatrix} \text{ is PSD.}
\end{aligned}
$$

Again, by Remark 11.7, when we are modeling some problem, we need not restrict ourselves to symmetric matrices only.

The matrix that we obtain by plugging into SP the values of vector x will also sometimes be written as $SP(x)$.

Given an ε and a semidefinite program which has some finite, polynomially bounded solution, the program can be solved in polynomial time, up to an error term of ε. In general, an error term cannot be avoided since the optimal solution to a semidefinite program can contain irrational numbers.

Semidefinite programming is a generalization of linear programming. This can be seen as follows. Let the linear inequalities of the linear program be of the form $a_1 x_1 + \ldots + a_m x_m + b \geqslant 0$. Given r linear inequalities $IN_i \geqslant 0$, $(1 \leqslant i \leqslant r)$, we can define the matrix SP for the semidefinite program to be of the following diagonal form:

$$\begin{pmatrix} IN_1 & 0 & \cdots & 0 \\ 0 & IN_2 & \cdots & 0 \\ \cdots & \cdots & \cdots & \cdots \\ 0 & 0 & \cdots & IN_r \end{pmatrix}$$

Since a diagonal matrix is PSD if and only if all of its entries are nonnegative, we have that the feasible solution spaces for the linear program and the semidefinite program are equal. Hence, linear programming can be seen as the special case of semidefinite programming where the given symmetric matrix SP is restricted to be diagonal.

The feasible solution space which is defined by a semidefinite program is convex. This can be seen as follows:

Given two feasible vectors x and y and its corresponding matrices $SP(x)$ and $SP(y)$, all vectors "between" x and y are also feasible, i.e., for all $0 \leqslant \lambda \leqslant 1$,

$$SP((1 - \lambda)x + \lambda y) = (1 - \lambda)SP(x) + \lambda SP(y)$$

is also PSD, as can easily be seen: For all v,

$$v^T \cdot SP((1 - \lambda)x + \lambda y) \cdot v = v^T \cdot (1 - \lambda) \cdot SP(x) \cdot v + v^T \cdot \lambda \cdot SP(y) \cdot v \geqslant 0.$$

This means that the feasible solution space which we can describe with the help of a semidefinite program is a convex space. As a consequence, we are not able to express a condition like "$x \geqslant 1$ or $x \leqslant -1$" in one dimension.

For a beginner, it is hard to get a feeling for what can and what cannot be formulated as a semidefinite program. One of the reasons is that - contrary to linear programs - it makes a big difference whether an inequality is of the type "$\leqslant$" or "$\geqslant$", as we shall soon see.

In the following, we show that a large class of quadratic constraints can be expressed in semidefinite programs.

11.3.1 Quadratically Constrained Quadratic Programming

Quadratically constrained quadratic programming can be solved using semidefinite programming, as we shall see now. Nevertheless, Vandenberghe and Boyd [VB96] state that as far as efficiency in the algorithms is concerned, one should better use interior-point methods particularly designed for this type of problems. A quadratically constrained quadratic program can be written as

$$\begin{aligned} \text{minimize} \quad & f_0(x) \\ \text{such that} \quad & f_i(x) \leqslant 0, \quad i = 1, \ldots, L \end{aligned}$$

where the f_i are convex quadratic functions $f_i(x) = (A_i x + b_i)^T (A_i x + b_i) - c_i^T x - d_i$. The corresponding semidefinite program looks as follows:

$$\begin{aligned} \text{minimize} \quad & t \\ \text{such that} \quad & \begin{pmatrix} I & A_0 x + b_0 \\ (A_0 x + b_0)^T & c_o^T x + d_0 + t \end{pmatrix} \text{ is PSD and} \\ & \forall i : \begin{pmatrix} I & A_i x + b_i \\ (A_i x + b_i)^T & c_i^T x + d_i \end{pmatrix} \text{ is PSD.} \end{aligned}$$

As we have seen earlier, the AND-condition of matrices being PSD can easily be translated into a semidefinite program.

As an example, consider the following problem in two dimensions:

$$\text{minimize} \quad -x + \frac{y}{2} \tag{11.1}$$

$$\text{such that} \quad y \geqslant x^2 \text{ and } y \leqslant \frac{x}{3} + \frac{1}{2}$$

The space of feasible solutions consists of all points located "between" the parabola and the straight line shown in Figure 11.1. We leave it as an exercise to the reader to compute the optimal solution (Exercise 11.2).

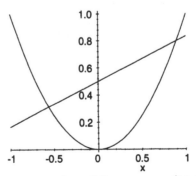

Fig. 11.1. Graphical representation of the program (11.1).

Since the feasible space is convex, we are not able to express a condition like $y \leqslant x^2$ in a semidefinite program. This shows that one has to be careful with the direction of the inequalities, in contrast to linear programs.

As another example, consider the symmetric 2×2-matrix

$$\begin{pmatrix} a & b \\ b & c \end{pmatrix}.$$

This matrix is PSD if and only if $a \geqslant 0, c \geqslant 0$, and $b^2 \leqslant ac$. Consequently,

$$\begin{pmatrix} x_1 & 2 \\ 2 & x_2 \end{pmatrix}$$

allows us to describe the feasible solution space $x_1 \cdot x_2 \geqslant 4, x_1 \geqslant 0$.

Finally, we remark that in some papers, a semidefinite program is defined by

minimize $C \bullet X$ **such that** $A_i \bullet X = b_i$ $(1 \leqslant i \leqslant m)$ and X is PSD

where X, C, A_i are all symmetric matrices. In fact, this is the dual (see Section 11.4) of the definition we gave.

11.3.2 Eigenvalues, Graph Theory and Semidefinite Programming

There are interesting connections between eigenvalues and graph theory, let us for example state without proof the "Fundamental Theorem of Algebraic Graph Theory" (see, e.g., [MR95b, p. 144]).

Theorem 11.9. *Let* $G = (V, E)$ *be an undirected (multi)graph with* n *vertices and maximum degree* Δ. *Then, under the canonical numbering of its eigenvalues* λ_i *for the adjacency matrix* $A(G)$, *the following holds:*

1. *If* G *is connected, then* $\lambda_2 < \lambda_1$.

2. *For* $1 \leqslant i \leqslant n, |\lambda_i| \leqslant \Delta$.

3. Δ *is an eigenvalue if and only if* G *is regular.*

4. G *is bipartite if and only if for every eigenvalue* λ *there is an eigenvalue* $-\lambda$ *of the same multiplicity.*

5. *Suppose that* G *is connected. Then,* G *is bipartite if and only if* $-\lambda_1$ *is an eigenvalue.*

Another connection is described by the so-called *Lovász number*. Given an adjacency matrix A, we obtain the matrix A' by changing A as follows: Put 1's on the main diagonal, and replace every 0 at a position (i, j) by the variable $x_{i,j}$.

The Lovász number of a graph is defined by

$$\theta(G) = \min_{\underline{x} \in \mathbb{R}^n} \lambda_1(A'(\underline{x}))$$

The Lovász number of a graph can be used to obtain an upper bound for the maximum clique-size $\omega(G)$ in a graph G, namely: $\omega(G) \leqslant \theta(G)$.

We sketch the proof of this property: Without loss of generality, the largest clique of size k is on the first k vertices. This means that the k-th "principal submatrix" contains ones only. From matrix theory, it is known that the largest eigenvalue of a matrix is at least the value of the largest eigenvalue of any principal submatrix. Since k is an eigenvalue of the k-th principal submatrix, we obtain

$$\theta(G) \geqslant \lambda_1(A') \geqslant k = \omega(G).$$

The largest eigenvalue of a symmetric matrix can be computed with the help of a semidefinite program. Let $\lambda_1, \ldots, \lambda_n$ be the eigenvalues of a matrix A. Then the eigenvalues of $z \cdot Id - A$ are $z - \lambda_1, \ldots, z - \lambda_n$. Hence, the solution to the following semidefinite program yields the largest eigenvalue of A:

minimize z **such that** $z \cdot Id - A$ is PSD

It is now obvious that we can also compute the Lovász number of a graph with the help of semidefinite programming.

11.4 Duality and an Interior-Point Method

Duality is an important concept in mathematical programming. It allows one to prove optimality of a given solution in a simple fashion. Furthermore, it plays a crucial role in algorithms which employ the interior-point method to solve such mathematical programs.

In this section, we will sketch very roughly the interior-point method by Ye which can be employed in solving a semidefinite program. The reader interested in details is referred to [Ye90].

Let us first recall the concept of duality from linear programming. A linear program (LP) consists (without loss of generality) of the following:

minimize $c^T x$ **such that** $Ax = b$ **and** $x \geqslant 0$.

As an example,

minimize $x_1 + 3x_2 + x_3$ **such that** $x_1 + x_2 = 3$, $x_3 - x_1 = 4$, $x_1, x_2, x_3 \geqslant 0$.

Without solving the linear program explicitly, we can get bounds from the linear equalities. For example, since the objective function $x_1 + 3x_2 + x_3$ is at least $x_2 + x_3$, and adding the first two linear equalities yields the condition $x_2 + x_3 = 7$, we know that the optimal solution cannot be smaller than seven. One can now see that taking any linear combination of the equalities yields a bound whenever it is smaller than the objective function. Of course, we want to obtain a bound which is as large as possible. This yields the following problem.

maximize $b^T \cdot y$ **such that** $A^T \cdot y \leqslant c$.

This is another linear program which is called the *dual* of the original one, which is called the *primal*. Our arguments above can be used to show that any solution to the dual yields a lower bound for the primal.

This principle is known as *weak duality*: The optimum value of the dual is not larger than the optimum value of the primal. Even more can be shown, namely that for linear programming, *strong* duality holds, i.e., both values agree:

If the primal or the dual is feasible, then their optima are the same.

This shows that duality can be used to prove optimality of some solution easily: We just provide two vectors x and y which are feasible points in the primal and dual, respectively. If $b^T y = c^T x$, we know that x must be an optimal point in the primal and dual problem.

It also makes sense to measure for an arbitrary point x how far away it is from the optimum. For this purpose, we solve the primal and dual simultaneously. Given a primal-dual pair x, y in between, one defines the *duality gap* as $c^T \cdot x - b^T \cdot y$. The smaller the gap, the closer we are to the optimum.

Similar concepts hold for semidefinite programming. Before exhibiting what the correspondences are, we want to sketch how duality can be used in an interior-point method for solving a linear program.

Khachian was the first to come up with a polynomial-time algorithm for solving linear programs. His method is known as the *ellipsoid algorithm*. Later, Karmarkar found another, more practical, polynomial-time algorithm which founded a family of algorithms known as *the interior-point method*.

On a very abstract level, an interior-point algorithm proceeds as follows: Given a point x in the interior of the polyhedron defined by the linear inequalities, it maps the polyhedron into another one in which x is "not very close to the boundaries." Then, the algorithm moves from x to some x' in the transformed space, and it maps x' back to some point in the original space. This step is repeated until some potential function tells us that we are very close to the optimal solution. Then, we can either stop and be happy with the good enough approximation, or we can apply another procedure which from this point obtains an optimal solution.

In defining the potential function, the size L of a linear program is needed which measures how many bits we need to store the program. We do not want to give a formal definition, since it is very close to the intuition. As one consequence, it is valid that every vertex of a linear program has rational coordinates where the numerator and denominator can be written using L bits only. (A vertex is a feasible point x such that there is no vector $y \neq 0$ with $x + y$ and $x - y$ feasible.)

There is the following lemma:

Lemma 11.10. *If x_1 and x_2 are vertices of $Ax = b, x \geqslant 0$, then either $c^T x_1 - c^T x_2 = 0$ or $c^T x_1 - c^T x_2 > 2^{-2L}$.*

This means that we only need to evaluate the objective function with an error less than 2^{-2L} since whenever we have reached a point which has distance less than 2^{-2L} to the optimum, then it must be the optimal point.

The first step in Ye's interior-point algorithm consists of *affine scaling*. Given a current solution pair x, y for the primal and dual, a scaling transformation (which depends on x and y) is applied to the LP. This affine scaling step does not change the duality gap.

The potential function is defined as follows:

$$G(x, y) := q \cdot \ln(x^T \cdot y) - \sum_{i=1}^{n} \ln(x_i y_i) \quad \text{for} \quad q = n + \lceil \sqrt{n} \rceil.$$

From the definition of the potential function G, it follows that if $G(x, y) \leqslant -k\sqrt{n}L$, then $x^T \cdot y$ is small enough to imply that x is an optimal solution to the primal problem. The high-level description of Ye's algorithm now looks as follows:

YE'S ALGORITHM
while $G(x, y) > -2\sqrt{n}L$
    do affine scaling
    do change either x or y according to some rule
end

It can be shown that the "rule" above can be chosen such that in every step of the while-loop, the potential function decreases by a constant amount, say $\frac{7}{120}$. It can also be shown that one can obtain an initial solution which has a potential of size $\mathcal{O}(\sqrt{n}L)$.

Altogether, this guarantees that the while loop will be executed $\mathcal{O}(\sqrt{n}L)$ many times, and it can be shown that every step of the operation can be performed in time $\mathcal{O}(n^3)$.

Let us now try to sketch very roughly what this algorithm looks like in the context of semidefinite programs: The primal and dual problems look as follows:

Primal	**Dual**
maximize $b^T y$	minimize $C \bullet X$
such that $C - \sum_{i=1}^{m} y_i A_i$ is PSD.	such that $A_i \bullet X = b_i, i = 1 \ldots m$
	and X is PSD.

Here is a proof of weak duality:

Lemma 11.11. *Let X be a feasible matrix for the dual and y be any feasible vector for the primal. Then $C \bullet X \geqslant b^T y$.*

Proof. $C \bullet X - \sum_{i=1}^{m} b_i y_i = C \bullet X - \sum_{i=1}^{m} (A_i \bullet X) y_i = (C - \sum_{i=1}^{m} y_i A_i) \bullet X \geqslant 0.$

The last inequality holds since the inner product of two PSD matrices is nonnegative. ∎

One of the things that make semidefinite programming more complex than linear programming is that the strong duality of a problem pair cannot be proved in a fashion as simple as for linear programming.

Nevertheless, it can be shown that whenever a polynomial *a priori bound* on the size of the primal and dual feasible sets are known, then the primal-dual pair of problems can be transformed into an equivalent pair for which strong duality holds.

A primal-dual solution pair now is a pair of matrices X and Y, and the potential function is defined as follows:

$$G(X, Y) = q \cdot \ln(X \bullet Y) - \ln \det(XY).$$

Like in Ye's algorithm, one now proceeds with a while-loop where in every single step the potential function is reduced by a constant amount, and after at most $\mathcal{O}(\sqrt{n} \, |\log \varepsilon|)$ executions of the while-loop a solution with duality gap at most ε can be found. Nevertheless, the details are much more difficult and beyond the scope of this survey. The interested reader may find some of those details in [Ali95] or in [Ye90].

As Alizadeh [Ali95] notes, the remarkable similarity between Ye's algorithm for linear programming and its version for semidefinite programming suggests that other LP interior-point methods, too, can be turned into algorithms for semidefinite programming rather mechanically.

Why an a priori bound makes sense. Whereas for a linear program, we have a bound of 2^L on the size of the solutions given the size L of the linear program, the situation is different for semidefinite programs. Consider the following example from Alizadeh:

$$\text{minimize} \quad x_n \quad \text{such that} \quad x_1 = 2 \text{ and } x_i \geqslant x_{i-1}^2$$

The optimum value of this program is of course 2^{2^n}. This is a semidefinite program since we have already seen earlier that the condition $x_i \geqslant x_{i-1}^2$ can be expressed. Just outputting this solution would take us time $\Omega(2^n)$.

11.5 Approximation Algorithms for MAXCUT

We recall the definition of MAXCUT.

MAXCUT
Instance: Given an undirected graph $G = (V, E)$.
Problem: Find a partition $V = V_1 \cup V_2$ such that the number of edges between V_1 and V_2 is maximized.

MAXCUT is an $\mathcal{NP}$-hard problem. There is also a weighted version in which there is a positive weight for every edge and we are asked to compute a partition where the sum of the edge weights in the cut is maximized.

11.5.1 MAXCUT and Classical Methods

It is known that the weight of the largest cut is at least 50 percent of the graph weight. A simple probabilistic argument works as follows: Scan the vertices v_1 to v_n and put each independently with probability $1/2$ into V_1 and with probability $1/2$ into V_2. The probability of an edge being in the cut is $1/2$. By linearity of expectation, the expected value is at least 50% of the graph weight, hence there exists a cut of this weight. As one might guess, this simple argument can be derandomized, leading to a simple greedy strategy which is a 2-approximation algorithm.

It has also been shown by Ngoc and Tuza (see, e.g., the survey paper by Poljak and Tuza [PT95]) that for every $0 < \varepsilon < 1/2$, it is $\mathcal{NP}$-complete to decide whether the largest cut of a graph has size at least $(1/2 + \varepsilon) \cdot |E|$.

Before the work of Goemans and Williamson, progress has only been on improving additional terms. We sketch one of those results from [HL96] here:

Theorem 11.12. *Every graph G has a cut of size at least $\frac{w(G)+w(M)}{2}$, where M is a matching in G. (Here, $w(G)$ and $w(M)$ denote the sum of weights of the edges in those subgraphs.) Given M, such a cut can be computed in linear time.*

We give a proof sketch of the existence. We process the edges of the matching consecutively. For an edge $e = \{v, w\}$ of the matching, we either add v to V_1 and w to V_2, or we add v to V_2 and w to V_1, each with probability $1/2$. Remaining vertices are distributed independently to either V_1 or V_2, each with probability $1/2$. It can now be seen that the edges in the matching appear in the cut, and other edges appear with probability $1/2$, which means that the expected value of the cut is $\frac{w(G)+w(M)}{2}$. (The reader is asked to derandomize this experiment in Exercise 11.3).

By Vizing's Theorem, the edges of every graph can be partitioned into at most $\Delta + 1$ matchings (where Δ denotes the maximum degree in a graph). Thus, one of them has size at least $|E|/(\Delta + 1)$ yielding that every graph has a cut of size at least $(|E|/2) \cdot (1 + 1/(\Delta + 1))$.

11.5.2 MAXCUT as a Semidefinite Program

We first observe that the optimum value of MAXCUT can be obtained through the following mathematical program:

$$\textbf{maximize} \quad \sum_{\{i,j\}\in E} \frac{1 - y_i y_j}{2} \quad \textbf{such that} \quad y_i \in \{-1, 1\} \text{ for all } 1 \leqslant i \leqslant n.$$

The idea is that $V_1 = \{i \mid y_i = 1\}$ and $V_2 = \{i \mid y_i = -1\}$ constitute the two classes of the partition for the cut. The term $(1 - y_i y_j)/2$ contributes 1 to the sum iff $y_i \neq y_j$ and 0 otherwise.

We want to relax this program into such a form that we can use semidefinite programming. If we relax the solution space from one dimension to n dimensions, we obtain the following mathematical program:

$$\textbf{maximize} \quad \sum_{\{i,j\}\in E} \frac{1 - \underline{y_i}\underline{y_j}}{2} \quad \textbf{such that} \quad \|\underline{y_i}\| = 1, \underline{y_i} \in \mathbb{R}^n.$$

(Note that in this section only, vector variables are underlined in order to distinguish them more clearly from integer variables.)

This is not yet a semidefinite program, but by introducing new variables, we can cast it as a semidefinite program:

$$\textbf{max} \quad \sum_{\{i,j\}\in E} \frac{1 - y_{i,j}}{2} \quad \textbf{such that} \quad \begin{pmatrix} 1 & y_{1,2} & y_{1,3} & \cdots & y_{1,n} \\ y_{1,2} & 1 & y_{2,3} & \cdots & y_{2,n} \\ y_{1,3} & y_{2,3} & 1 & \cdots & y_{3,n} \\ \vdots & \vdots & \vdots & \ddots & \vdots \\ y_{1,n} & y_{2,n} & y_{3,n} & \cdots & 1 \end{pmatrix} \text{ is PSD.}$$

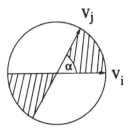

Fig. 11.2. The angle $\alpha = \arccos(\underline{v_i} \cdot \underline{v_j})$ between the vectors $\underline{v_i}$ and $\underline{v_j}$ is invariant under rotation.

The reason is the following: By the third equivalence condition in Lemma 11.6, the matrix defines a solution space in which every variable $y_{i,j}$ can be written as the product of some $\underline{v_i} \cdot \underline{v_j}$. The diagonal guarantees that $\|\underline{v_i}\| = 1$ for all i.

This is a relaxation since any solution in one dimension (with the other components equal to zero) yields a feasible solution of this semidefinite program with the same objective function value.

Now, assume that we are given an (optimal) solution to this semidefinite program. We proceed as follows in order to obtain a partition :

- Use Cholesky decomposition to compute $\underline{v_i}$, $i = 1, \ldots, n$ such that $y_{i,j} = \underline{v_i} \cdot \underline{v_j}$. This can be done in time $\mathcal{O}(n^3)$.

- Choose randomly a hyperplane. (This can be done by choosing some random vector $\underline{r}$ as the normal of the hyperplane.) For this purpose, use the rotationally symmetric distribution.

- Choose V_1 to consist of all vertices whose vectors are on one side of the hyperplane (i.e., $\underline{r} \cdot \underline{v_i} \leqslant 0$) and let $V_2 := V \setminus V_1$.

The process of turning the vectors $\underline{v_i}$ into elements from $\{-1, 1\}$ by choosing a hyperplane is also called the "rounding procedure." It remains to show that the partition so obtained leads to a cutsize which is at least 87% of the optimum cutsize. For this purpose, we consider the expected value of the cutsize obtained. Because of linearity, we only need to consider for two given vectors $\underline{v_i}$ and $\underline{v_j}$ the probability that they are on different sides of the hyperplane. Since the product $\underline{v_i} \cdot \underline{v_j}$ is invariant under rotation, we can consider the plane defined by the two vectors, depicted in Figure 11.2.

First, we note that the probability that the hyperplane H chosen randomly is equal to the considered plane is equal to zero.

In the other cases, H and the plane defined by $\underline{v_i}$ and $\underline{v_j}$ intersect in a straight line. If the two vectors are to be on different sides of the hyperplane H, then this intersection line has to lie "between" $\underline{v}_i$ and $\underline{v}_j$ (i.e., fall into the shaded part of the figure) which happens with probability

$$\frac{2\alpha}{2\pi} = \frac{\alpha}{\pi} = \frac{\arccos(\underline{v}_i \cdot \underline{v}_j)}{\pi}.$$

Hence, the expected value of the cut is equal to

$$\mathbf{E}[cutsize] = \sum_{\{i,j\}\in E} \frac{\arccos(\underline{v}_i \cdot \underline{v}_j)}{\pi}. \tag{11.2}$$

We remark that one can also arrive at this expression by observing that

$$\mathbf{E}[\operatorname{sgn}(\underline{r} \cdot \underline{v}_i) \cdot \operatorname{sgn}(\underline{r} \cdot \underline{v}_j)] = \frac{1}{2\pi} \cdot \int_{\varphi=0}^{2\pi} \operatorname{sgn}(\cos\varphi) \cdot \operatorname{sgn}(\cos(\varphi - \alpha))\, d\varphi = 1 - 2 \cdot \frac{\alpha}{\pi}.$$

Whenever we solve a concrete semidefinite relaxation of the MaxCut problem, we obtain an upper bound on the MaxCut solution. The quality of the outcome of the rounding procedure can then directly be measured by comparing it with this upper bound.

Nevertheless, one is of course interested in how good one can guarantee the expected solution to be in the worst case. The next two subsections will analyze this worst case behavior. One should keep in mind, though, that experiments indicate that in "most" cases, the quality of the solutions produced is higher than in the worst case.

Analyzing the Quality Coarsely. We compare the expected value (11.2) with the value of the relaxed program. One way to do so is to compare every single term $\arccos(\underline{v}_i \cdot \underline{v}_j)/\pi$ with $(1 - \underline{v}_i \cdot \underline{v}_j)/2$.
Solving a simple calculus exercise, one can show that in the range $-1 \leqslant y \leqslant 1$,

$$\frac{\arccos(y)}{\pi} \geqslant 0.87856 \cdot \frac{1-y}{2}.$$

(Let us denote by $\alpha_{GW} = 0.87856\ldots$ the maximum constant we could have used.)
A sketch of the involved functions may also be quite helpful. The following figure shows the function $\arccos(y)/\pi$ as well as the function $(1-y)/2$.

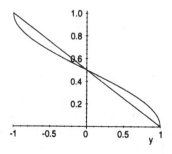

The following two figures show the quotient $\frac{\arccos(y)}{\pi}/\frac{1-y}{2}$ in the range $-1 \leqslant y \leqslant 1$ and $-1 \leqslant y \leqslant 0$, respectively. (At first sight, it is rather surprising that the quotient is larger than 1 for $y > 0$, since our randomized procedure then yields terms which are larger than the terms in the relaxed program.)

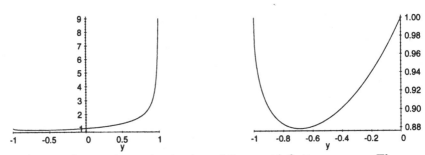

Let $Relax_{opt}$ denote the optimal value of the semidefinite program. The conclusion of the analysis above is that the expected value of the cutsize produced by the rounding procedure is at least $0.878 \cdot Relax_{opt}$, hence the quality of the solution is also not larger than $1/0.878 \approx 1.139$.

We also see that the quality of the solution obtained could be improved if we could achieve that $\underline{v_i} \cdot \underline{v_j}$ is "far away" from approximately -0.689.

Thus, one might hope that extra constraints could improve the quality of the solution. In this direction, Karloff [Kar96] has shown that the quality of the rounding procedure cannot be improved by adding *linear* inequality constraints to the semidefinite program. For MAXCUT, the ratio can be improved for some families of graphs, but not in the general worst case. However, additional linear inequality constraints can improve the approximation ratio in the case of MAXDICUT and MAXE2SAT, see Section 11.8.

Analyzing the Quality More Precisely. In the above subsection, we have analyzed the quality of our solution term by term. We can improve our estimations by taking into account the value of the solution our semidefinite program has produced.

For this purpose assume that the optimum is assumed for some values of $y_{i,j}$, and let $Y_{sum} := \sum_{\{i,j\} \in E} y_{i,j}$. Then,

$$Relax_{opt} := \sum_{\{i,j\} \in E} \frac{1 - y_{i,j}}{2} = \frac{|E|}{2} - \frac{1}{2} \cdot Y_{sum}$$

depends only on Y_{sum} and not on each individual value $y_{i,j}$.

If $Relax_{opt}$ is known, then $Y_{sum} = |E| - 2 \cdot Relax_{opt} \leqslant 0$ is known. We can ask for which individual values of $y_{i,j}$ summing to Y_{sum}, $S := \sum_{i,j} \arccos(y_{i,j})$ attains a minimum. Observe that $\arccos(y)$ is convex on $[-1, 0]$ and concave on $[0, 1]$ (the derivative of $\arccos y$ is $-1/\sqrt{1 - y^2}$).

In Exercise 11.4, we ask the reader to verify the following: If Y_{sum} is fixed, then the $y_{i,j}$ for which $\sum_{i,j} \arccos(y_{i,j})$ attains a minimum, fulfill the following property:

There is a value $Y \leqslant 0$ and an r with $|E|/2 \leqslant r \leqslant |E|$ such that r of the $y_{i,j}$ are equal to Y and $|E| - r$ of the $y_{i,j}$ are equal to one.

Then, $\sum_{\{i,j\} \in E} \arccos(y_{i,j}) = r \cdot \arccos(\frac{Y_{sum} - |E| + r}{r})$ and the quotient $\frac{\sum_{\{i,j\}} \arccos(y_{i,j})}{Relax_{opt}}$ can be estimated. We obtain that

$$
\begin{aligned}
\frac{\mathbf{E}[\text{cutsize}]}{Relax_{opt}} &\geqslant \min_{|E|/2 \leqslant r \leqslant |E|} \frac{r \cdot \arccos(\frac{Y_{sum} - |E| + r}{r})}{\pi \cdot Relax_{opt}} \\
&= \min_{r} \frac{\arccos(\frac{r - 2 \cdot Relax_{opt}}{r})}{\pi \cdot Relax_{opt}/r} \\
&= \frac{\arccos(1 - 2x)}{\pi \cdot x} =: q(x),
\end{aligned}
$$

where we have substituted $x := Relax_{opt}/r$ for the r where the minimum is attained. We have obtained the following: Given a graph G with optimal cutsize MC, and "relative" cutsize $y := MC/|E|$, it holds that $y \leqslant x$ and thus the "rounding quality" $\mathbf{E}[\text{cutsize}]/Relax_{opt}$ obtained for graph G is at least $\min_{x \geqslant y} q(x)$. Below, we give a plot of $Q(y) := \min_{x \geqslant y} q(x)$.

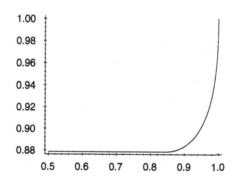

As we can see, the quality of the rounding procedure becomes better when the relative maximum cutsize of the input graph becomes larger.

The above analysis only considers the quality of the rounding procedure. Karloff has shown in [Kar96] that there is a family of graphs for which the quality of the rounding procedure is arbitrarily close to α_{GW}. Nevertheless, it might be possible to obtain a better approximation ratio than $1/\alpha_{GW} \approx 1.139$ by using a different rounding procedure, since for the family of graphs constructed by Karloff, the optimum cutsize is equal to $Relax_{opt}$.

Thus, it is also a natural question to ask how much larger $Relax_{opt}$ can be compared to the optimum MAXCUT solution MAX_{opt}.

Karloff [Kar96] calls this the "integrality ratio", and remarks that from his paper, nothing can be concluded about this "integrality ratio."

On the other hand, Goemans and Williamson mention that for the input graph "5-cycle", $MAX_{opt}/Relax_{opt} = 0.88445\ldots$ holds which gives a graph where the relaxation is relatively far away from the original best solution.

It is clear that for these graphs, even a better rounding procedure would not lead to better results.

The above considerations suggest that it would be nice if we could keep a substantial subset of all products $v_i \cdot v_j$ away from ≈ -0.689, where the worst case approximation $0.878\ldots$ is attained. One might hope that adding extra constraints might lead to better approximation ratios.

Implementation Remarks. For practical purposes, it is probably not a good idea to derandomize the probabilistic algorithm given by Goemans and Williamson, since it seems very likely that after only a few rounds of choosing random hyperplanes, one should find a good enough approximation, and the quality of the approximation can also be controlled by comparing with the upper bound of the semidefinite program.

Nevertheless, it is an interesting theoretical problem to show that semidefinite programming also yields a deterministic approximation algorithm.

In the original proceedings paper by Goemans and Williamson, a derandomization procedure was suggested which later turned out to have a flaw. A new suggestion was made by Mahajan and Ramesh [MR95a], but their arguments are rather involved and technical which is why we omit them in this survey. One can only hope that a simpler deterministic procedure will be found.

Just one little remark remains as far as the implementation of the randomized algorithm is concerned. How do we draw the vector r, i.e., how do we obtain the rotationally symmetric distribution? For this purpose, one can draw n values r_1 to r_n independently, using the standard normal distribution. Unit length of the vector can be achieved by a normalization.

For the purposes of the MAXCUT-algorithm, a normalization is not necessary since we are only interested in the sign of $r \cdot v_i$. The standard normal distribution can be simulated using the uniform distribution between 0 and 1, for details see [Knu81, pp. 117 and 130].

11.6 Modeling Asymmetric Problems

In the following, we describe approximation algorithms for MAXDICUT and MAX2SAT/MAXE2SAT. The algorithms are based on semidefinite programming. All results in this section (as well as many in the following section) are due to Goemans and Williamson [GW95].

The general approach which yields good randomized approximation algorithms for MAXDICUT and MAXE2SAT is very similar to the approach used to approximate MAXCUT with ratio 1.139, cf. Section 11.5.2. We summarize the three main steps of this approach:

1. **Modeling.** First, model the problem as an integer quadratic problem over some set $Y = \{y_1, \ldots, y_n\}$ of variables. The objective function is linear in $\{y_i y_j : y_i, y_j \in Y\}$. The only restrictions on the variables are $y_i \in \{-1, 1\}$, $y_i \in Y$.

2. **Relaxation.** Consider the semidefinite relaxation of the integer quadratic problem. The relaxed problem has $|Y|^2$ variables $y_{i,j}$. The objective function is linear in these variables. In the relaxation, the restriction is that the matrix $(y_{i,j})$ is PSD and that all entries on the main diagonal of this matrix are one.

   An optimal solution of the semidefinite program can be computed in polynomial time with any desired precision ε.

3. **Rounding.** By a Cholesky decomposition, the solution of the relaxation can be expressed by n vectors $v_i \in \mathbb{R}^n$. Use these vectors in a probabilistic experiment, i.e., choose a hyperplane at random and assign $+1$ and -1 to the variables y_i. The result is a (suboptimal) solution of the integer quadratic problem.

In order to approximate MAXDICUT and MAXE2SAT, step 2 will be exactly the same as for MAXCUT. Step 3, the rounding, will be almost the same. However, the modeling in step 1 is different.

The difference in the modeling stems from an asymmetry which is inherent in the problems MAXDICUT and MAXE2SAT. MAXCUT is a symmetric problem, since switching the role of the "left" and the "right" set of the partition does not change the size of the cut. Thus, the objective function of the integer quadratic program of MAXCUT models only whether the vertices of an edge are in different sets:

$$\text{maximize} \quad \sum_{\{i,j\} \in E} \frac{1 - y_i \cdot y_j}{2}$$
$$\text{such that} \quad y_i \in \{-1, 1\} \quad \forall i \in \{1, \ldots, n\}$$

Of course, for any set S of vertices, the value of the objective function for the cuts $(S, \overline{S})$ and $(\overline{S}, S)$ is the same. In contrast, for a directed cut the direction of an edge does make a difference. The objective function must distinguish the directed cuts $(S, \overline{S})$ and $(\overline{S}, S)$.

We describe an approach to model these problems by a semidefinite program.

11.6.1 Approximating MaxDiCut

As before, we want the values of the variables y_i of the integer quadratic program to be restricted to 1 or -1. The reason is that this simplifies the relaxation. The objective function must be linear in the product $y_i \cdot y_j$ of any pair (y_i, y_j) of variables. However, we cannot distinguish edge (i, j) from edge (j, i) by just looking at the product $y_i \cdot y_j$.

The trick Goemans and Williamson use to model the direction of an edge is to introduce a new variable y_0. Like other variables, y_0 may be -1 or 1. Now, the graph-theoretical property

$$i \in S$$

is expressed in the integer quadratic program by the property

$$y_i \cdot y_0 = 1.$$

In this way, the term

$$(1 + y_i y_0) \cdot (1 - y_j y_0) \, / \, 4 \;\; = \;\; (1 + y_i y_0 - y_j y_0 - y_i y_j) \, / \, 4 \qquad (11.3)$$

is equal to 1 if $y_i = y_0 = -y_j$. Otherwise, it is 0. Thus, (11.3) represents the contribution of edge (i, j) to the objective function. The value of the whole directed cut, i.e., the objective function for MaxDiCut, is the sum of term (11.3) over all edges in the graph. In this way, we get

$$\textbf{maximize} \qquad \sum_{(i,j) \in E} \frac{1 + y_i y_0 - y_j y_0 - y_i y_j}{4} \qquad (11.4)$$

$$\textbf{such that} \qquad y_i \in \{-1, 1\} \;\; \forall i \in \{0, \ldots, n\}$$

as an integer quadratic program modeling the problem MaxDiCut.

Remark 11.13. An equivalent view of (11.4) is that an isolated vertex $0 \notin V$ has been added to the graph. For any cut, the objective function counts exactly those edges (i, j) where the tail i is in the same set as 0 and where the head j is in the other set.

As in the MAXCUT approximation, step 2 of the MAXDICUT approximation consists of relaxing to

$$\text{maximize} \quad \sum_{(i,j)\in E} \frac{1 + y_i y_0 - y_j y_0 - y_i y_j}{4} \quad (11.5)$$

$$\text{such that} \quad y_i \in \mathbb{R}^{n+1}, \|y_i\| = 1 \ \forall i \in \{0,\ldots,n\}.$$

Note that in this case the relaxation is to the $(n+1)$-dimensional vector space.

Letting $y_{i,j} = y_i \cdot y_j$, we obtain the semidefinite program

$$\text{maximize} \quad \sum_{(i,j)\in E} \frac{1 + y_{i,0} - y_{j,0} - y_{i,j}}{4} \quad (11.6)$$

$$\text{such that} \quad \text{the matrix } (y_{i,j}) \text{ is PSD,}$$

$$y_{i,i} = 1 \ \forall i \in \{0,\ldots,n\}.$$

By solving (11.6) and computing the Cholesky decomposition, we get vectors $v_0, v_1, \ldots, v_n$ such that $(y_{i,j}) = (v_0, v_1, \ldots, v_n)^T (v_0, v_1, \ldots, v_n)$ maximizes the value of the objective function (within the desired precision).

The next step consists of rounding the vectors $v_0, v_1, \ldots, v_n$. Because of the special role of v_0, this step slightly differs from the rounding step for MAXCUT. In a certain sense, the inner product $v_i \cdot v_0$ measures the probability of vertex i being in cut S of the output. If $v_i \cdot v_0 = 1$, then $i \in S$ with probability 1, if $v_i \cdot v_0 = -1$, then $i \in S$ with probability 0. More formally, with uniform distribution, choose a unit vector $r \in \mathbb{R}^{n+1}$ and let H_r be the hyperplane through the origin which has normal r. Let

$$S := \{i : \text{sgn}(v_i \cdot r) = \text{sgn}(v_0 \cdot r)\}$$

be the set of those vertices i where v_i and v_0 are on the same side of H_r. The randomized procedure outputs this set S.

The probability of edge (i, j) occurring in the directed cut is exactly

$$\text{Prob}_{r\in\mathbb{R}^{n+1}, \|r\|=1}[\text{sgn}(v_i \cdot r) = \text{sgn}(v_0 \cdot r) \neq \text{sgn}(v_j \cdot r)].$$

By arguments from spherical geometry (see [GW95] for details), for any unit vectors $v_i, v_j, v_0 \in \mathbb{R}^{n+1}$, this probability is at least

$$0.796 \cdot \frac{1 + v_i v_0 - v_j v_0 - v_i v_j}{4}.$$

This yields

Theorem 11.14 ([GW95]). MAXDICUT *can be approximated by a randomized polynomial-time algorithm with approximation ratio 1.257, i.e., the expected value of the computed solution is at least 0.796 times the optimal solution.*

11.6.2 Approximating MAX2SAT

We turn to the approximation of MAX2SAT. Of course, the results in this section also hold for MAXE2SAT.

As MAXDICUT, MAX2SAT is an asymmetric problem: In general, switching the truth values in an assignment for the variables in a Boolean formula will change the number of clauses which are satisfied.

As before, we will use an additional variable y_0 when modeling MAX2SAT. ("The value of y_0 will determine whether -1 or 1 will correspond to 'true' in the MAX2SAT instance" [GW95].) Let $v(C) \in \{0,1\}$ denote the contribution of clause C to the objective function. Thus, the contribution of a singleton clause x_i (or $\overline{x}_i$) is

$$v(x_i) = \frac{1 + y_i \cdot y_0}{2} \qquad (11.7)$$

(or

$$v(\overline{x}_i) = \frac{1 - y_i \cdot y_0}{2}, \qquad (11.8)$$

respectively). For clauses of length two, say $x_i \vee \overline{x}_j$, we have

$$
\begin{aligned}
v(x_i \vee \overline{x}_j) &= 1 - v(\overline{x}_i) \cdot v(x_j) \\
&= 1 - \frac{1 - y_i \cdot y_0}{2} \cdot \frac{1 + y_j \cdot y_0}{2} \\
&= \frac{1 + y_i \cdot y_0}{4} + \frac{1 - y_j \cdot y_0}{4} + \frac{1 + y_i \cdot y_j}{4}. \qquad (11.9)
\end{aligned}
$$

In this way, the value of all clauses of length up to two can be expressed as nonnegative linear combinations of $1 + y_i \cdot y_j$ and $1 - y_i \cdot y_j$. Thus, we can compute nonnegative $a_{i,j}$, $b_{i,j}$ such that

$$\sum_{i=1}^{m} v(C_i) = \sum_{i,j} a_{i,j}(1 + y_i \cdot y_j) + b_{i,j}(1 - y_i \cdot y_j).$$

The MAX2SAT instance is thus modelled by the integer quadratic program

$$
\begin{aligned}
\textbf{maximize} \quad & \sum_{i,j} a_{i,j}(1 + y_i \cdot y_j) + b_{i,j}(1 - y_i \cdot y_j) \qquad (11.10) \\
\textbf{such that} \quad & y_i \in \{-1, 1\} \quad \forall i \in \{0, \dots, n\}
\end{aligned}
$$

and the semidefinite formulation of the relaxation is

$$
\begin{aligned}
\textbf{maximize} \quad & \sum_{i,j} a_{i,j}(1 + y_{i,j}) + b_{i,j}(1 - y_{i,j}) \qquad (11.11) \\
\textbf{such that} \quad & \text{the matrix } (y_{i,j}) \text{ is PSD,} \\
& y_{i,i} = 1 \quad \forall i \in \{0, \dots, n\}.
\end{aligned}
$$

In the rounding procedure, those variables x_i are set to true where the vectors v_i and v_0 of the Cholesky decomposition are on the same side of the random hyperplane.

By an argument that is similar to the one used in the analysis of MAXCUT, the expected value of the computed solution is at least 0.878 times the value of an optimal solution. I.e., we have

Theorem 11.15 ([GW95]). MAX2SAT *(and, thus,* MAXE2SAT*) can be approximated by a randomized polynomial-time algorithm with approximation ratio 1.139.*

11.7 Combining Semidefinite Programming with Classical Approximation Algorithms

Prior to the work in [GW95], at least three different approaches to approximate MAXEkSAT/MAXSAT have been proposed: Johnson's algorithm, an algorithm based on network flow, and an algorithm based on linear programming. To distinguish these approaches from the semidefinite programming technique, we call them *classical* algorithms.

Johnson's algorithm. Johnson's algorithm [Joh74] is the derandomization of the probabilistic algorithm where the truth assignment is uniformly chosen among all possible assignments. This algorithm is described in Chapter 2 and derandomized in Chapter 3. MAXEkSAT is solved with approximation ratio $1/(1 - 2^{-k})$. The approximation ratio for MAXSAT is at least 2. (Poljak and Turzík describe another MAXSAT algorithm that achieves the same approximation ratio.)

Network flow. Yannakakis [Yan92] describes an approximation algorithm for MAXSAT which is based on computing maximal flows in networks. Yannakakis' algorithm is rather involved. Asano et al. [AHOH96] give an outline of this algorithm. The main idea is to construct several networks. The maximum flow in these networks is used to partition the set of variables in the Boolean formula into three classes. The variables in the first (second, third) class are independently set to be true with probability 0.5 (0.555, 0.75, respectively). The expected value of the approximation ratio of this algorithm is shown to be 1.334. The algorithm can be derandomized.

Linear programming. Chapter 2 contains a description of a MAXSAT-algorithm due to Goemans and Williamson [GW94b] which solves the linear relaxation of an integer program modeling MAXSAT. The solution of the linear program determines for each variable the probability with which this variable is set to true. This algorithm solves MAXkSAT with approximation ratio

k	Johnson's algorithm	Network flow	Linear programming
1	0.5	0.75	1.0
2	0.75	0.75	0.75
3	0.875	0.75	0.703
4	0.937	0.765	0.683
5	0.968	0.762	0.672
6	0.984	0.822	0.665
7	0.992	0.866	0.660

Table 11.1. For three basic MAXSAT algorithms, a lower bound on the probability that the computed solution satisfies a fixed clause of length k is given, $k = 1, \ldots, 7$.

$1/(1 - (1 - 1/k)^k)$ and MAXSAT with approximation ratio $e/(e - 1) = 1.582$. It can be derandomized.

For all of these algorithms, the approximation ratio depends on the concrete MAXSAT instance. If the clauses in the formula are rather long, then Johnson's algorithm is very good. If almost all of the clauses are very short, then the algorithm based on linear programming might be better. However, the algorithm based on network flows is the only one of these algorithms where the analysis guarantees approximation ratio 1.334. Thus, among these algorithms there is no "best" one. Table 11.1 summarizes (lower bounds on) the approximation ratio of these algorithms on instances of MAXEkSAT for different k, $1 \leqslant k \leqslant 7$.

In Chapter 2, we have seen that the merits of different MAXSAT algorithms can be combined. The combination of two different algorithms may yield a better approximation of MAXSAT. Namely, another 1.334-approximation is obtained by combining Johnson's algorithm and the algorithm based on linear programming. On any input, the combined algorithm calls both of these algorithms. This yields two truth assignments. The combined algorithm outputs that truth assignment which satisfies the larger number of clauses.

In the rest of this section we use a variant of this technique where the algorithms are combined in a probabilistic way. In our examples, three different algorithms will be combined.

First, we describe MAXSAT approximations that are based on semidefinite programming (Section 11.7.1). Then, these algorithms are combined with a classical algorithm (Section 11.7.2).

11.7.1 Handling Long Clauses

The algorithm described in Section 11.6.2 handles only clauses of length at most two. In the following, we describe how Goemans and Williamson treat longer clauses.

Let $\Phi = C_1 \wedge \ldots \wedge C_m$ be a Boolean formula in conjunctive normal form. For any clause C_j let $\ell(C_j)$ denote the length of the clause and let I_j^+ (or I_j^-) denote the set of non-negated (negated, respectively) variables in C_j.

Recall Goemans and Williamson's formulation of Φ as a *linear program* ((2.1) in Chapter 2):

$$\textbf{maximize} \quad \sum_{j=1}^m z_j \tag{11.12}$$
$$\textbf{such that} \quad \sum_{i \in I_j^+} y_i + \sum_{i \in I_j^-}(1 - y_i) \geqslant z_j \quad \forall j \in \{1, \ldots, m\}$$
$$0 \leqslant y_i \leqslant 1 \quad \forall i \in \{1, \ldots, n\}$$
$$0 \leqslant z_j \leqslant 1 \quad \forall j \in \{1, \ldots, m\}$$

In a very similar way, by using the function v introduced in (11.7)–(11.9), MAXSAT is modelled by

$$\textbf{maximize} \quad \sum_{j=1}^m z_j \tag{11.13}$$
$$\textbf{such that} \quad \sum_{i \in I_j^+} v(x_i) + \sum_{i \in I_j^-} v(\overline{x}_i) \geqslant z_j \quad \forall j \in \{1, \ldots, m\}$$
$$v(C_j) \geqslant z_j \quad \forall j \in \{1, \ldots, m\}, \ell(C_j) = 2$$
$$y_i \cdot y_i = 1 \quad \forall i \in \{0, \ldots, n\}$$
$$0 \leqslant z_j \leqslant 1 \quad \forall j \in \{1, \ldots, m\}.$$

To obtain the corresponding semidefinite formulation, replace each vector product $y_i \cdot y_k$ by a new variable $y_{i,k}$. As restrictions, we get that the matrix $A_1 := (y_{i,k})_{i,k \in \{0,\ldots,n\}}$ is PSD and that a system of linear inequalities in the variables $y_{i,k}$, $i, k \in \{0, \ldots, n\}$, and z_j, $j \in \{1, \ldots, m\}$ is satisfied. As noted in Section 11.3, the linear inequalities can be transformed into a matrix $A_2 := SP(y_{0,0}, \ldots, y_{n,n}, z_1, \ldots, z_m)$ which is PSD if and only if all inequalities are satisfied. Let A be the matrix

$$\begin{pmatrix} A_1 & 0 \\ 0 & A_2 \end{pmatrix}$$

which is PSD if and only if A_1 and A_2 are PSD. Then

$$\textbf{maximize} \quad \sum_{j=1}^m z_j \tag{11.14}$$
$$\textbf{such that} \quad \text{matrix } A \text{ is PSD}$$
$$y_{i,i} = 1 \quad \forall i \in \{0, \ldots, n\}$$

is the corresponding semidefinite relaxation.

Let $\hat{y}_{i,k}$ and $\hat{z}_j$ denote the value of the variables in a solution of (11.14).

By **A1**, we denote the algorithm which computes the Cholesky decomposition of the matrix $(\hat{y}_{i,k})$ and applies Goemans-Williamson's rounding procedure to the solution (as described in Section 11.6.2). For each clause C_j of length one or two,

$$\sum_{i \in I_j^+} v(x_i) + \sum_{i \in I_j^-} v(\overline{x}_i) \;\geqslant\; v(C_j),$$

i.e., $v(C_j) \geqslant z_j$ implies $\sum_{i \in I_j^+} v(x_i) + \sum_{i \in I_j^-} v(\overline{x}_i) \geqslant z_j$. Thus, algorithm **A1** satisfies instances of MAX2SAT with probability at least 0.878.

By **A2** we denote the following algorithm: Independently for each variable, set x_i to be true with probability $(1 + \hat{y}_{i,0})/2$. Then, by the same argument as in the analysis of (11.12) (cf. Chapter 2), the probability that clause C_j is satisfied is at least

$$\left(1 - \left(1 - \frac{1}{\ell(C_j)}\right)^{\ell(C_j)}\right) \hat{z}_j.$$

11.7.2 Approximating MAXSAT

Let (p_1, p_2, p_3) be a probability vector, i.e., $0 \leqslant p_j \leqslant 1$ and $p_1 + p_2 + p_3 = 1$. Consider the following combination of three algorithms:

1. With probability p_1, execute algorithm **A1**.

2. With probability p_2, execute algorithm **A2**.

3. With probability p_3, execute Johnson's algorithm.

The expected value of the solution of these algorithms is

$$\sum_{j:\ell(C_j)\leqslant 2} 0.878\, \hat{z}_j, \quad \sum_{j=1}^{k}\left(1 - \left(1 - \frac{1}{\ell(C_j)}\right)^{\ell(C_j)}\right)\hat{z}_j, \quad \text{or} \quad \sum_{j=1}^{k}\left(1 - \frac{1}{2^{\ell(C_j)}}\right),$$

respectively. Thus, the solution of the combined algorithm has expected value

$$p_1 \sum_{j:\ell(C_j)\leqslant 2} 0.878\, \hat{z}_j + p_2 \sum_{j=1}^{k}\left(1 - \left(1 - \frac{1}{\ell(C_j)}\right)^{\ell(C_j)}\right)\hat{z}_j + p_3 \sum_{j=1}^{k}\left(1 - \frac{1}{2^{\ell(C_j)}}\right).$$

If we choose $p_1 = 0.0430$ and $p_2 = p_3 = 0.4785$, then by a numerical computation it can be checked that this term is at least 0.755 times the optimum of (11.14). I.e., we have

Theorem 11.16 ([GW95]). MAXSAT *can be approximated by a randomized polynomial-time algorithm with approximation ratio 1.324.*

Note that in a practical implementation we would use a deterministic combination of **A1**, **A2**, and Johnson's algorithm. The combined algorithm outputs that truth assignment which satisfies the larger number of clauses. The approximation ratio of this algorithm is also at least 1.324.

Using a similar approach, where long clauses are handled in a slightly different way, Goemans and Williamson also achieve an approximation ratio of 1.319 for MAXSAT.

Asano et al. have proposed another method for handling long clauses by semidefinite programs [AOH96]. They obtain an algorithm with approximation ratio 1.307. Tuning the latter method, Asano et al. [AHOH96] have obtained an 1.304-approximation algorithm for MAXSAT. However, the best known approximation for MAXSAT is an algorithm by Asano [Asa97] which is based on a refinement of Yannakakis' algorithm. This algorithm has approximation ratio 1.299.

11.8 Improving the Approximation Ratio

In this section we describe two techniques due to Feige and Goemans [FG95] and to Karloff [Kar96] which improve the approximation ratio of the semidefinite programs for MAXE2SAT and MAXDICUT. The idea is to add linear restrictions to the semidefinite program and to modify the rounding procedure. This improves the approximation ratio for MAXE2SAT to 1.075 and the approximation ratio for MAXDICUT to 1.165.

11.8.1 Adding Constraints

Note that we can add any (polynomial) number of linear restrictions to a semidefinite program whilst keeping the running time of the approximation algorithm polynomial. The constraints considered here are valid for all truth assignments and all cuts, respectively. (In particular, the constraints do not depend on the instance of the problem.) However, they do not hold for all the vectors in the relaxation. Thus, they may change the solution of the relaxation. This may improve the approximation ratio of the algorithm.

Feige and Goemans [FG95] discuss the use of two sets of linear inequality constraints which sharpen the restrictions of MAXE2SAT and MAXDICUT. Their first set contains for all $i, j, k \in \{0, \ldots, n\}$ the inequalities

$$
\begin{aligned}
y_i \cdot y_j + y_j \cdot y_k + y_k \cdot y_i &\geqslant -1 \\
-y_i \cdot y_j - y_j \cdot y_k + y_k \cdot y_i &\geqslant -1 \\
-y_i \cdot y_j + y_j \cdot y_k - y_k \cdot y_i &\geqslant -1 \\
y_i \cdot y_j - y_j \cdot y_k - y_k \cdot y_i &\geqslant -1,
\end{aligned}
$$

These inequalities are valid, since they hold for any assignment of y_i, y_j, y_k with -1 and 1. For MAXE2SAT it expresses the *tertium non datur*. For MAXDICUT and MAXCUT it expresses the fact that any cut partitions the set of vertices into two sets.

The second set of constraints is a subset of the first one. It contains for all $i, j \in \{0, \ldots, n\}$ the inequalities

$$y_i \cdot y_0 + y_j \cdot y_0 + y_i \cdot y_j \geqslant -1$$
$$-y_i \cdot y_0 - y_j \cdot y_0 + y_i \cdot y_j \geqslant -1$$
$$-y_i \cdot y_0 + y_j \cdot y_0 - y_i \cdot y_j \geqslant -1$$
$$y_i \cdot y_0 - y_j \cdot y_0 - y_i \cdot y_j \geqslant -1.$$

Consider Goemans-Williamson's formulation of MAXE2SAT described in Section 11.6.2. For an instance with one clause, say $x_1 \vee x_2$, the objective function is $(3 + y_1 \cdot y_0 + y_2 \cdot y_0 - y_1 \cdot y_2)/4$. In the relaxation of the original formulation, there are vectors v_0, v_1, v_2 where $v_1 \cdot v_0 = v_2 \cdot v_0 = -v_1 \cdot v_2 = 0.5$, i.e., where the objective function is $9/8$. Thus, there is an instance where the approximation ratio is at most 0.888. If the second set of constraints is added to the program, then the value of the objective function is at most 1 for any feasible solution of this instance.

11.8.2 Nonuniform Rounding

Feige and Goemans [FG95] introduced the concept of *nonuniform rounding*, a technique which improves Goemans-Williamson's MAXDICUT and MAXE2SAT approximation.

The idea is to modify step 3, the rounding, in a way that takes advantage of the special role of vector v_0.

First, Feige and Goemans consider those outputs of the semidefinite program which give rise to the worst case of the approximation ratio. Consider a clause in a MAXE2SAT instance, say $\bar{x}_i \vee \bar{x}_j$. Its contribution to the objective function is $(3 - y_i \cdot y_0 - y_j \cdot y_0 - y_i \cdot y_j)/4$. Let v_0, v_i, v_j be the vectors computed by the approximation algorithm. In Section 11.6.2 we have seen that this clause is satisfied with probability at least 0.878 times its contribution in the objective function, if the "uniform rounding" procedure of Goemans and Williamson is used. It can be shown that the worst case ≈ 0.878 is attained exactly if two of the inner products $v_i \cdot v_0, v_j \cdot v_0, v_i \cdot v_j$ are approximately -0.689 and the other product is 1, cf. also the analysis of the MAXCUT algorithm in Section 11.5.2. Thus, in the worst case, either $v_i \cdot v_0 \leqslant 0$ or $v_j \cdot v_0 \leqslant 0$. (Note that the triple of worst case vectors is not excluded by the additional constraints given in Section 11.8.1.)

Consider the algorithm which assigns x_i to true if $v_i \cdot v_0 \geqslant 0$ and to false otherwise. (Call this *crude rounding*.) If the triple of vectors v_0, v_i, v_j is a worst

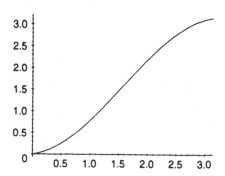

Fig. 11.3. Plot of the function f used for nonuniform rounding in the approximation of MaxE2Sat.

case for the uniform rounding step, then crude rounding ensures that the clause $\overline{x}_i \vee \overline{x}_j$ is always satisfied. Thus, for the worst case vectors of the uniform rounding procedure, crude rounding is better. Of course, it is not a good idea to use crude rounding, since it is much worse in many other cases.

We want to skew the uniform rounding procedure a little bit. We would like to use a rounding procedure which combines uniform rounding with a pinch of crude rounding. That is, if $v_i \cdot v_0 > 0$, then the probability of x_i being true should be increased. If $v_i \cdot v_0 < 0$, then the probability of x_i being true should be decreased.

Technically, this effect is obtained in the following way. Let ϑ be the angle between v_0 and v_i. Fix some function $f : [0, \pi] \to [0, \pi]$. Then, map vector v_i to that vector v_i' which is on the same hyperplane as the vectors v_i and v_0 and where the angle between v_0 and v_i' is $f(\vartheta)$. Finally, perform the uniform rounding procedure using the vectors v_0 and $v_1', \ldots, v_n'$.

The choice of $f(\vartheta) = \vartheta$ corresponds to uniform rounding. We require $f(\pi - \vartheta) = \pi - f(\vartheta)$ in order to ensure that negated and unnegated literals are handled in the same way.

Among other functions, Feige and Goemans study the effect of

$$f(\vartheta) = \vartheta + 0.806765 \left[\frac{\pi}{2}(1 - \cos \vartheta) - \vartheta \right] \tag{11.15}$$

on the approximation ratio in nonuniform rounding, cf. Figure 11.3. Using this function, they obtain the following approximation ratio.

Theorem 11.17 ([FG95]). MaxE2Sat *can be approximated by a randomized polynomial-time algorithm with approximation ratio 1.075, i.e., the expected value of the computed solution is at least 0.931 times the optimal solution.*

Sketch of Proof. By numerical computation, analyze the approximation ratio of the following modification of the algorithm described in Section 11.6.2:

- Add the constraints listed in Section 11.8.1.

- Use function (11.15) for nonuniform rounding.

In the analysis, first discretize the space of all vectors v_0, v_i, v_j that satisfy the additional constraints. In fact, it is sufficient to discretize the space of all possible angles between v_0, v_i and v_j. Then, numerically compute the ratio between the probability that a clause is satisfied and the contribution of that clause to the objective function. Check that this ratio is at least 0.931 for all these vectors. ∎

The importance of the additional constraints is stressed by the fact that, for nonuniform rounding using this function f, the worst case of the performance ratio is attained for a triple of vectors where $v_i \cdot v_0 + v_j \cdot v_0 + v_i \cdot v_j = -1$.

Using the same method with a different function f, Feige and Goemans obtain

Theorem 11.18 ([FG95]). MAXDICUT *can be approximated by a randomized polynomial-time algorithm with approximation ratio 0.859.*

11.9 Modeling Maximum Independent Set as a Semidefinite Program?

Just for didactical purposes, we consider how one might turn MAXIMUMINDEPENDENTSET into a semidefinite program. We leave to the interested reader the investigation whether this can be used to obtain better approximation algorithms for MAXIMUMINDEPENDENTSET than are currently known.

MAXIMUMINDEPENDENTSET
Instance: Given an undirected graph $G = (V, E)$.
Problem: Find a subset $V_1 \subseteq V$ such that V_1 does not contain any edge and $|V_1|$ is maximized.

We first claim that the size of a maximum independent set in a graph (and the corresponding independent set) can be obtained by finding a solution to the following mathematical program:

$$\max \ x_1 + \cdots + x_n - \sum_{\{i,j\} \in E} x_i x_j \quad \text{such that } \forall i: \ 0 \leqslant x_i \leqslant 1 \quad (11.16)$$

Then, the following mathematical program also yields the optimum:

$$\textbf{max } x_1^2 + \cdots + x_n^2 - \sum_{\{i,j\}\in E} x_i x_j \quad \textbf{such that } \forall i : \; 0 \leqslant x_i \leqslant 1 \quad (11.17)$$

This follows from the observation that the objective function considered in one variable only (others fixed) is a parabola which is open into the positive y-direction. Hence, we have a guarantee that for either $x_i = 0$ or $x_i = 1$, the optimum is obtained.

Now, let us show that (11.16) represents MAXIMUMINDEPENDENTSET. Given an independent set I and setting $x_i = 1$ for $i \in I$ (and $x_i = 0$ otherwise), we see that the optimal solution of (11.16) is at least the size of a maximum independent set.

Assume that we are given a solution of (11.16). Since the objective function is linear in every variable, we can assume that all x_i are either 0 or 1. We construct an independent set I with the same value as the objective function as follows: If the set $I := \{i \mid x_i = 1\}$ is not an independent set, we proceed as follows: If $x_i = x_j = 1$ and $\{i, j\}$ is an edge in the graph, then we set $x_i = 0$ which means that we have not decreased the objective function. (One vertex was eliminated, but at least one edge was also removed.)

Hence, the optimal solution of (11.16) or (11.17) yields the optimum cardinality of an independent set. Since by Theorem 4.13 MAXIMUMINDEPENDENTSET cannot be approximated within a factor n^ε (unless $\mathcal{P} = \mathcal{NP}$), it is unlikely that we can solve this mathematical program with the help of semidefinite programming, although it looks very much like the problem formulation for MAXCUT.

Consider for example the following semidefinite program P:

$$\textbf{maximize } y_{1,1} + \cdots + y_{n,n} - \sum_{\{i,j\}\in E} y_{i,j} \quad \textbf{such that } Y \text{ is PSD}$$

We may also add the linear inequalities $0 \leqslant y_{i,j} \leqslant 1$ to P. Since this semidefinite program is a relaxation of (11.17), we have obtained a method to compute an upper bound, unfortunately, this upper bound is useless, as the value of the semidefinite program always is n, since one can choose Y to be the identity matrix. The question remains whether there are more suitable relaxation approaches for MAXIMUMINDEPENDENTSET.

Exercises

Exercise 11.1. Given the following sets M_i,

- $M_1 = \{(x_1, x_2) : x_1 \geqslant x_2, x_1 \geqslant 1\}$,
- $M_2 = \{(x_1, x_2) : x_1 x_2 \geqslant 1, x_1 \geqslant 0, x_2 \geqslant 0\}$,

– $M_3 = \{(x_1, x_2) : x_1^2 + x_2^2 \leqslant 1\}$

– $M_4 = \{(x_1, x_2) : x_1^4 + x_2^4 \leqslant 1\}$

construct matrices F_i^0, F_i^1 and F_i^2 such that

$$M_i = \{(x_1, x_2) : \text{The matrix } F_i(x_1, x_2) := F_i^0 + x_1 \cdot F_i^1 + x_2 \cdot F_i^2 \text{ is PSD}\}.$$

Exercise 11.2. Find an optimal solution to the semidefinite program (11.1) (parabola and line).

Exercise 11.3. Derandomize the approximation algorithm for MAXCUT which is described in Section 11.5.1.

Exercise 11.4. Let $-1 \leqslant y_i \leqslant 1, i = 1, \ldots, T$ and let $Y_{sum} := \sum_{i=1}^{T} y_i \leqslant 0$. Prove the following: If Y_{sum} is fixed, then the y_i for which $\sum_i \arccos(y_i)$ attains a minimum, fulfill the following property:
There is a value $Y \leqslant 0$ and an r with $T/2 \leqslant r \leqslant T$ such that r of the y_i are equal to Y and $T - r$ of the y_i are equal to one.

12. Dense Instances of Hard Optimization Problems

Katja Wolf

12.1 Introduction

Some years ago only very few optimization problems like KNAPSACK were known to have polynomial time approximation schemes. Since the PCP-Theorem implies that all $\mathcal{APX}$-hard problems are unlikely to permit a PTAS, a consequent approach is to look for those subclasses which are easier to handle with respect to approximation. Baker [Bak94] shows that several $\mathcal{NP}$-hard graph problems have a PTAS on planar graphs. Chapter 13 is devoted to the construction of PTASs for the TRAVELING SALESMAN problem and the k-MST problem in Euclidean metric spaces. Recently, research has been extended to the approximability of a large class of problems, including MAXCUT and MAXEkSAT, on dense instances. The topic of this chapter is to survey the approximation results that have been achieved in this area.

We give a definition of density which depends on the type of the problem we consider.

Definition 12.1. *A graph with n vertices is* dense *if it has $\Omega(n^2)$ edges. It is* everywhere dense *if the minimum degree of the graph is $\Omega(n)$. A kSAT formula in n Boolean variables is* dense *if it consists of $\Omega(n^k)$ clauses.*

We shall follow the work of Arora, Karger and Karpinski [AKK95a], who combine randomized sampling and rounding techniques with linear programming to design a scheme with additive approximation guarantee for smooth integer programs.

Definition 12.2. *Let $p \in \mathbb{Z}[x_1, \ldots, x_n]$ be a degree d polynomial with integer coefficients. p is called* smooth, *if the coefficients preceding each term of degree i are bounded by a constant factor times n^{d-i}.*
A smooth degree d integer program has the objective

$$\begin{array}{ll} \text{maximize (minimize)} & p(x_1, \ldots, x_n) \\ \text{subject to} & x_i \in \{0, 1\} \quad \text{for} \quad 1 \leqslant i \leqslant n. \end{array}$$

The central result is the following theorem.

Theorem 12.3. *Let* **Opt** *denote the optimal value of the objective function of a smooth degree d integer program. For each $\varepsilon > 0$ there is an $n^{\mathcal{O}(1/\varepsilon^2)}$ time algorithm computing a 0-1 assignment to the variables which satisfies*

$$p(x_1, \ldots, x_n) \geqslant \mathbf{Opt} - \varepsilon\, n^d \quad \text{for a maximization problem, and}$$
$$p(x_1, \ldots, x_n) \leqslant \mathbf{Opt} + \varepsilon\, n^d \quad \text{for a minimization problem.}$$

As we will see, the approximation scheme with additive performance guarantee yields a (multiplicative) PTAS when applied to dense instances of the optimization problems we are concerned with. But first, let us describe the underlying ideas, treating MAXCUT as an example. We defer the proof of the theorem until Section 12.3.

12.2 Motivation and Preliminaries

Let $G = (V, E)$ be an undirected graph with vertex set V, ($|V| = n$), and edge set E. $\Gamma(v) := \{u \in V \mid (u, v) \in E\}$ denotes the set of vertices adjacent to vertex v. The MAXCUT problem is to determine a set $W \subset V$ such that the number of edges having one endpoint in W and the other endpoint in $V - W$ is maximized. We can easily formulate MAXCUT as a smooth quadratic integer program if we introduce a variable x_v for each vertex $v \in V$:

$$\text{maximize} \quad \sum_{v \in V} \left((1 - x_v) \cdot \sum_{u \in \Gamma(v)} x_u \right)$$
$$\text{subject to} \quad x_v \in \{0, 1\}, \quad v \in V.$$

Let us interpret $x_v = 0$ as "v is on the left-hand side W" and $x_v = 1$ as "v is on the right-hand side $V - W$" of the cut. In the above sum we count for each left vertex the number of its neighbors on the right.

In a maximum cut each vertex must be situated opposite to the majority of its neighbors, because otherwise one could move a vertex and increase the number of edges crossing the cut. So, we could determine the position of a vertex, if we knew to which side most of its neighbors belong. It seems as if nothing is won, since we do not know the neighbors' positions either.

A useful trick in order to cope with this circularity is taking a sample and fixing the positions of the sampled vertices arbitrarily. For the size of the sample we choose $k = \Theta(\log n)$.

Once we have arranged the sampled vertices, we could place each unsampled vertex opposite the majority of its sampled neighbors. This will not be successful, unless the numbers of right and left sampled neighbors differ a great deal. But a different approach is more promising: By inspecting the sample, we can for each vertex estimate the *fraction* of its neighbors that should be part of the right-hand side.

In order to explain the progress we have made, we need some notations.

$$r_v \quad := \quad \sum_{u \in \Gamma(v)} x_u$$

r_v is the number of v's neighbors on the right-hand side of the cut.

s_v :   an estimate for r_v. Count v's right-hand side neighbors in the sample and multiply the result by n/k.

The Sampling Lemma (which we will prove in Section 12.2.3) guarantees that, if we choose a sample of size $k = \Theta(\log n / \varepsilon^2)$, with high probability s_v and r_v will differ by an additive error of at most εn. We can restate the objective function as

$$\text{maximize} \quad \sum_{v \in V} (1 - x_v) \cdot r_v$$

and search for a solution to the following approximate system with $2n$ linear inequalities:

$$\text{maximize} \quad \sum_{v \in V} (1 - x_v) \cdot s_v$$

$$\sum_{u \in \Gamma(v)} x_u \quad \leqslant \quad s_v + \varepsilon n$$

$$\sum_{u \in \Gamma(v)} x_u \quad \geqslant \quad s_v - \varepsilon n$$

$$x_v \in \{0, 1\}, \quad v \in V.$$

Note that we now have converted the original quadratic MAXCUT program into a linear program. The reduction of the degree is a benefit of the sampling. We obtain a fractional solution $0 \leqslant x_v \leqslant 1$ ($v \in V$) by linear programming and afterwards apply Raghavan and Thompson's randomized rounding method [RT87] to produce a 0-1 assignment.

Thus far, we have described a single run for an arbitrary allocation of the sample. We repeat the same procedure for each of the $2^k = \text{poly}(n)$ possible arrangements of the vertices in the sample. Of course, we implicitly consider the optimal constellation of the sampled vertices among our trials. This technique is often called *exhaustive sampling*. In the end, we take the assignment that yields the largest cutsize.

It remains to be shown that the resulting approximation is within an additive error of εn^2 of the optimum. The proof is given in Section 12.3.

12.2.1 Chernoff Bounds

Before we go into the details of randomized rounding and random sampling, let us briefly review Chernoff bounds to gain insight into the technical background of the method. The analysis of the rounding procedure is quite intricate, because we allow negative coefficients and must derive suitable tail bounds for the deviation of a random variable from its mean.

Lemma 12.4. *Let $X_1, \ldots, X_n$ be independent Bernoulli random variables,*

$$\text{Prob}[X_i = 1] \; = \; p_i \; = \; 1 - \text{Prob}[X_i = 0].$$

and let $w_1, \ldots, w_n \in [0, 1]$. X denotes the weighted sum of the Bernoulli trials, $X = \sum_{i=1}^{n} w_i X_i$, and $\mu = \mathbf{E}[X]$. Then the following inequality holds

$$\text{Prob}\left[\, |X - \mu| > t \,\right] \leqslant 2\,e^{-2t^2/n} \qquad \text{for } t > 0.$$

For a proof of the lemma we direct the reader to [Hoe63] or to [MR95b]. It is conducted in a similar way as the proof of Theorem 3.5.

Corollary 12.5. *For each $f > 0$ there is a constant $c > 0$ such that under the assumptions of Lemma 12.4*

$$\text{Prob}\left[\, |X - \mu| > c\,\sqrt{n \log n}\, \right] \leqslant 2\,n^{-f}.$$

Proof.   We apply Lemma 12.4 with $t := c\,\sqrt{n \log n}$ and $c = \sqrt{f/2}$ and obtain

$$\text{Prob}\left[\, |X - \mu| > c\,\sqrt{n \log n}\, \right] \leqslant 2\,e^{-2c^2 \log n} \leqslant 2\,n^{-2c^2} \leqslant 2\,n^{-f}.$$

∎

Lemma 12.6 (Positive and Negative Variables).
Let $X_1, \ldots, X_n$ be independent Bernoulli random variables such that $\{1, \ldots, n\}$ is the disjoint union of two sets I and J, and $0 \leqslant w_i \leqslant 1$ for $i \in I$, and $-1 \leqslant w_j \leqslant 0$ for $j \in J$.

Then for each $f > 0$ there are constants $c_1 > 0$ and $c_2 > 0$ such that

$$\text{Prob}\left[\, |X - \mu| > c_1\,\sqrt{n \log n}\, \right] \leqslant c_2 n^{-f}.$$

Sketch of Proof.   We define

$$X^{(1)} = \sum_{i \in I} w_i X_i, \quad \mu^{(1)} = \mathbf{E}[X^{(1)}], \quad X^{(2)} = \sum_{j \in J} w_j X_j, \quad \mu^{(2)} = \mathbf{E}[X^{(2)}].$$

Note that $|X - \mu| \leqslant |X^{(1)} - \mu^{(1)}| + |-X^{(2)} - (-\mu^{(2)})|$, so we can apply Corollary 12.5 to $X^{(1)}$ and to $-X^{(2)}$ and combine the resulting inequalities appropriately. ∎

Remark 12.7. As common in the literature, we use the term *with high probability* to express that the probability is $1 - n^{-\Omega(1)}$ when n tends to infinity.

12.2.2 Randomized Rounding

After this technical intermezzo, we are now well-prepared for the rounding procedure.

Lemma 12.8 (Randomized Rounding).
Assume that $x = (x_1, \ldots, x_n)^T$ is a fractional solution to the linear program

$$A x = b$$
$$0 \leqslant x_i \leqslant 1 \quad (i = 1, \ldots, n),$$

where the coefficients of the matrix A are $\mathcal{O}(1)$.
Randomly construct a 0-1 vector $y = (y_1, \ldots, y_n)^T$ by setting $y_i = 1$ or $y_i = 0$ for all $i \in \{1, \ldots, n\}$ independently according to

$$\mathrm{Prob}[y_i = 1] = x_i = 1 - \mathrm{Prob}[y_i = 0].$$

Then, with high probability, y satisfies

$$A y = \left(b_1 + \mathcal{O}(\sqrt{n \log n}), \ldots, b_n + \mathcal{O}(\sqrt{n \log n}) \right)^T.$$

Proof. We have imposed that the coefficients of A are bounded, say $a_{ij} \in [-c, c]$ for $1 \leqslant i, j \leqslant n$. If we define $\hat{a}_{ij} := a_{ij}/c$, $\hat{b}_i := b_i/c$ and notice that

$$\mu_i := \mathbf{E}\left[\sum_{j=1}^n \hat{a}_{ij} y_j \right] = \sum_{j=1}^n \hat{a}_{ij} \, \mathbf{E}[y_j] = \sum_{j=1}^n \hat{a}_{ij} x_j = \hat{b}_i,$$

we can apply Lemma 12.6 to each row i of the system. So we have

$$\left| \sum_{j=1}^n \hat{a}_{ij} y_j - \hat{b}_i \right| \leqslant c_1 \sqrt{n \log n}$$

with probability at least $1 - c_2 n^{-f}$, respectively

$$\left| \sum_{j=1}^n a_{ij} y_j - b_i \right| \leqslant c c_1 \sqrt{n \log n} = \mathcal{O}(\sqrt{n \log n})$$

with probability at least $1 - c_2 n^{-f}$. Now choose $f \geqslant 2$ and enforce that the assertion holds for the whole system with probability $1 - c_2 n^{-f+1}$. ∎

12.2.3 Random Sampling

Lemma 12.9 (Random Sampling with Replacement).
Let r be the sum of n numbers $a_1, \ldots, a_n \in [-c, c]$. By sampling with replacement we choose $k = \Theta(\log n/\varepsilon^2)$ (not necessarily distinct) elements and compute their sum s. Then, with high probability, we have

$$\frac{sn}{k} \in \left[r - \varepsilon n, \, r + \varepsilon n \right].$$

Thus, we can take sn/k as an estimate for r, but we have to tolerate a linear error term.

For convenience, we cite an inequality for continuous random variables due to Hoeffding ([Hoe63], Theorem 2):

If $Y_1, \ldots, Y_k$ are independent random variables and $a \leqslant Y_i \leqslant b$ for $i = 1, \ldots, k$, let $\overline{Y} = 1/k \sum_{i=1}^{k} Y_i$ and $\mu = \mathbf{E}[\overline{Y}]$, then

$$\mathrm{Prob}\left[\,|\,\overline{Y} - \mu\,| \geqslant t\,\right] \leqslant e^{-2kt^2/(b-a)^2} \quad \text{for} \quad t > 0.$$

Proof of Lemma 12.9. Introduce a random variable X_i for the element drawn in the i-th turn, and let $\overline{X} = 1/k \sum_{i=1}^{k} X_i$. We have $\mathbf{E}[X_i] = 1/n \sum_{j=1}^{n} a_j$ for each i, and since we draw with replacement, the X_i are independent and identically distributed.
Observing $\mathbf{E}[\overline{X}] = 1/n \sum_{j=1}^{n} a_j$, we can use the above inequality and get

$$\mathrm{Prob}\left[\,|\,\overline{X} - \frac{1}{n}\sum_{j=1}^{n} a_j\,| \geqslant \varepsilon\,\right] = \mathrm{Prob}\left[\,|\,\frac{ns}{k} - r\,| \geqslant \varepsilon n\,\right] \leqslant 2e^{-k\varepsilon^2/(2c^2)}.$$

Choose a suitable $k = \Theta(\log n/\varepsilon^2)$, and the lemma follows. Note that the probability can be influenced by varying the size of the sample within a constant factor. ∎

12.3 Approximating Smooth Integer Programs

In this section we are going to prove Theorem 12.3 for quadratic programs, and we will indicate how the idea can be generalized to degree d objective functions. We only address the maximization version, minimization is similar.

12.3.1 The Quadratic Case

Theorem 12.10. *Consider the following quadratic integer program*

$$\begin{aligned} \text{maximize} \quad & x^T A x + b^T x \\ \text{subject to} \quad & x = (x_1, \ldots, x_n)^T \in \{0,1\}^n \end{aligned}$$

and let $A = (a_{ij})$ be an $(n \times n)$ matrix with coefficients $a_{ij} = \mathcal{O}(1)$, $b = (b_1, \ldots, b_n)^T$, $b_j = \mathcal{O}(n)$, and let $\mathbf{Opt}$ denote the maximum value of the objective function.
There is an algorithm which for arbitrary $\varepsilon > 0$ constructs a 0-1 vector $y = (y_1, \ldots, y_n)^T$ for which

$$y^T A y + b^T y \geqslant \mathbf{Opt} - \varepsilon n^2.$$

y is computed in time $n^{\mathcal{O}(1/\varepsilon^2)}$.

We describe a randomized algorithm that involves linear programming and the sampling and rounding methods we have provided in the previous section. Later on, in Section 12.3.2, we will sketch how the algorithm can be derandomized in order to obtain a deterministic scheme. Throughout the proof we imitate the approach we have already outlined for MAXCUT.

Proof. The first step is to transform the quadratic program into a linear program. Let $x^* = (x_1^*, \ldots, x_n^*)^T$ be a solution that maximizes the value of the objective function. Of course, we do not really know such a vector, we are just about to demonstrate how such a vector or at least an approximate one can be constructed. But to explain the idea of the algorithm and for its analysis, we assume for the moment that we already know x^*. We define $r^T = (r_1, \ldots, r_n)$ via

$$r^T = x^{*T} A + b^T,$$

that is, $r_i = \sum_{j=1}^n x_j^* a_{ji} + b_i$, the inner product of x^* and the i-th column of A plus b_i, and obtain a new linear system

$$
\begin{aligned}
\text{maximize} \quad & r^T x \\
& x^T A + b^T = r^T \\
& x \in \{0,1\}^n.
\end{aligned}
\tag{12.1}
$$

Clearly, x^* is an optimal solution to this system.

Our preliminary goal now is to find an estimate vector $s = (s_1, \ldots, s_n)^T$, which is a good approximation for r. s is allowed to differ from r in every coordinate by a linear error

$$|r_i - s_i| \leqslant \varepsilon n \quad \text{for } 1 \leqslant i \leqslant n. \tag{$*$}$$

Since we do not know a direct way to determine s without knowing x^*, we establish a set of candidate vectors for s, one of which is guaranteed to satisfy ($*$). Like in MAXCUT we choose a (multi-)set Z of $k = \Theta(\log n / \varepsilon^2)$ indices by random sampling. We process each of the 2^k possible 0-1 assignments for the x_j, $j \in Z$, as an outer loop for the algorithm and for each assignment we compute a vector s given by

$$s_i = \frac{n}{k} \sum_{j \in Z} a_{ji} x_j + b_i \quad \text{for } 1 \leqslant i \leqslant n.$$

This yields a set S of 2^k candidate vectors for s. We know that one of the assignments equals x^* with respect to the drawn indices in Z

$$s_i^* = \frac{n}{k} \sum_{j \in Z} a_{ji} x_j^* + b_i \quad \text{for } 1 \leqslant i \leqslant n.$$

Since b_i is fixed and the a_{ji} are $\mathcal{O}(1)$, we can apply the Sampling Lemma to $\sum_{j \in Z} a_{ji} x_j^*$ which ensures

$$|r_i - s_i^*| \leqslant \varepsilon n,$$

say, with probability at least $1 - 2n^{-2}$ for one constraint i, respectively with probability at least $1 - 2n^{-1}$ for all n sums.

Let us pause here and recapitulate the proof up to this point. We have obtained a set of vectors S and know that this set contains a certain vector s, namely s^*, which with high probability approximates r within the required accuracy. All this was done without actually involving x^* in the computation. The remainder of the algorithm is executed for each of the candidate vectors in S. Eventually, we take the one with the best result, i.e., the largest value of the objective function. For the analysis let us concentrate on that vector s which satisfies $(*)$.

Look at the following program

$$\text{maximize} \quad s^T x \tag{12.2}$$

$$x^T A + b^T \geqslant (s_1 - \varepsilon n, \dots, s_n - \varepsilon n)$$
$$x^T A + b^T \leqslant (s_1 + \varepsilon n, \dots, s_n + \varepsilon n)$$
$$x \in \{0, 1\}^n.$$

(Here the inequalities are supposed to hold for every coordinate of a vector.) x^* is a feasible 0-1 solution to the above system, and

$$s^T x^* = r^T x^* - (r^T - s^T) x^* = x^{*T} A x^* + b^T x^* - (r^T - s^T) x^* \geqslant \mathbf{Opt} - \varepsilon n^2.$$

If we relax the integrality condition and replace it by $x \in [0, 1]^n$, we can use linear programming and find an optimal fractional solution x in polynomial time. Because of the relaxation we have $s^T x \geqslant s^T x^*$. Then we apply the rounding procedure we saw in Lemma 12.8. The resulting 0-1 vector y satisfies

$$y^T A + b^T \geqslant \left(s_1 - \varepsilon n - \mathcal{O}(\sqrt{n \log n}), \dots, s_n - \varepsilon n - \mathcal{O}(\sqrt{n \log n}) \right)$$
$$y^T A + b^T \leqslant \left(s_1 + \varepsilon n + \mathcal{O}(\sqrt{n \log n}), \dots, s_n + \varepsilon n + \mathcal{O}(\sqrt{n \log n}) \right)$$

with high probability. The value of $s^T y$ can be bounded from below with the help of the following inequality

$$s^T y \geqslant s^T x - \mathcal{O}(n\sqrt{n \log n})$$
$$\geqslant s^T x^* - \mathcal{O}(n\sqrt{n \log n})$$
$$\geqslant \mathbf{Opt} - \varepsilon n^2 - \mathcal{O}(n\sqrt{n \log n}).$$

Define an error vector δ via $\delta^T = x^T A + b^T - s^T$. As for each coordinate $|\delta_i| \leqslant \varepsilon n$ holds, we get $|\delta^T y| \leqslant \varepsilon n^2$.

Finally, let us combine the inequalities in order to bound the possible deviation from the value of the objective function of the original program.

$$
\begin{aligned}
y^T A y + b^T y &= (y^T A + b^T) y \\
&= (y^T A + b^T - (x^T A + b^T)) y + \delta^T y + s^T y \\
&\geqslant \mathcal{O}\left(n\sqrt{n\log n}\right) - \varepsilon\, n^2 + \left(\mathbf{Opt} - \varepsilon\, n^2 - \mathcal{O}\left(n\sqrt{n\log n}\right)\right) \\
&= \mathbf{Opt} - \Big(2\varepsilon + o(1)\Big)\, n^2
\end{aligned}
$$

The running time of the algorithm is dominated by solving linear programs and subsequent randomized rounding of the solution vectors which can be done in polynomial time. This is repeated for every possible 0-1 assignment to the sampled indices, that is $2^k = 2^{\mathcal{O}(\log n/\varepsilon^2)} = n^{\mathcal{O}(1/\varepsilon^2)}$ times. The algorithm outputs the vector with the largest value of the objective function. ∎

Remark 12.11. The error caused by the sampling is significantly larger than the error produced by randomized rounding. So if the sampling could be replaced by a better method, this would improve the approximation guarantee.

12.3.2 Derandomization and Generalization of the Scheme

In the proof for the quadratic case we demonstrated how a quadratic program is converted into a program with a linear objective function and linear constraints (see (12.1) and (12.2)). In the same fashion, we can express the optimum of a degree d polynomial with the help of degree $d-1$ polynomials. Again by random sampling we recursively transform these constraints into constraints of a lower degree, ending up with a system of linear constraints. Theorem 12.3 is obtained by an induction on the degree. An immediate consequence is that the same algorithm works if the original program has smooth degree $d-1$ constraints. The following generalization holds.

Corollary 12.12. *Consider a given smooth degree d integer program which is, in addition, subject to a polynomial number of constraints*

$$
q_j(x_1, \ldots, x_n) \leqslant b_j\,,
$$

each q_j being a smooth polynomial of degree $d_j < d$. Then, for each $\varepsilon > 0$, the algorithm constructs a 0-1 solution vector y within an additive error of $\varepsilon\, n^d$ of the optimum, and satisfies

$$
q_j(y_1, \ldots, y_n) \leqslant b_j + \varepsilon\, n^{d_j}\,.
$$

So far, we have only described a randomized scheme. A deterministic procedure can be derived by derandomizing the rounding algorithm and the sampling procedure.

Raghavan [Rag88] coined the term "method of conditional probabilities" for the derandomization of randomized rounding. A detailed description of the method of conditional probabilities is presented in Chapter 3.

Derandomizations of the Sampling Lemma are published in [BGG93] and in [BR94]. Gillman [Gil93] shows that choosing $\Theta(\log n/\varepsilon^2)$ indices independently can be simulated by taking the indices reached along a random walk of length $\Theta(\log n/\varepsilon^2)$ on an expander graph with constant expansion and bounded degree. Because of the bounded degree, there are only $n^{\mathcal{O}(1/\varepsilon^2)}$ different random walks. So it remains polynomial, even if we process the algorithm for all samples produced from these random walks and take the best result.

12.4 Polynomial Time Approximation Schemes for Dense MaxCut and MaxEkSat Problems

In this section we will apply the approximation procedure developed in Section 12.3 and realize that it turns out to be a PTAS on dense MaxCut and MaxEkSat instances. Notice that for both problems exact optimization remains $\mathcal{NP}$-hard when we restrict ourselves to dense instances.

To understand this for MaxCut, we reduce a general graph to a dense graph by adding a complete bipartite graph on $2n$ vertices. The resulting graph is dense because the average degree is linear in the size of the graph. The initial graph has a cut of at least k edges if and only if the dense graph has a cut of at least $k + n^2$ edges.

A reduction from MaxEkSat to the dense MaxEkSat problem is performed by adding dummy clauses that are always satisfied, regardless of the assignment to the Boolean variables.

12.4.1 MaxCut

We have already given a smooth quadratic formulation for MaxCut. By Theorem 12.3 for α and $\varepsilon > 0$, we can compute an assignment to the variables, in $n^{\mathcal{O}(1/(\varepsilon\alpha)^2)}$ steps, such that the size c of the cut satisfies

$$c \geqslant \mathbf{Opt} - \frac{\varepsilon\alpha}{2}\, n^2\,.$$

For a dense graph, having αn^2 edges, one can randomly produce a cut with expected cutsize $\mathrm{E}[\mathrm{Cut}]$ of at least $\alpha n^2/2$ edges by putting each vertex on the left or on the right side with probability $1/2$.

$$\mathrm{E}[\mathrm{Cut}] \;=\; \sum_{e\in E}\mathrm{Prob}[e\in\mathrm{Cut}] \;=\; \frac{1}{2}|E| \;\geqslant\; \frac{1}{2}\alpha n^2$$

This implies $\mathbf{Opt} \geqslant \alpha n^2/2$. Summarizing, we have

$$c \geqslant \mathbf{Opt} - \frac{\varepsilon\,\alpha}{2}\,n^2 \geqslant \mathbf{Opt} - \varepsilon\,\mathbf{Opt} = (1 - \varepsilon)\,\mathbf{Opt}.$$

In a similar way one can devise a formulation for MAXDICUT as a smooth quadratic program and achieve a PTAS for dense directed graphs. We are not going to elaborate on this further. The reader who is interested in approximation algorithms for general MAXCUT problems is directed to Chapter 11. Semidefinite programming tools and randomized rounding are applied to a relaxation of MAXCUT. For recent non-approximability results concerning MAXCUT and MAXDICUT see Chapter 8.

12.4.2 MAXEkSAT

Let us now consider a dense EkSAT formula with Boolean variables $y_1, \ldots, y_n$ and $m \geqslant \alpha\,n^k$ clauses. For each Boolean variable y_i we introduce a 0-1 variable x_i ($i = 1, \ldots, n$). Then we transform each positive literal y_i into $1 - x_i$ and each negative literal $\overline{y_j}$ into x_j. The Boolean operator $\vee$ is replaced by the integer multiplication and for the whole clause we get the final expression by subtracting the product from 1.

For instance: $\overline{y_1} \vee y_2 \vee y_3 \quad\longrightarrow\quad 1 - x_1(1 - x_2)(1 - x_3).$

A satisfying Boolean assignment to the y-variables of a clause corresponds to a 0-1 assignment to the x-variables that evaluates to 1. True is translated into 1, false into 0.

Denote the resulting term of the j-th clause by $p_j(x_1, \ldots, x_n)$. We have turned the original kSAT formula into the following smooth degree k program.

$$\text{maximize} \quad \sum_{j=1}^{m} p_j(x_1, \ldots, x_n)$$
$$\text{subject to} \quad x_i \in \{0, 1\} \quad \text{for} \quad 1 \leqslant i \leqslant n.$$

Let **Opt** be the maximum number of clauses of the formula any Boolean assignment can satisfy and let t be the number of satisfied clauses the algorithm produces. For a random true/false assignment a clause of length k is false with probability 2^{-k}, true with probability $1 - 2^{-k}$. So we deduce

$$\mathbf{Opt} \geqslant (1 - 2^{-k})\,m \geqslant (1 - 2^{-k})\,\alpha\,n^k.$$

For $\varepsilon > 0$, in time $n^{\mathcal{O}(2^{2k}/(\varepsilon\,\alpha)^2)}$, the scheme produces a solution with

$$t \geqslant \mathbf{Opt} - \frac{\varepsilon\,\alpha}{2^k}\,n^k \geqslant \mathbf{Opt}\left(1 - \varepsilon\,\frac{1}{2^k - 1}\right) \geqslant (1 - \varepsilon)\,\mathbf{Opt}$$

true clauses.

12.5 Related Work

Apart from the approximation schemes for the problems we have presented here, Arora, Karger and Karpinski have designed PTASs for GRAPHBISECTION of everywhere dense graphs and for DENSESTSUBGRAPH, as well as an exact algorithm for 3-COLORING of 3-colorable graphs. (This will be considered in the Exercises). They also generalize the notion of density to MAX-SNP problems, representing them as MAX-k-FUNCTIONSAT instances for a suitable k, which depends on the special problem, and obtain similar PTASs as in Section 12.4.2 for the resulting instances, if they are dense. In MAX-k-FUNCTIONSAT the input consists of m Boolean functions in n variables, each function depending only on k variables. The objective is to assign truth values to the variables so as to satisfy the maximum possible number of these functions. An instance is dense, if it consists of $\Omega(n^k)$ functions. (For the connection to MAX-SNP see also [Pap94].)

Fernandez de la Vega [Fer96] has devised randomized approximation schemes for MAXCUT and MAXACYCLICSUBGRAPH. In his placement procedure he applies sampling techniques to obtain a subset W of the vertices, which completely belongs to one side of the maximum cut. The remaining vertices are allocated according to their adjacency relations to W. The running time of the algorithm is $\mathcal{O}(2^{1/\varepsilon^{2+o(1)}} n^2)$.

A paper by Arora, Frieze and Kaplan [AFK96] deals with assignment problems with additional linear constraints. They present a new rounding procedure for fractional bipartite perfect matchings, different from the Raghavan-Thompson rounding method described in Lemma 12.8. Here, the variables are not rounded independently. They give an additive approximation algorithm for the smooth QUADRATICASSIGNMENTPROBLEM, which uses random sampling to reduce the degree by one. The algorithm is similar to the presented scheme, except for the new rounding technique. The running time of the algorithm is only quasi-polynomial: $n^{\mathcal{O}(\log n/\varepsilon^2)}$, but they give PTASs for dense instances of MINLINEAR-ARRANGEMENT, MAXACYCLICSUBGRAPH and several other problems.

Karpinski and Zelikovsky [KZ97] study STEINERTREE problems where each terminal must be adjacent to $\Omega(n)$ non-terminal vertices and give a PTAS for this situation. For the SETCOVER problem, they call an instance with a family of m subsets of a set of n elements ε-dense, if each element belongs to εm subsets. They show that SETCOVER on dense instances is not $\mathcal{NP}$-hard, unless $\mathcal{NP} \subseteq \text{DTIME}(n^{\log n})$, and provide a greedy algorithm with approximation ratio $c \log n$ for every $c > 0$. Open questions are whether there is a constant-factor approximation or even an exact polynomial time algorithm for dense SETCOVER and whether the dense STEINERTREE problem is in $\mathcal{P}$.

Exercises

The results discussed in the exercises are due to Arora, Karger and Karpinski. (See [AKK95b].)

Exercise 12.1. GRAPHBISECTION
Instance: Given an undirected graph $G = (V, E)$ with $2n$ vertices.
Problem: Partition the vertex set into two disjoint subsets V_1 and V_2 of size n such that the number of edges having one endpoint in V_1 and the other endpoint in V_2 is minimized.

Show that the existence of a PTAS for dense instances of the GRAPHBISECTION problem implies the existence of a PTAS on general instances.

Exercise 12.2. DENSESTSUBGRAPH
Instance: Given an undirected graph $G = (V, E)$ and a positive integer k.
Problem: Determine a subset V' of the vertices, $|V'| = k$, such that the graph $G' = (V', E(V'))$ has the most edges among all induced subgraphs on k vertices.

Construct a PTAS for DENSESTSUBGRAPH on everywhere dense graphs for $k = \Omega(n)$.

Exercise 12.3. Sketch a randomized polynomial algorithm for 3-COLORING of everywhere dense 3-colorable graphs based on a random sampling approach.
(**Hint:** If the random sample is large enough, one can show that with high probability every vertex has a sampled neighbor. How can this observation be used to extend a proper coloring of the sample to a coloring of the whole vertex set?)

13. Polynomial Time Approximation Schemes for Geometric Optimization Problems in Euclidean Metric Spaces

Richard Mayr, Annette Schelten

13.1 Introduction

Many geometric optimization problems which are interesting and important in practice have been shown to be $\mathcal{NP}$-hard. However, for some cases it is possible to compute approximate solutions in polynomial time.

This chapter describes approaches to find polynomial time algorithms that compute approximate solutions to various geometric optimization problems in Euclidean metric spaces. It is mainly based on the recent work of Arora [Aro97], which significantly improved previous works by Arora [Aro96] and Mitchell [Mit96]. First let us define the problems we will consider:

TRAVELING SALESMAN PROBLEM (TSP)

Instance: A graph $G = (V, E)$ on n nodes with edge weights defined by $c : E \longrightarrow \mathbb{R}$.

Problem: Find a cycle in G with $|V| = n$ edges that contains all nodes and has minimal cost.

Such a cycle is called the *salesman tour* for G. A variant of this problem is k-TSP where for a given $k < n$ the aim is to find the k vertices with the shortest traveling salesman tour, together with this tour. Both problems are $\mathcal{NP}$-hard.

In practice the graph is often embedded in the Euclidean space $\mathbb{R}^d$ with dimension $d \in \mathbb{N}$. If additionally the distances are given by the Euclidean norm (l_2-norm) we speak of *Euclidean* TSP:

EUCLIDEAN TSP

Instance: A graph $G = (V, E)$ in $\mathbb{R}^d$ on n nodes with edge weights defined by the Euclidean norm.

Problem: Find a cycle in G with $|V| = n$ edges that contains all nodes and has minimal cost.

Since the distances between the nodes are given by a norm, we are in particular dealing with complete graphs. The more general case that the graph lies in an arbitrary metric space is called *metric* TSP.

It has long been known that TSP is $\mathcal{NP}$-hard [Kar72]. Later it was shown that this even holds for the special case of Euclidean TSP [Pap77, GGJ76]. While – under the assumption that $\mathcal{P} \neq \mathcal{NP}$ – metric TSP and many other problems do not have a polynomial time approximation scheme (proved by Arora, Lund, Motwani, Sudan and Szegedy [ALM+92]), there is a polynomial time approximation scheme (PTAS) for Euclidean TSP.

In the following we will describe such a PTAS where the graph lies in the Euclidean space $\mathbb{R}^2$. It can be generalized to the d-dimensional Euclidean space $\mathbb{R}^d$, as well as to other related problems (see Section 13.5).

Arora's [Aro97] new algorithm is a randomized PTAS, which finds a $(1 + \frac{1}{c})$-approximation to Euclidean TSP in time $\mathcal{O}(n(\log n)^{\mathcal{O}(c)})$ where $c > 0$ and n denotes the number of given nodes. In earlier papers Arora [Aro96] and Mitchell [Mit96] independently obtained PTAS that need $\mathcal{O}(n^{\mathcal{O}(c)})$ time.

Quoting the author, the main idea of the PTAS relies "on the fact that the plane can be recursively partitioned such that some $(1+\frac{1}{c})$-approximate salesman tour crosses each line of the partition at most $t = \mathcal{O}(c \log n)$ times. Such a tour can be found by dynamic programming." [Aro97].

13.2 Definitions and Notations

In the following we use the convention that rectangles are axis-aligned. The *size* of a rectangle is the length of its longer edge. For a given set of n nodes the *bounding box* is the smallest square enclosing them. We will always denote the size of any bounding box by L and without loss of generality let $L \in \mathbb{N}$.

In the following we assume that there is a grid of granularity 1, such that all nodes lie on grid points. Moreover, we assume that the bounding box has size $L = nc$. This is no restriction, since any instance of the problem can be transformed into one that satisfies these conditions with small extra cost: Place a grid of granularity $L/(nc)$ over the instance and move every node to the nearest grid point. (It is possible that two nodes now lie on the same grid point.) If we find a solution for this new instance, then we also have a solution for the original instance. Let p be a salesman path for the new instance. We can construct a salesman path p' for the original instance as follows: For every node add two edges, from the node to the nearest grid point and back. The total length of these additional edges is $\leqslant 2n \cdot \sqrt{2}/2 \cdot L/(nc)$. As the length **Opt** of the optimal tour is at least L, this is $\leqslant \sqrt{2}\,\mathbf{Opt}/c = \mathcal{O}(\mathbf{Opt}/c)$. This additional cost is acceptable, since we are only interested in $(1 + 1/c)$-approximations. So far the minimal internode distance in the new instance is $L/(nc)$. Now we scale all distances by $L/(nc)$ and obtain a minimal internode distance of 1. The new size of the bounding box is now nc. The transformation only requires $\mathcal{O}(n \log n)$ time.

The definitions of a *dissection* and a *quadtree* are essential to this PTAS. Both are recursive partitionings of the bounding box into smaller squares. To get a dissection of the bounding box we partition any square into four squares of equal size until each square has length at most 1. These four squares are called the children of the bigger one. Similarly we define a quadtree. Also this time each square is divided into four equal squares, but we already stop when a square contains at most one node.

Instead of "normal" dissections or quadtrees we will consider shifted ones. For $a, b \in \mathbb{N}$ the (a, b)-*shift* of the dissection is obtained by shifting all x- and y-coordinates of all lines by a and b respectively, and then reducing them modulo L. This can lead to cases where previous rectangles are now wrapped around. Consider a simple dissection of depth one. There is only the bounding box and its four children. If it is shifted by shifts $a, b \leqslant L/2$, then three of the original children will be wrapped around. The only child that is not wrapped around is the child-square that used to reside in the lower left corner; it is now approximately in the center of the bounding box.

Let there be a shifted dissection. To construct the corresponding *quadtree with shift* (a, b) we cut off the partitioning at each square that contains at most one node.

In a dissection or in a quadtree a child-square has exactly half the size of its parent-square (the size is defined as the length of the longer edge). The size of the bounding box is nc and c is a constant, thus at a depth of $\mathcal{O}(\log n)$ the rectangles in the quadtree have size 1. As the minimal internode distance is 1, the construction stops there, because there is at most one node in the square. Thus any quadtree – shifted or not – has depth $\mathcal{O}(\log n)$ and consists of $\mathcal{O}$(number of nodes · depth) = $\mathcal{O}(n \log n)$ squares. Hence the quadtree can be obtained in time $\mathcal{O}(n \log^2 n)$.

Finally we define the concept of a *light salesman tour*, which is crucial for the following.

Definition 13.1. *Let $m, t \in \mathbb{N}$. For each square we introduce an additional node at each corner and m further equally-spaced nodes on each of its four edges. These points are called portals.*

We say that a salesman tour is (m, t)-light with respect to the shifted dissection if it crosses each edge of each square in the dissection at most t times and all crossings go through one of the portals. A tour can use a portal more than once.

To find a nearly optimal salesman tour by dynamic programming means to find an approximate tour in each subsquare and to put these tours together. Therefore we have to consider all possibilities where a segment of a tour can enter a square. In a (m, t)-light salesman tour these points are limited by the discrete set of portals. Since the distances are given by the Euclidean norm, the triangle-inequality holds, which means that the direct way between two nodes is always at most as long as the way through an additional point. Hence removing these additional points (portals) never increases the cost of the tour.

13.3 The Structure Theorem and Its Proof

The algorithm, which we will describe in Section 13.4, mainly relies on the following Structure Theorem:

Theorem 13.2. *(Structure Theorem) Consider an Euclidean TSP instance with bounding box of size L and the corresponding dissection with shift (a, b) where $a, b \in \mathbb{N}$ with $0 \leqslant a, b \leqslant L$ are picked randomly (every choice of a, b has the same probability).*

Then with probability at least $1/2$ this dissection has a salesman tour of cost at most $(1 + 1/c)\,\mathbf{Opt}$ that is (m, t)-light where $m = \mathcal{O}(c \, \log L)$, $t = \mathcal{O}(c)$ and $m \geqslant t \log L$.

In the proof of this theorem we assume that we have an optimal tour p and randomly chosen shifts $a, b \in \mathbb{N}$. Then we modify this tour until we get a new approximate salesman tour – probably through additional nodes – in the dissection with shift (a, b) which is (m, t)-light. The following lemma helps us to reduce the number of crossings between the salesman tour and the edges of each square.

Lemma 13.3. *(Patching Lemma) Let p be a closed path. Consider a line segment S of length l and assume that p crosses S at least three times. Then there exist line segments on S of total length less than or equal to $6l$ such that the addition of these segments to p changes p into a closed path p' of length $|p| + 6l$ that crosses S at most twice.*

Proof. We assume that p (of length $|p|$) is a closed path that crosses S $2k + 2$ ($2k + 1$, respectively) times. Let $x_1, \ldots, x_{2k+2}$ ($x_1, \ldots, x_{2k+1}$, resp.) be the corresponding crossing points. For $i = 1, \ldots, 2k$ break the path p at the points x_i. This implies that p breaks into $2k$ subpaths $P_1, \ldots, P_{2k}$. Also we double all these points x_i, $i = 1, \ldots, 2k$ and call these copies x'_i. Now we are adding the line segments of a minimal cost salesman tour through $x_1, \ldots, x_{2k}$, a minimal cost salesman tour through $x'_1, \ldots, x'_{2k}$ and also the line segments of minimal cost perfect matchings among $x_1, \ldots, x_{2k}$ and among $x'_1, \ldots, x'_{2k}$. All added line segments are lying on S which yields an Eulerian graph and hence there is a tour p' through $x_1, \ldots, x_{2k+2}, x'_1, \ldots, x'_{2k+2}$ using these line segments and the paths $P_1, \ldots, P_{2k}$. (An Eulerian graph is a connected graph which has a tour through all edges. These graphs are characterized by the fact that they are connected and that each point has even degree.) The new path p' is now using line-segments of S, nevertheless there are at most two crossing points. Also the added segments are of length at most $6l$ and hence $|p'| \leq |p| + 6l$.   ∎

There is a longer proof [Aro97] that shows that one can also find such a tour p' where the length is only increased by $3l$, and probably even this can be improved. In the proof of the Structure Theorem we just need the fact that p only increases by a constant factor of the length of S.

The next lemma shows an upper bound on how often a salesman tour crosses grid lines. It also will be used in the proof of the Structure Theorem. We defined that all nodes lie on grid points of a grid of granularity 1 (see Section 13.2). So the minimal internode distance is at least 1 and thus all line segments in the salesman path have a length of at least 1.

Lemma 13.4. *Let p be a path consisting of segments of length at least 1 and $t(p, l)$ the number of times that p crosses a horizontal or vertical line l of the grid, then*

$$\sum_{l:vertical} t(p, l) \; + \sum_{l:horizontal} t(p, l) \; \leq \; \left(\sqrt{2} + 2 \right) length(p).$$

Proof. Regard an edge of p of length s and let x and y be the lengths of the horizontal and vertical projections of this edge, thus $x^2 + y^2 = s^2$. The edge contributes at most $(x + 1) + (y + 1)$ to the left hand side. As $s \geq 1$ we have

$$x + y + 2 \; \leq \; \sqrt{2(x^2 + y^2)} + 2 \; = \; \sqrt{2}s + 2 \; \leq \; \left(\sqrt{2} + \frac{2}{s} \right) s \leq (\sqrt{2} + 2)s$$

Taking the sum over all edges of p yields the result.   ∎

Definition 13.5. *The squares in a (shifted) dissection can be assigned a level. The bounding box is at level 0, its four children at level 1, and so on. Consider a grid line through the bounding box that contains line-segments of the dissection. If such a line l contains an edge of a square of level i in the dissection we say that l also has level i. For each line l there is a minimal $i \in \mathbb{N}$ such that l has level i. We call i the* minimal level *of l. (Note that l also has all levels $j \geq i$.)*

Proof of the Structure Theorem. Assume that there is an optimal tour p and shifts $a, b \in \mathbb{N}$ which are choosen randomly. Now we consider the dissection with shift (a, b) and show how the tour is modified into an approximate salesman tour which is (m, t)-light. In particular we determine how much this modification increases the cost.

As defined in Section 13.2 all nodes lie on grid points of a grid of granularity 1. Without loss of generality we assume that L is a power of 2. This ensures that all lines used in the dissection are grid lines. And since the shift values a and b are integers, the lines of the shifted dissection are still grid lines.

In the following we only consider the vertical lines of the dissection, for horizontal ones the modification works analogously. Let i be the minimal level of such a vertical line l. By definition the property to be (m, t)-light consists of two components.

(1) The salesman tour crosses each edge of a square with level i at most t times.

To meet this first condition we use the Patching Lemma. We charge the cost increases to the line l to which it is applied.

The edges on the vertical line l correspond to the line segments which lie between the y-coordinates $b + r \cdot \frac{L}{2^i} \bmod L$ and $b + (r+1) \cdot \frac{L}{2^i} \bmod L$ $(r = 0, 1, 2, \ldots, 2^i - 1)$. For each $j \geq i$ let $l'_{j,r}$ denote the line segment between the y-coordinates $b + r \cdot \frac{L}{2^j} \bmod L$ and $b + (r+1) \cdot \frac{L}{2^j} \bmod L$. In order to minimize the cost of the reducing of crossings we do not apply the lemma directly, but do the patching bottom-up for every j with $\log L \geq j \geq i$. Therefore we use the following procedure $CHANGE(l, i, b)$.

$CHANGE(l, i, b)$
for $j = \log L$ **down to** i **do**
    for $r = 0, 1, \ldots, 2^j - 1$ **do**
        if $l'_{j,r}$ is crossed by the current tour more often than t times, then
        reduce the number of crossings by applying the Patching Lemma
    end
end

Obviously, for every u, all modifications for $j = \log L, \ldots, u + 1$ influence the modification for $j = u$. In each application of the Patching Lemma to a vertical line l of the dissection we add some vertical line segments of l to the tour p which implies that patching a vertical line l possibly increases the number of times the tour crosses a horizontal line h of the dissection. Fortunately, we can assume that the increase in the number of crossings at h due to the patching on l is at most two. Otherwise we would again apply the Patching Lemma to reduce them to two. This additional application does not increase the cost, because all added segments lie on l and have no horizontal separation. These two extra crossings have to be taken into account for both horizontal and vertical lines. So altogether we need 4 extra crossings.

It may happen that a line segment is wrapped around and the Patching Lemma has to be applied separately for each part. In this case we can just ensure that the number of crossings after this procedure is four, not two as in the normal case.

Now we are considering how much this procedure increases the cost of the tour. The additional cost is influenced by the length of the bounding box L, the length of the new line segments introduced by the Patching Lemma and by the number of segments to which the Patching Lemma has to be applied. For $j \geqslant i$ let $c_{l,j}(b)$ denote the number of segments to which we apply the Patching Lemma in the j-th iteration. This implies that the additional cost due to $CHANGE(l, i, b)$ is less than or equal to

$$\sum_{j \geqslant i} c_{l,j}(b) \cdot 6 \cdot \frac{L}{2^j}.$$

Whenever the Patching Lemma is applied at least $t+1$ crossings are replaced by at most 4. Thus every application of the Patching Lemma decreases the number of crossings by at least $t - 3$.

Since the original tour p crosses l only $t(p, l)$ times, we conclude $\sum_{j \geqslant 1} c_{l,j}(b) \leqslant \frac{t(p,l)}{t-3}$ for each $b \in \mathbb{N}$.

The shifts $a, b \in \mathbb{N}$ are chosen randomly. Hence the probability that a line l has level i is $2^i / L$ for all $i \leqslant \log L$. Thus for every vertical line l the expected cost which is charged to the line l is

$$= \sum_{i \geqslant 1} \frac{2^i}{L} \cdot \text{cost increase due to } CHANGE(l, i, b)$$

$$\leqslant \sum_{i \geqslant 1} \frac{2^i}{L} \cdot \sum_{j \geqslant i} c_{l,j}(b) \cdot 6 \cdot \frac{L}{2^j}$$

$$= 6 \cdot \sum_{j \geqslant 1} \frac{c_{l,j}(b)}{2^j} \cdot \sum_{i \leqslant j} 2^i$$

$$\leqslant 6 \cdot \sum_{j \geqslant 1} 2 \cdot c_{l,j}(b)$$

$$\leqslant \frac{12\, t(p, l)}{t - 3}.$$

Since l is vertical this inequality holds for all $b \in \{0, 1, \ldots, L - 1\}$.

(2) All these crossings have to be at portals.

This time we simply move each crossing to the nearest portal which implies the following cost:

Again consider a line l of the dissection with minimal level i. There are at most $t(p, l)$ crossings which have to be moved and for each of it the cost to move it to the nearest portal is at most $\frac{L}{2^i m}$. As we assumed $m \geqslant t \log L$, the cost for this step is upperbounded by

$$\sum_{i=1}^{\log L} \frac{2^i}{L} \cdot t(p,l) \cdot \frac{L}{2^i m} = \frac{t(p,l)\log L}{m} \leqslant t(p,l)/t.$$

Adding the cost in (1) and (2) we find that the expected cost for modifying an optimal salesman tour p into a salesman tour which is (m,t)-light at line l is at most:

$$\frac{12 \cdot t(p,l)}{t-3} + \frac{t(p,l)}{t} = \frac{t(p,l)}{t}(13 + \frac{36}{t-3}).$$

For sufficiently large t ($t \geqslant 12$) this is upperbounded by

$$\frac{17\, t(p,l)}{t}.$$

By taking the sum over all lines of the dissection and using Lemma 13.4 (remember that all lines of the dissection are grid lines) we get

$$\sum_{l:vertical} \frac{17t(p,l)}{t} + \sum_{l:horizontal} \frac{17t(p,l)}{t} \leqslant \left(\sqrt{2}+2\right)\frac{17\,\mathbf{Opt}}{t}.$$

Thus when $t \geqslant 34c\left(\sqrt{2}+2\right)$ the tour cost increases by at most $\mathbf{Opt}/(2c)$. It follows from the Markov inequality that with probability $\geqslant 1/2$ the increase is not more than $\mathbf{Opt}/c$ and hence the cost of the best (m,t)-light salesman tour corresponding to the shifted dissection is at most $(1 + 1/c)\,\mathbf{Opt}$.

$t \geqslant 34c\left(\sqrt{2}+2\right) \approx 116.08 \cdot c$ implies that $t = 117c$ suffices. However, in step (1) we saw that we sometimes need 4 extra crossings. Since $c \geqslant 1$, a value of $t = 121c$ suffices. This satisfies our previous assumption that $t \geqslant 12$.

Arora [Aro97] showed that the Patching Lemma also holds for a constant 3 instead of 6. By using this lower constant and several other optimizations he achieves smaller values for t. So far the best known is about $t = 10c$. ∎

13.4 The Algorithm

Our aim is to find an approximate salesman tour through n given nodes in $\mathcal{O}(n\log^{\mathcal{O}(c)} n)$-time. In the following we will simply speak of a solution when we mean an approximate solution in contrast to an optimal one.

Like in Section 13.2 we assume that all nodes are on grid points of a grid of granularity 1 and that the size L of the bounding box is $L = nc$.

Let $0 \leqslant a, b \leqslant L$ with $a, b \in \mathbb{N}$ be two randomly chosen shifts. Using dynamic programming the algorithm computes the best (m, t)-light salesman tour with respect to the corresponding quadtree with shift (a, b). (Remember that constructing this shifted quadtree needs at most $\mathcal{O}(n \log^2 n)$ time.) The Structure Theorem guarantees that with probability at least $1/2$ this salesman tour has a cost of at most $(1 + 1/c)\,\mathbf{Opt}$.

The algorithm has to find a solution for all nontrivial squares. As the depth of the quadtree is $\mathcal{O}(\log n)$ these are $\mathcal{O}(\text{number of squares with a node} \cdot depth) = \mathcal{O}(n \log n)$ squares. For each side of each square there are at most $\mathcal{O}(m^t)$ possibilities where the salesman tour could cross the boundary of the square (these are $\mathcal{O}(m^{4t})$ per square). Each considered tour enters and leaves the square at one of these portals, altogether there are at most $(4t)!$ different pairs of crossings.

Each of the resulting subproblems is specified by the following parameters:

(X) again $\mathcal{O}(m^{4t})$ possibilities to choose at most t portals on the edges of the subsquare.

(Y) $(4t)^{4t} \cdot (4t)!$ possible orderings of these portals.

Recursively this procedure solves the subproblems and keeps the choice that leads to the lowest cost in (X) and (Y). The algorithm uses a lookup table to store these solutions, thus it never solves the same subproblem twice. Hence the running time is bounded by $\mathcal{O}(T \cdot (m^{4t})^2 \cdot (4t)^{4t} \cdot (4t)!^2)$ where T denotes the number of these squares. Since $T = \mathcal{O}(n \log n)$, $m = \mathcal{O}(c \log L)$ and $L = nc$ we obtain a running time of $\mathcal{O}(n \log n \cdot (c\, \log(nc))^{\mathcal{O}(c)}) = \mathcal{O}(n(\log n)^{\mathcal{O}(c)})$.

13.5 Applications to Some Related Problems

In this section we want to give a short overview to which optimization problems this algorithm or a similar one can be applied.

A possible generalization is to regard spaces with a higher dimension. As there is a R^d version of the Patching Lemma [Aro97], the Structure Theorem and the algorithm can be generalized to higher dimensions. For any fixed dimension d this modified algorithm runs in polynomial time. In time $\mathcal{O}(n(\log n)^{(\mathcal{O}(dc))^{d-1}})$ it computes an (m, t)-light approximate salesman tour where $m = \mathcal{O}((\sqrt{d}c \log n)^{d-1})$ and $t = \mathcal{O}((\sqrt{d}c)^{d-1})$.

The Patching Lemma and Lemma 13.4 hold independently of the considered problem if these are similar problems like k-TSP, k-MST and so on. And so does the basic structure of the Structure Theorem. For Euclidean k-TSP – the more general version of Euclidean TSP – there are two main changes. First $L < \mathbf{Opt}$ can happen and hence it is more complicated to fulfill the assumption that there is a grid of granularity one such that all nodes lie on grid points (section 13.2). To construct an approximation P to **Opt** with $P \leqslant \mathcal{O}(\textbf{Opt} \cdot n)$ helps [Aro97]. Additionally we are confronted with the problem which k nodes out of the given n the salesman tour should use. Nevertheless the algorithm works similar and the running time only grows by a factor of $\mathcal{O}(k)$. The same happens for the Euclidean case of k-MST with $k < n$ where k-MST is defined as follows:

k-MINIMAL SPANNING TREE PROBLEM (k-MST)

Instance: A graph $G = (V, E)$ with edge weights defined by $c : E \longrightarrow R$ and an integer $k < n$.

Problem: Find a spanning tree on k nodes that has a minimal sum of weights.

While this problem is $\mathcal{NP}$-hard, there are polynomial time solutions in the special case of $k = n$.

If we are not searching for a minimal tour but for a minimal spanning tree through all nodes of a graph, then we can possibly minimize the lengths of the tree by introducing additional intermediate points. These points are called *Steiner points* and the connected tree is a *Steiner tree*.

The algorithm also has to be modified a little bit to find such a minimal Steiner tree. This time there possibly occur Steiner points in trivial squares. (Trivial squares are squares that do not contain any node of the input.) But since we are interested in computing an (m, t)-light Steiner tree the number of these points in any trivial square is bounded by $\mathcal{O}(t)$. If we use a regular grid of $\mathcal{O}(t \log n)$ vertical and $\mathcal{O}(t \log n)$ horizontal lines in such a square then we get $\mathcal{O}((t \log n)^2)$ grid points. Choosing the additional Steiner points out of this set neither increases the cost of the Steiner tree too much, nor are there too many possibilities ($\mathcal{O}((t \log n)^{\mathcal{O}(t)})$). Hence the running time does not change.

13.6 The Earlier PTAS

In Arora's PTAS the Euclidean space is partitioned recursively into smaller parts such that all have a common simple structure. The important point is that there is always a near optimal solution to the optimization problem that "agrees closely" with the partitioning. This means that as the depth of the tiling increases there is still a solution that "agrees" with this tiling and whose quality differs from the previous best approximation only by a small constant factor.

Both earlier PTAS – the one due to Arora [Aro96] and the one due to Mitchell [Mit96] – are based on the same concept, except that these algorithms are not randomized. Arora used an extremely simple partitioning technique called $1/3 : 2/3$-tiling, while [MBCV97] Mitchell, Blum, Chalasani and Vempala introduced a new partitioning technique (guillotine subdivisions) which leads to an algorithm that achieves a constant factor approximation for the Euclidean k-MINIMAL SPANNING TREE PROBLEM. By generalizing this idea to m-guillotine subdivisions Mitchell [Mit96] obtained PTAS for k-MST, $k \leqslant n$ and TSP.

An m-guillotine subdivision is a recursively defined partition of the bounding box into smaller instances. In these algorithms there are, at each step, $\mathcal{O}(n)$ different possibilities to partition a problem into two subproblems. Also the number of so called boundary information is limited. All these possibilities are considered one by one which implies a running time of $\mathcal{O}(n^{\mathcal{O}(c)})$.

In contrast to this Arora now considers exactly one partition and looks at all levels at once. There are $\log n$ levels, hence this gives a running time of $\log^{\mathcal{O}(c)} n$ for each subproblem. Since the number of these subproblems is polynomial in n, the whole running time is bounded by $\mathcal{O}(n \log^{\mathcal{O}(c)} n)$.

Finally, the question occurs if these algorithms are applicable in practice. Unfortunately this is not yet the case. Until Arora and Mitchell showed that the Euclidean TSP belongs to the $\mathcal{NP}$-hard problems which have a PTAS, Christofides' approximation algorithm was the best known. Nevertheless, these PTAS are mainly of theoretical interest. In polynomial time the algorithm of Christofides [Chr76] computes a tour that has at most $3/2$ times the optimal cost. And still in this special case this algorithm is more practical. While the Christofides approximation runs in time $\mathcal{O}(n^{2.5})$, Arora's current version is much slower. So far in the best implementation the $\mathrm{poly}(\log n)$ term still grows like $(\log n)^{40c}$ [Aro97].

Bibliography

[ABI86]   N. Alon, L. Babai, and A. Itai. A fast and simple randomized parallel algorithm for the maximal independent set problem. *Journal of Algorithms*, 7:567–583, 1986.

[AFK96]   S. Arora, A. Frieze, and H. Kaplan. A new rounding procedure for the assignment problem with applications to dense graph arrangement problems. In *Proceedings of the 37th Annual IEEE Symposium on Foundations of Computer Science*, pages 21–30, 1996.

[AHOH96]   T. Asano, K. Hori, T. Ono, and T. Hirata. Approximation algorithms for MAXSAT: Semidefinite programming and network flows approach. Technical report, 1996.

[AKK95a]   S. Arora, D. Karger, and M. Karpinski. Polynomial time approximation schemes for dense instances of $\mathcal{NP}$-hard problems. In *Proceedings of the 27th Annual ACM Symposium on Theory of Computing*, pages 284–293, 1995.

[AKK95b]   S. Arora, D. Karger, and M. Karpinski. Polynomial time approximation schemes for dense instances of $\mathcal{NP}$-hard problems. Unpublished manuscript, November 1995.

[AL96]   S. Arora and C. Lund. Hardness of approximations. In D.S. Hochbaum, editor, *Approximation Algorithms for $\mathcal{NP}$-hard Problems*, pages 399–446. PWS Publishing Company, 1996.

[Ali95]   F. Alizadeh. Interior point methods in semidefinite programming with applications to combinatorial optimization. *SIAM Journal on Optimization*, 5(1):13–51, 1995.

[ALM+92]   S. Arora, C. Lund, R. Motwani, M. Sudan, and M. Szegedy. Proof verification and hardness of approximation problems. In *Proceedings of the 33rd Annual IEEE Symposium on Foundations of Computer Science*, pages 14–23, 1992.

[AOH96]   T. Asano, T. Ono, and T. Hirata. Approximation algorithms for the maximum satisfiability problem. *Nordic Journal of Computing*, 3:388–404, 1996.

[Aro94]   S. Arora. *Probabilistic checking of proofs and hardness of approxi-mation problems*. PhD thesis, Department of Computer Science, Berkeley, 1994.

[Aro96]   S. Arora. Polynomial time approximation schemes for Euclidean TSP and other geometric problems. In *Proceedings of the 37th An-nual IEEE Symposium on Foundations of Computer Science*, pages 2–11, 1996.

[Aro97]   S. Arora. Nearly linear time approximation schemes for Euclidean TSP and other geometric problems. In *Proceedings of the 38th An-nual IEEE Symposium on Foundations of Computer Science*, 1997. To appear.

[AS92]   S. Arora and S. Safra. Probabilistic checking of proofs: A new char-acterization of $\mathcal{NP}$. In *Proceedings of the 33rd Annual IEEE Sym-posium on Foundations of Computer Science*, pages 2–13, 1992.

[Asa97]   T. Asano. Approximation algorithms for MAXSAT: Yannakakis vs. Goemans-Williamson. In *Proceedings of the 5th Israel Symposium on the Theory of Computing and Systems*, 1997. To appear.

[Bak94]   B.S. Baker. Approximation algorithms for $\mathcal{NP}$-complete problems on planar graphs. *Journal of the ACM*, 41(1):153–180, 1994.

[BC93]   D.P. Bovet and P. Crescenzi. *Introduction to the Theory of Com-plexity*. Prentice-Hall, 1993.

[BCW80]   M. Blum, A.K. Chandra, and M.W. Wegman. Equivalence of free Boolean graphs can be decided probabilistically in polynomial time. *Information Processing Letters*, 10:80–82, 1980.

[Bel96]   M. Bellare. Proof checking and approximation: Towards tight re-sults. *Complexity Theory Column of Sigact News*, 27, 1996.

[BFL91]   L. Babai, L. Fortnow, and C. Lund. Non-deterministic exponential time has two-prover interactive protocols. *Computational Complex-ity*, 1:3–40, 1991.

[BFLS91]   L. Babai, L. Fortnow, L. Levin, and M. Szegedy. Checking compu-tations in polylogarithmic time. In *Proceedings of the 23rd Annual ACM Symposium on Theory of Computing*, pages 21–31, 1991.

[BGG93]   M. Bellare, O. Goldreich, and S. Goldwasser. Randomness in inter-active proofs. *Computational Complexity*, 3:319–354, 1993.

[BGKW88]   M. Ben-Or, S. Goldwasser, J. Kilian, and A. Wigderson. Multi-prover interactive proofs: How to remove intractability assumptions. In *Proceedings of the 20th Annual ACM Symposium on Theory of Computing*, pages 113–131, 1988.

[BGLR93]   M. Bellare, S. Goldwasser, C. Lund, and A. Russell. Efficient prob-abilistically checkable proofs and applications to approximation. In

Proceedings of the 25th Annual ACM Symposium on Theory of Computing, 1993.

[BGS95]   M. Bellare, O. Goldreich, and M. Sudan. Free bits, PCPs and non-approximability. Technical Report TR95-024, Electronic Colloquium on Computational Complexity, 1995.

[BH92]   R. Boppana and M.M. Halldórsson. Approximating maximum independent set by excluding subgraphs. *BIT*, 32:180–196, 1992.

[BM95]   R. Bacik and S. Mahajan. Semidefinite programming and its applications to $\mathcal{NP}$ problems. In *Proceedings of the 1st Computing and Combinatorics Conference (COCOON)*. Lecture Notes in Computer Science 959, Springer Verlag, 1995.

[BR94]   M. Bellare and J. Rompel. Randomness-efficient oblivious sampling. In *Proceedings of the 35th Annual IEEE Symposium on Foundations of Computer Science*, pages 276–287, 1994.

[BS94]   M. Bellare and M. Sudan. Improved non-approximability results. In *Proceedings of the 26th Annual ACM Symposium on Theory of Computing*, pages 184–193, 1994.

[BW86]   E. Berlekamp and L. Welch. Error correction of algebraic block codes. US Patent Number 4,633,470, 1986.

[Chr76]   N. Christofides. Worst-case analysis of a new heuristic for the traveling salesman problem. Technical Report 388, Graduate School of Industrial Administration, Carnegie-Mellon University, Pittsburgh, 1976.

[CK81]   I. Csiszar and J. Korner. *Information Theory: Coding Theorems for Discrete Memoryless Systems*. Academic Press, 1981.

[CKST95]   P. Crescenzi, V. Kann, R. Silvestri, and L. Trevisan. Structure in approximation classes. In *Proceedings of the 1st Computing and Combinatorics Conference (COCOON)*, pages 539–548. Lecture Notes in Computer Science 959, Springer Verlag, 1995.

[Coo71]   S.A. Cook. The complexity of theorem proving procedures. In *Proceedings of the 3rd Annual ACM Symposium on Theory of Computing*, pages 151–158, 1971.

[CST96]   P. Crescenzi, R. Silvestri, and L. Trevisan. To weight or not to weight: Where is the question? In *Proceedings of the 4th Israel Symposium on the Theory of Computing and Systems*, pages 68–77, 1996.

[Dan63]   G.B. Dantzig. *Linear Programming and Extensions*. Princeton University Press, 1963.

[DF85]   M.E. Dyer and A.M. Frieze. A simple heuristic for the p-center problem. *Operations Research Letters*, 3:285–288, 1985.

[Fag74]   R. Fagin. Generalized first-order spectra and polynomial-time rec-
           ognizable sets. In *SIAM-AMS Proceedings*, pages 43–73, 1974.

[Fei91]   U. Feige. On the success probability of the two provers in one-
           round proof systems. In *Proceedings of the 6th IEEE Symposium on
           Structure in Complexity Theory*, pages 116–123, 1991.

[Fei95]   U. Feige. Error reduction by parallel repetition – the state of the
           art. Technical Report CS95-32, WISDOM Technical reports in Com-
           puter Science, 1995.

[Fei96]   U. Feige. A threshold of ln n for approximating set cover. In *Proceed-
           ings of the 28th Annual ACM Symposium on Theory of Computing*,
           pages 314–318, 1996.

[Fer96]   W. Fernandez de la Vega. MAXCUT has a randomized approxima-
           tion scheme in dense graphs. *Random Structures and Algorithms*,
           8(3):187–198, 1996.

[FG95]   U. Feige and M. Goemans. Approximating the value of two prover
           proof systems, with applications to MAX2SAT and MAXDICUT. In
           *Proceedings of the 3rd Israel Symposium on the Theory of Computing
           and Systems*, pages 182–189, 1995.

[FGL⁺91]   U. Feige, S. Goldwasser, L. Lovász, S. Safra, and M. Szegedy. Ap-
           proximating clique is almost $\mathcal{NP}$-complete. In *Proceedings of the
           32nd Annual IEEE Symposium on Foundations of Computer Sci-
           ence*, pages 2–12, 1991.

[FK91]   H. Furstenberg and Y. Katznelson. A density version of the Hales-
           Jewett theorem. *Journal d'Analyse Mathematique*, 57:64–119, 1991.

[FK94]   U. Feige and J. Kilian. Two-prover protocols – low error at afford-
           able rates. In *Proceedings of the 26th Annual ACM Symposium on
           Theory of Computing*, pages 172–183, 1994.

[FL92]   U. Feige and L. Lovász. Two-prover one-round proof systems: Their
           power and their problems. In *Proceedings of the 24th Annual ACM
           Symposium on Theory of Computing*, pages 733–744, 1992.

[Fri91]   J. Friedman. On the second eigenvalue and random walks in random
           d-regular graphs. *Combinatorica*, 11(4):331–362, 1991.

[FRS88]   L. Fortnow, J. Rompel, and M. Sipser. On the power of multi-prover
           interactive protocols. In *Proceedings of the 3rd IEEE Symposium on
           Structure in Complexity Theory*, pages 156–161, 1988.

[FRS90]   L. Fortnow, J. Rompel, and M. Sipser. Errata for on the power
           of multi-prover interactive protocols. In *Proceedings of the 5th
           IEEE Symposium on Structure in Complexity Theory*, pages 318–
           319, 1990.

[FV96]   U. Feige and O. Verbitsky. Error reduction by parallel repetition – A
           negative result. In *Proceedings of the 11th Annual IEEE Conference
           on Computational Complexity Theory*, 1996.

[FW91]   M. Formann and F. Wagner. A packing problem with applications to
           lettering of maps. In *Proceedings of the 7th Annual ACM Symposium
           on Computational Geometry*, pages 281–288, 1991.

[GG81]   O. Gabber and Z. Galil. Explicit constructions of linear-sized super-
           concentrators. *Journal of Computer and System Sciences*, 22:407–
           420, 1981.

[GGJ76]   M.R. Garey, R.L. Graham, and D.S. Johnson. Some $\mathcal{NP}$-complete
           geometric problems. In *Proceedings of the 8th Annual ACM Sym-
           posium on Theory of Computing*, pages 10–22, 1976.

[Gil93]   D. Gillman. A Chernoff bound for random walks on expanders. In
           *Proceedings of the 34th Annual IEEE Symposium on Foundations
           of Computer Science*, pages 680–691, 1993.

[GJ79]   M.R. Garey and D.S. Johnson. *Computers and Intractability. A
           Guide to the Theory of $\mathcal{NP}$-Completeness*. W.H. Freeman and Com-
           pany, 1979.

[Goe97]   M.X. Goemans. Semidefinite programming in combinatorial opti-
           mization. In *Proceedings of the 16th International Symposium on
           Mathematical Programming*, 1997.

[Gol95]   O. Goldreich. Probabilistic proof systems (survey). Technical re-
           port, Department of Computer Science and Applied Mathematics,
           Weizmann Institute of Science, 1995.

[Gra90]   R.M. Gray. *Entropy and Information Theory*. Springer-Verlag, 1990.

[GS92]   P. Gemmell and M. Sudan. Highly resilient correctors for polyno-
           mials. *Information Processing Letters*, 43:169–174, 1992.

[GvL86]   G.H. Golub and C.F. van Loan. *Matrix Computations*. North Oxford
           Academic, 1986.

[GW94a]   M.X. Goemans and D.P. Williamson. .878 approximation algorithms
           for MAXCUT and MAX2SAT. In *Proceedings of the 26th Annual
           ACM Symposium on Theory of Computing*, pages 422–431, 1994.

[GW94b]   M.X. Goemans and D.P. Williamson. New 3/4-approximation algo-
           rithms for the maximum satisfiability problem. *SIAM Journal on
           Discrete Mathematics*, 7:656–666, 1994.

[GW95]   M.X. Goemans and D.P. Williamson. Improved approximation algo-
           rithms for maximum cut and satisfiability problems using semidefi-
           nite programming. *Journal of the ACM*, 42:1115–1145, 1995.

[Hås96a]   J. Håstad. Clique is hard to approximate within $n^{1-\epsilon}$. In *Proceedings of the 37th Annual IEEE Symposium on Foundations of Computer Science*, pages 627–636, 1996.

[Hås96b]   J. Håstad. Testing of the long code and hardness for clique. In *Proceedings of the 28th Annual ACM Symposium on Theory of Computing*, pages 11–19, 1996.

[Hås97a]   J. Håstad. Clique is hard to approximate within $n^{1-\epsilon}$. Technical Report TR97-038, Electronic Colloquium on Computational Complexity, 1997. An earlier version was presented in *Proceedings of the 37th Annual IEEE Symposium on Foundations of Computer Science* 1996, pp. 627-636.

[Hås97b]   J. Håstad. Some optimal inapproximability results. In *Proceedings of the 29th Annual ACM Symposium on Theory of Computing*, pages 1–10, 1997. Also appeared as Technical Report TR97-037, Electronic Colloquium on Computational Complexity.

[HL96]   T. Hofmeister and H. Lefmann. A combinatorial design approach to MAXCUT. *Random Structures and Algorithms*, 9(1-2):163–175, 1996.

[HN79]   W. Hsu and G. Nemhauser. Easy and hard bottleneck location problems. *Discrete Applied Mathematics*, 1:209–216, 1979.

[Hoe63]   W. Hoeffding. Probability inequalities for sums of bounded random variables. *Journal of the American Statistical Association*, 58:13–30, 1963.

[HPS94]   S. Hougardy, H.J. Prömel, and A. Steger. Probabilistically checkable proofs and their consequences for approximation algorithms. *Discrete Mathematics*, 136:175–223, 1994.

[HS85]   D.S. Hochbaum and D. Shmoys. A best possible heuristic for the k-center problem. *Mathematics of Operations Research*, 10:180–184, 1985.

[IK75]   O.H. Ibarra and C.E. Kim. Fast approximation algorithms for the knapsack and sum of subsets problems. *Journal of the ACM*, 22:463–468, 1975.

[Joh74]   D.S. Johnson. Approximation algorithms for combinatorial problems. *Journal of Computer and System Sciences*, 9:256–278, 1974.

[Kar72]   R.M. Karp. Reducibility among combinatorial problems. In R.E. Miller and J.W. Thatcher, editors, *Complexity of Computer Computations*, pages 85–103. Plenum Press, 1972.

[Kar84]   N. Karmarkar. A new polynomial-time algorithm for linear programming. *Combinatorica*, 4(4):373–395, 1984.

[Kar91]   R.M. Karp. An introduction to randomized algorithms. *Discrete Applied Mathematics*, 34:165–201, 1991.

[Kar96]   H.J. Karloff. How good is the Goemans-Williamson MAXCUT algorithm? In *Proceedings of the 28th Annual ACM Symposium on Theory of Computing*, pages 427–434, 1996.

[Kha79]   L.G. Khachian. A polynomial algorithm in linear programming. *Soviet Mathematics Doklady*, 20:191–194, 1979.

[KK82]   N. Karmarkar and R.M. Karp. An efficient approximation scheme for the one-dimensional bin packing problem. In *Proceedings of the 23rd Annual IEEE Symposium on Foundations of Computer Science*, pages 312–320, 1982.

[KL96]   P. Klein and H.-I. Lu. Efficient approximation algorithms for semidefinite programs arising from MAXCUT and COLORING. In *Proceedings of the 28th Annual ACM Symposium on Theory of Computing*, pages 338–347, 1996.

[KMS94]   D. Karger, R. Motwani, and M. Sudan. Approximate graph coloring by semidefinite programming. In *Proceedings of the 35th Annual IEEE Symposium on Foundations of Computer Science*, pages 2–13, 1994.

[KMSV94]   S. Khanna, R. Motwani, M. Sudan, and U. Vazirani. On syntactic versus computational views of approximability. In *Proceedings of the 26th Annual ACM Symposium on Theory of Computing*, pages 819–830, 1994.

[Knu81]   D.E. Knuth. *The Art of Computer Programming, Vol. 2, Seminumerical Algorithms*. Addison-Wesley, 2nd edition, 1981.

[KR90]   R.M. Karp and V. Ramachandran. Parallel algorithms for shared-memory machines. In J. van Leeuwen, editor, *Handbook of Theoretical Computer Science, Vol. A, Algorithms and Complexity*, pages 869–941. Elsevier, 1990.

[KZ97]   M. Karpinski and A. Zelikovsky. Approximating dense cases of covering problems. Technical Report TR97-004, Electronic Colloquium on Computational Complexity, 1997.

[LFKN92]   C. Lund, L. Fortnow, H. Karloff, and N. Nisan. Algebraic methods for interactive proof systems. *Journal of the ACM*, 39(4):859–868, 1992.

[LMN93]   N. Linial, Y. Mansour, and N. Nisan. Constant depth circuits, Fourier transform and learnability. *Journal of the ACM*, 40(3):607–620, 1993.

[LN87]   R. Lidl and H. Niederreiter. *Finite Fields*. Cambridge University Press, 1987.

[Lov75]   L. Lovász. On the ratio of the optimal integral and fractional covers. *Discrete Mathematics*, 13:383–390, 1975.

[LPS86]   A. Lubotzky, R. Phillips, and P. Sarnak. Explicit expanders and the Ramanujan conjectures. In *Proceedings of the 18th Annual ACM Symposium on Theory of Computing*, 1986.

[Lub85]   M. Luby. A simple parallel algorithm for the maximal independent set problem. In *Proceedings of the 17th Annual ACM Symposium on Theory of Computing*, pages 1–10, 1985.

[Lub86]   M. Luby. A simple parallel algorithm for the maximal independent set problem. *SIAM Journal on Computing*, 15(4):1036–1053, 1986.

[LY93]   C. Lund and M. Yannakakis. On the hardness of approximating minimization problems. In *Proceedings of the 25th Annual ACM Symposium on Theory of Computing*, pages 286–293, 1993. Also appears in: *Journal of the ACM* 41: 960-981, 1994.

[Mar75]   G.A. Margulis. Explicit construction of concentrators. *Problems of Information Transmission*, 10:325–332, 1975.

[MBCV97]   J.S.B. Mitchell, A. Blum, P. Chalasani, and S. Vempala. A constant-factor approximation algorithm for the geometric k-MST problem in the plane. *SIAM Journal on Computing*, pages 402–408, 1997. To appear. Journal version of: J.S.B. Mitchell. Guillotine subdivisions approximate polygonal subdivisions: A simple new method for the geometric k-MST problem. In *Proceedings of the 7th Annual ACM-SIAM Symposium on Discrete Algorithms.* 1996.

[MBG⁺93]   A. Menezes, I. Blake, X. Gao, R. Mullin, S. Vanstone, and T. Yaghoobian. *Applications of Finite Fields*. Kluwer Academic Publishers, 1993.

[Mit96]   J.S.B. Mitchell. Guillotine subdivisions approximate polygonal subdivisions: Part II – a simple polynomial-time approximation scheme for geometric k-MST, TSP, and related problems. Technical report, Department of Applied Mathematics and Statistics, Stony Brook, 1996.

[MNR96]   R. Motwani, J. Naor, and P. Raghavan. Randomized approximation algorithms in combinatorial optimization. In D.S. Hochbaum, editor, *Approximation Algorithms for NP-hard Problems*, pages 447–481. PWS Publishing Company, 1996.

[MR95a]   S. Mahajan and H. Ramesh. Derandomizing semidefinite programming based approximation algorithms. In *Proceedings of the 36th Annual IEEE Symposium on Foundations of Computer Science*, pages 162–169, 1995.

[MR95b]   R. Motwani and P. Raghavan. *Randomized Algorithms*. Cambridge University Press, 1995.

[MS77]   F.J. MacWilliams and N.J.A. Sloane. *The Theory of Error-Correct-ing Codes.* North-Holland, 1977.

[Nis96]   N. Nisan. Extracting randomness: How and why – A survey. In *Proceedings of the 11th IEEE Symposium on Structure in Complexity Theory*, 1996.

[NSS95]   M. Naor, L. Schulman, and A. Srinivasan. Splitters and near-optimal derandomization. In *Proceedings of the 36th Annual IEEE Symposium of Foundations of Computer Science*, pages 182–191, 1995.

[Pap77]   C.H. Papadimitriou. The Euclidean traveling salesman problem is $\mathcal{NP}$-complete. *Theoretical Computer Science*, 4:237–244, 1977.

[Pap94]   C.H. Papadimitriou. *Computational Complexity.* Addison-Wesley, 1994.

[PM81]   A. Paz and S. Moran. Non deterministic polynomial optimization problems and their approximation. *Theoretical Computer Science*, 15:251–277, 1981.

[PS82]   C.H. Papadimitriou and K. Steiglitz. *Combinatorial Optimization: Algorithms and Complexity.* Prentice-Hall, 1982.

[PS94]   A. Polishchuk and D.A. Spielman. Nearly linear-size holographic proofs. In *Proceedings of the 26th Annual ACM Symposium on Theory of Computing*, pages 194–203, 1994.

[PT95]   S. Poljak and Z. Tuza. Maximum cuts and large bipartite subgraphs. In *Combinatorial Optimization*, volume 20 of *DIMACS Series in Discrete Mathematics and Theoretical Computer Science*, pages 181–244, 1995.

[PY91]   C.H. Papadimitriou and M. Yannakakis. Optimization, approximation, and complexity classes. *Journal of Computer and System Sciences*, 43:425–440, 1991.

[Rag88]   P. Raghavan. Probabilistic construction of deterministic algorithms: approximating packing integer programs. *Journal of Computer and System Sciences*, 37:130–143, 1988.

[Raz95]   R. Raz. A parallel repetition theorem. In *Proceedings of the 27th Annual ACM Symposium on Theory of Computing*, pages 447–456, 1995.

[Raz97]   R. Raz. A parallel repetition theorem (full version). Unpublished manuscript, 1997.

[RS92]   R. Rubinfeld and M. Sudan. Self-testing polynomial functions efficiently and over rational domains. In *Proceedings of the 3rd Annual ACM-SIAM Symposium on Discrete Algorithms*, pages 23–32, 1992.

[RT87]   P. Raghavan and C.D. Thompson. Randomized rounding: a technique for provably good algorithms and algorithmic proofs. *Combinatorica*, 7:365–374, 1987.

[Spe94]   J. Spencer. Randomization, derandomization and antirandomization: three games. *Theoretical Computer Science*, 131:415–429, 1994.

[SS77]   R. Solovay and V. Strassen. A fast Monte-Carlo test for primality. *SIAM Journal on Computing*, 6(1):84–85, 1977. An erratum appeared in *SIAM Journal on Computing* 7(1): 118, 1978.

[Sud92]   M. Sudan. *Efficient checking of polynomials and proofs and the hardness of approximation problems*. PhD thesis, Department of Computer Science, Berkeley, 1992.

[SW97]   M. Sudan and D.P. Williamson. Some notes on Håstad's proof. Unpublished manuscript, 1997.

[Tre97]   L. Trevisan. *Reductions and (Non)-Approximability*. PhD thesis, Computer Science Department, University of Rome"La Sapienza", 1997.

[TSSW96]   L. Trevisan, G.B. Sorkin, M. Sudan, and D.P. Williamson. Gadgets, approximation, and linear programming. In *Proceedings of the 37th Annual IEEE Symposium on Foundations of Computer Science*, pages 617–626, 1996.

[VB96]   L. Vandenberghe and S. Boyd. Semidefinite programming. *SIAM Review*, 38:49–95, 1996.

[Ver94]   O. Verbitsky. Towards the parallel repetition conjecture. In *Proceedings of the 9th IEEE Symposium on Structure in Complexity Theory*, pages 304–307, 1994.

[Ver96]   O. Verbitsky. Towards the parallel repetition conjecture. *Theoretical Computer Science*, 157(2):277–282, 1996.

[Weg93]   I. Wegener. *Theoretische Informatik: eine algorithmenorientierte Einführung*. B.G. Teubner, 1993.

[Yan92]   M. Yannakakis. On the approximation of maximum satisfiability. In *Proceedings of the 3rd Annual ACM-SIAM Symposium on Discrete Algorithms*, pages 1–9, 1992.

[Ye90]   Y. Ye. An $\mathcal{O}(n^3 L)$ potential reduction algorithm for linear programming. In *Contemporary Mathematics*, volume 114, pages 91–107, 1990.

[Zuc93]   D. Zuckerman. $\mathcal{NP}$-complete problems have a version that's hard to approximate. In *Proceedings of the 8th IEEE Symposium on Structure in Complexity Theory*, pages 305–312, 1993. Journal version, entitled: On unapproximable versions of $\mathcal{NP}$-complete problems. In *SIAM Journal on Computing*, 25: 1293-1304, 1996.

Author Index

Subject Index

List of Contributors

Artur Andrzejak
Departement Informatik
ETH Zürich
CH-8092 Zürich
Switzerland
artur@inf.ethz.ch

Clemens Gröpl
Lehrstuhl Algorithmen und Komplexität
Institut für Informatik
Humboldt Universität zu Berlin
D-10099 Berlin
Germany
groepl@informatik.hu-berlin.de

Volker Heun
Lehrstuhl Effiziente Algorithmen
Institut für Informatik
Technische Universität München
D-80290 München
Germany
heun@informatik.tu-muenchen.de

Thomas Hofmeister
Lehrstuhl II
Fachbereich Informatik
Universität Dortmund
D-44221 Dortmund
Germany
hofmeist@ls2.informatik.uni-dortmund.de

Stefan Hougardy
Lehrstuhl Algorithmen und Komplexität
Institut für Informatik
Humboldt Universität zu Berlin
D-10099 Berlin
Germany
hougardy@informatik.hu-berlin.de

Martin Hühne
Lehrstuhl II
Fachbereich Informatik
Universität Dortmund
D-44221 Dortmund
Germany
huehne@ls2.informatik.uni-dortmund.de

Thomas Jansen
Lehrstuhl II
Fachbereich Informatik
Universität Dortmund
D-44221 Dortmund
Germany
jansen@ls2.informatik.uni-dortmund.de

Richard Mayr
Lehrstuhl Theoretische Informatik und
Grundlagen der KI
Institut für Informatik
Technische Universität München
D-80290 München
Germany
mayrri@informatik.tu-muenchen.de

Wolfgang Merkle
Mathematisches Institut
Ruprecht-Karls-Universität Heidelberg
Im Neuenheimer Feld 294
D-69120 Heidelberg
Germany
merkle@math.uni-heidelberg.de

Martin Mundhenk
Lehrstuhl für Theoretische Informatik
Fachbereich IV - Informatik
Universität Trier
D-54286 Trier
Germany
mundhenk@ti.uni-trier.de

Claus Rick
Institut für Informatik IV
Rheinische-Friedrich-Wilhelms-
Universität Bonn
Römerstr. 164
D-53117 Bonn
Germany
rick@cs.uni-bonn.de

Hein Röhrig
Bahnhofstr. 19
D-66125 Saarbücken-Dudweiler
Germany
roehrig@mpi-sb.mpg.de

Annette Schelten
Lehrstuhl für Diskrete Mathematik
und Grundlagen der Informatik
BTU Cottbus
Postfach 10 13 44
D-03013 Cottbus
Germany
schelten@math.tu-cottbus.de

Sebastian Seibert
Lehrstuhl für Informatik I
RWTH Aachen
D-52056 Aachen
Germany
seibert@i1.informatik.rwth-aachen.de

Detlef Sieling
Lehrstuhl II
Fachbereich Informatik
Universität Dortmund
D-44221 Dortmund
Germany
sieling@ls2.informatik.uni-dortmund.de

Martin Skutella
Fachbereich Mathematik
Sekr. MA 6-1
TU Berlin
Straße des 17. Juni 136
D-10623 Berlin
Germany
skutella@math.tu-berlin.de

Anna Slobodová
ITWM-Trier
Bahnhofstraße 30-32
D-54292 Trier
Germany
anna@ti.uni-trier.de

Ulrich Weigand
IMMD 1
Friedrich-Alexander-Universität
Erlangen-Nürnberg
Martensstraße
D-91058 Erlangen
Germany
weigand@informatik.uni-erlangen.de

Thomas Wilke
Institut für Informatik
und Praktische Mathematik
Christian-Albrechts-Universität zu Kiel
Olshausenstraße 40
D-24098 Kiel
Germany
tw@informatik.uni-kiel.de

Katja Wolf
Zentrum für Paralleles Rechnen
Universität zu Köln
Weyertal 80
D-50931 Köln
Germany
kwolf@informatik.uni-koeln.de

Alexander Wolff
Institut für Informatik
Freie Universität Berlin
Takustraße 9
D-14195 Berlin
Germany
awolff@inf.fu-berlin.de

Springer
and the
environment

At Springer we firmly believe that an
international science publisher has a
special obligation to the environment,
and our corporate policies consistently
reflect this conviction.
We also expect our business partners –
paper mills, printers, packaging
manufacturers, etc. – to commit
themselves to using materials and
production processes that do not harm
the environment. The paper in this
book is made from low- or no-chlorine
pulp and is acid free, in conformance
with international standards for paper
permanency.

Lecture Notes in Computer Science

For information about Vols. 1–1296

please contact your bookseller or Springer-Verlag